AF213567

CISM International Centre for Mechanical Sciences

Courses and Lectures

Volume 593

For more than 40 years the book series edited by CISM, "International Centre for Mechanical Sciences: Courses and Lectures", has presented groundbreaking developments in mechanics and computational engineering methods. It covers such fields as solid and fluid mechanics, mechanics of materials, micro- and nanomechanics, biomechanics, and mechatronics. The papers are written by international authorities in the field. The books are at graduate level but may include some introductory material.

More information about this series at http://www.springer.com/series/76

Marco Paggi · David Hills
Editors

Modeling and Simulation of Tribological Problems in Technology

Springer

Editors
Marco Paggi
IMT School for Advanced Studies Lucca
Lucca, Italy

David Hills
Department of Engineering Science
University of Oxford
Oxford, UK

ISSN 0254-1971 ISSN 2309-3706 (electronic)
CISM International Centre for Mechanical Sciences
ISBN 978-3-030-20376-4 ISBN 978-3-030-20377-1 (eBook)
https://doi.org/10.1007/978-3-030-20377-1

This Springer imprint is published by the registered company Springer Nature Switzerland AG
The registered company address is: Gewerbestrasse 11, 6330 Cham, Switzerland

Preface

Significant advances in contact mechanics have been achieved since the first theoretical derivations 130 years ago with Hertz, primarily associated with the solution of contact problems in statics and dynamics involving later on friction, adhesion, wear, roughness, heat or electric conduction, and also with materials not only linear elastic. Principles of contact mechanics can be applied in many traditional mechanical engineering areas such as locomotive wheel–rail contact, coupling devices, braking systems, tires, bearings, combustion engines, mechanical linkages, gasket seals, metal forming, ultrasonic welding, electrical contacts, and many others. Current challenges in the field regard to the extension of contact mechanics methodologies to the micro- and the nanoscale, to coupled multi-field problems, and to advanced mechanical engineering, microelectronics, and nanomechanics applications involving roughness, adhesion, friction, and wear.

With the goal to convey, in a self-contained manner, the fundamental concepts for the classification of the types of contact, the mathematical methods for the formulation of the contact problems, and the numerical methods required for their solution, we organized a course on "Modelling and simulation of tribological problems in technology" in the International Centre for Mechanical Sciences (CISM) in Udine, Italy, from May 28 to June 1, 2018. Such a course featured 5 days of lectures delivered by A. Almqvist (Luleå University of Technology, Sweden), J. R. Barber (University of Michigan, USA), D. Dini (Imperial College London, UK), D. A. Hills (University of Oxford, UK), and M. Paggi (IMT School for Advanced Studies Lucca, Italy) to an audience of more than 40 researchers from academia and industry.

As a consequence, this book and its seven chapters—based on the lectures of the aforementioned CISM course—aim at conveying a strong background on the theory and numerical methods for contact mechanics, with also the in-depth treatment of cutting-edge research topics and applications. The book is primarily tailored for doctoral students of applied mathematics, mechanics, engineering, and physics with a strong research interest in theoretical modeling, numerical simulation, and experimental characterization of contact problems in technology. It is also suited for young and senior researchers in the above-mentioned and neighboring fields working in academia or in private research and development centers, interested in gaining a compact yet comprehensive overview of contact mechanics from its fundamental mathematical background, to the computational methods and the experimental techniques available for the solution of contact problems.

As a start, Chap. 1 "Fundamentals of Elastic Contacts" provides a classification of contact problems and the half-space solutions for linear elasticity. For the class of complete contacts, asymptotic methods are formulated and applied to mechanical engineering problems. Chapter 2 "Contact Problems Involving Friction" further extends the discussion to contact problems with friction and partial slip, with attention to coupling between the normal and the tangential contact problems, also in elastodynamics. Chapter 3 "Nonequilibrium Molecular Dynamics Simulations of Tribological Systems" focuses on modeling contact problems at the nanoscale, exploiting nonequilibrium molecular dynamics methods. Chapter 4 "Computational Methods for Contact Problems with Roughness" introduces computational methods for the solution of normal and tangential contact problems at the microscale, with special attention to modeling of surface roughness. Chapter 5 "Emergent Properties from Contact Between Rough Interfaces", exploiting the methods presented in the previous chapter, focuses on the key research question of how nonlinear interactions between contact patches induced by roughness across different length scales influence the emergent physico-mechanical properties of an interface. Chapter 6 "Modelling Flows in Lubrication" introduces the reader to lubrication theory and describes the governing equations, models and methods that can be used to simulate various types of lubricated systems. Finally, Chap. 7 "Contact Mechanics of Rubber and Soft Matter" focuses on the role of viscoelasticity and adhesion in contact problems, the methods for their solution and also the characterization of the phenomenon of energy dissipation in tangential contacts.

We would like to thank all the colleagues for their great efforts and dedication to share their knowledge, and their engagement in the CISM lectures and the contributions to this book.

Lucca, Italy Marco Paggi
Oxford, UK David Hills

Contents

Chapter 1
Fundamentals of Elastic Contacts

David Hills and Hendrik Andresen

Abstract Contacts are classified into the fundamental types and their characteristics briefly explored. The formulation of incomplete contacts using a half-plane formulation is then developed, and used to obtain the plane solution for a Hertzian contact, while providing the framework for many other geometries. Williams' solution for a sharp infinite elastic wedge is described in detail, and it is shown how this may be applied to advantage in understanding complete contacts and especially their near-edge properties. We then go back to look at incomplete contacts and analyse how they respond when there is interfacial friction present and they remain stationary, but a partial slip state evolves. The chapter concludes by reviewing the other possible types of contact.

Introduction and Taxonomy

There are only two ways in which loads can be transferred into and out of any component in an engineering assembly—one is through body forces, which are often small, the main exception being centrifugal forces developed in rotating machinery—and the other is by contact with other components. It follows that loads which, in general terms, are diffusely carried in the majority of a component are localised where it is in contact with other things. Detailed consideration of contacts is therefore important, whether they are stationary (or notionally stationary), sliding or rolling. Here, we are going to consider only contacts made from elastic (usually linear elastic) material, and capable of analysis within linear elastic theory, so that it follows that displacements are small and also that conventional first-order definitions of strains apply, which may be an important consideration in certain indentation problems, as we shall see.

D. Hills (✉) · H. Andresen
Department of Engineering Science, University of Oxford,
Parks Road, Oxford OX1 3PJ, UK
e-mail: david.hills@eng.ox.ac.uk

M. Paggi and D. Hills (eds.), *Modeling and Simulation of Tribological Problems in Technology*, CISM International Centre for Mechanical Sciences 593,
https://doi.org/10.1007/978-3-030-20377-1_1

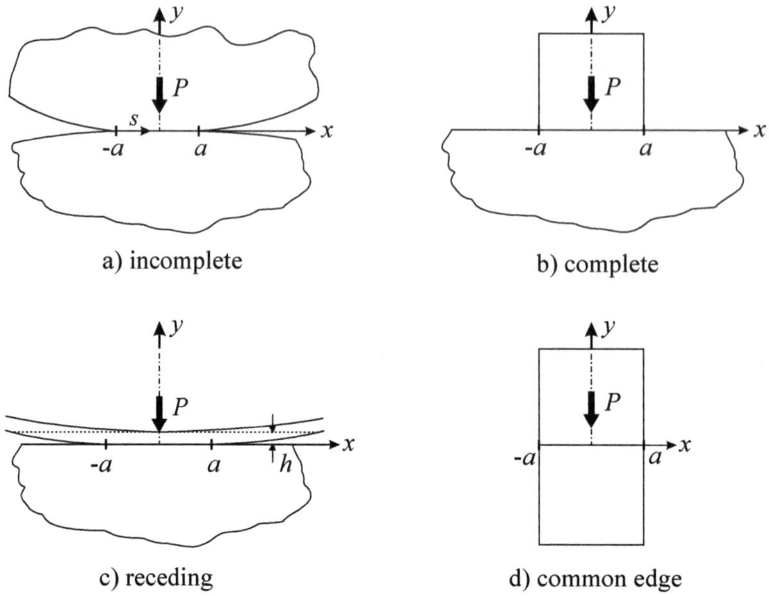

Fig. 1.1 Types of contact

It is very instructive to start a systematic study by looking at the kinds of contacts which arise, and to begin to think about their basic properties and what these imply. Figure 1.1 shows the four main kinds of contact. We will restrict ourselves throughout to two-dimensional contacts, and assume that their extent perpendicular to the plane of the page is large, so that plane strain conditions are obtained. One of the things we shall concentrate on is behaviour of the contact edges and here even three-dimensional contacts have properties which are locally 'plane'. We will, where appropriate, be able also to include out-of-plane deformation by looking at the corresponding anti-plane solution.

Incomplete, or non-conformal contacts, Fig. 1.1a, are those which arise when bodies having convex front surfaces are pressed into contact. They have the properties that the contact increases in size (or advances) as the applied load is increased, and that the contact pressure falls smoothly to zero at the edges in a locally square root bounded manner, so that $p(s) \sim \sqrt{s}$. Often, provided that the size of the contact is small compared with the extent of the body, we can approximate each to a 'half-plane'. The best known incomplete contact is that between two circular cylinders having their axes parallel, and this is the plane form of the configuration famously studied by Hertz over 130 years ago (Hertz 1881). It was he who made this sweeping simplification but which, it turns out, is quite well satisfied by a wide range of contacts. When we come to solve the contact problem in detail we will use a formulation appropriate to a half-plane for each body and it follows that, because the contact interface is a plane of symmetry, when a normal load is applied, surface particles

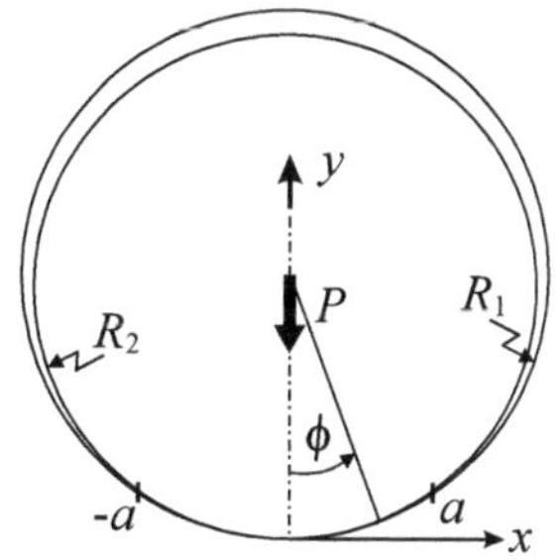

Fig. 1.2 Incomplete contact which is nearly conformal

in the contact surfaces will displace laterally by the same amount if the bodies are elastically similar. No interfacial shear tractions therefore arise. Equally, if a shear force is applied so that shear tractions develop along the interface, the surface normal displacement of each body will be the same so that no change in relative surface gradient arises, and therefore the contact pressure distribution is unmodified. Contacts of this kind, where the normal problem induces no shear traction and shear tractions cause no modification of the normal pressure, regardless of the value of the coefficient of friction, are said to be 'uncoupled'.

Some contacts which are advancing in character do not permit representation of the bodies by half-planes. A good example is shown in Fig. 1.2 which may be thought of as the traditional civil engineering bearing bolt. The small clearance between the pin and hole in which it is journalled ($\Delta = R_1 - R_2 > 0$ but $\Delta \ll R_2$) means that a formulation appropriate to an elastic disk and anti-disk is needed (Persson 1964). These contacts are also said to be 'nearly conformal'.

A wedge-shaped body pressed into the surface of a second body is another example of an incomplete contact and is mentioned because surface elements of the contacting bodies will have to rotate as they pass into the contact patch under increasing normal load, with a combined rotation equal to the external wedge angle. Recall that linear elasticity requires the rotation of elements of material to be small and so solutions to this problem using this underlying theory will be valid only when the wedge is very shallow (say with an external wedge angle of no more than 10° to 15°), and solutions for steeper wedges than this should be viewed with suspicion (Truman et al. 1995).

We turn, now, to contacts where the front faces of the bodies have the same profile in the unloaded configuration, and so as they are placed together they *conform*, such as the square block resting on the elastically similar half-plane, Fig. 1.1b. In these problems, the size of the contact is fixed by the points where there is a discontinuity in profile gradient, and the size of the contact is independent of the applied load. A half-plane idealisation of the square block is inappropriate because the side faces of the block must be traction free and so no general, simple, closed-form formulation is possible. However, if we are interested in mainly the contact edges then, as we shall see later, one possibility is to assume that the contact interface does not slip, and hence an observation point near the edge experiences a three-quarter plane domain or, more generally, if the block had inclined sides, then a wedge. The Williams analysis of these (section "Williams' Solution") show that $p(s) \sim s^{\lambda-1}$, where $\lambda < 1$, and is therefore singular. Even if the contact edge slips it will usually be the case that the pressure takes the same form and $\lambda < 1$, although there are cases (Karuppanan and Hills 2008) where $\lambda > 1$, and hence the contact pressure is bounded.

The problem of an elastic strip or layer resting on an elastically similar half-plane also conforms but when the extent of the loading on the upper surface is much less than the lateral extent of the strip—in Fig. 1.1c, we have the extreme case of a line load on what may be an infinitely long strip—the application of a finite normal load causes the contact to become much smaller, or to *recede*. In a problem of this kind where the basic geometry includes only a solitary length scale (here, the strip thickness, h) the contact 'snaps' to its final size upon application of an infinitesimal normal load. Increasing the force merely causes the state of stress and displacements to increase in proportion, but the separation points remain the same and, in all cases, $p(s) \sim \sqrt{s}$ (if the bodies are elastically similar). Returning to the 'nearly conformal' contact illustrated in Fig. 1.2, we note that, when $\Delta = 0$ the application of a radial force to the pin causes the contact to 'snap' to an included angle of $\phi = 87.46°$. This is also the angle to which an advancing contact extends when $R \gg \Delta > 0$ and to which it smoothly recedes when $-R \ll \Delta < 0$ (Ciavarella et al. 2006).

It will be apparent that, in all the contacts considered so far, the contact pressure adjacent to the contact edge can take one of only two values—$p(s) \to 0$ as $s \to 0$ in the cases of advancing or receding contacts, and $p(s) \to \infty$ as $s \to 0$ in the case of most stationary, complete contacts. The last type of contact to be considered, illustrated in Fig. 1.1d, is when the bodies conform over their front faces but where the length of those faces is the same in each body, terminated by an abrupt change in front-face gradient. If the angle between the side faces is not 180°, a local analysis would again follow the Williams' analysis for a wedge, but for the cases where the contact side faces are collinear, forming a straight line, and that face is traction free, the local contact pressure is finite ($p(s) \sim s^0$). Also, in the absence of tractions the only non-zero component of stress lies parallel with the free surface and, in the absence of out-of-plane loading but in the presence of some increasing in-plane loading, the edges must be the last point to slip. This is obvious in the case when the interface is perpendicular to the free face as the complement to the free-face shear traction lies along the contact interface. We might call problems of this kind 'common-edge'.

A general point which emerges from this taxonomy is that, in the cases of conformal contact (of whatever class—complete, receding, and common edge) a possible method of investigating what happens along the interface is simply to assume that it is in intimate contact and adhered, i.e. we model the contact pair as one body—as if they form a monolith—and then we look at the tractions along the interface. If the normal (direct) traction σ_n is everywhere negative then there is no tendency to separate (which forms one of the Signorini conditions—see Barber's chapter). If we also look at the shear traction, σ_t, and form the ratio $-|\sigma_t|/\sigma_n$ then, providing that ratio is everywhere less than the coefficient of friction, f, the interface will also be stuck and the initial assumptions justified. When the latter condition (alone) is not satisfied slip may be introduced by using an array of glide dislocations (Moore et al. 2018) and in an uncoupled contact such an array would not change the contact pressure distribution but in others it would, and might well cause separation.

Half-Plane Problems

The vast majority of solved contact problems—in the sense that there is a simple, clear algebraic expression of the contact law—the relationship between the applied load and contact size—and a simple description of the contact pressure distribution fall into the category of those where the bodies may be adequately idealised as half-planes.

The starting point of their solution is the Flamant solution using an Airy stress function description of the stress state within a wedge (Barber 2010), particularised to the case where the internal wedge angle is 180°, Fig. 1.3. If the normal load is P and shear force Q (in each case per unit depth into the page), the appropriate Airy function is

$$\phi = -\frac{r\theta}{\pi}\left(P\sin\theta + Q\cos\theta\right) \tag{1.1}$$

from which we may find the stress components by differentiation alone, using the relations

$$\sigma_{rr}\left(r,\theta\right) = \frac{1}{r}\frac{\partial\phi}{\partial r} + \frac{1}{r^2}\frac{\partial^2\phi}{\partial\theta^2} \tag{1.2}$$

$$\sigma_{\theta\theta}\left(r,\theta\right) = \frac{\partial^2\phi}{\partial r^2} \tag{1.3}$$

$$\sigma_{r\theta}\left(r,\theta\right) = -\frac{\partial}{\partial r}\left(\frac{1}{r}\frac{\partial\phi}{\partial\theta}\right), \tag{1.4}$$

giving $\sigma_{r\theta} = \sigma_{\theta\theta} = 0$ everywhere, and

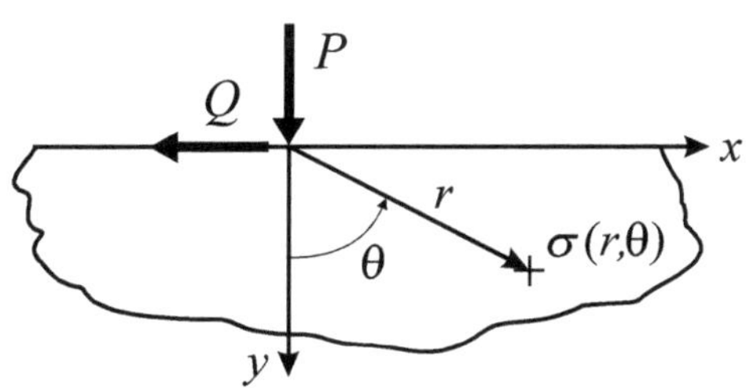

Fig. 1.3 Forces acting on a half-plane

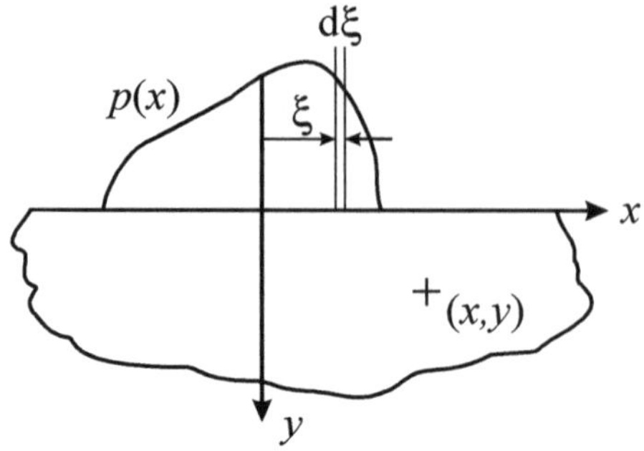

Fig. 1.4 Contact pressure distribution for Flamant's solution

$$\sigma_{rr} = -\frac{2}{\pi r}\left(P\cos\theta - Q\sin\theta\right). \tag{1.5}$$

So, the state of stress induced is very simple: there is only one non-zero component, a radial direct stress acting towards the point of application of the load. In the case of the normal load, P, this takes its maximum compressive value immediately beneath the load and falls smoothly to zero as the surface is approached. With the shear force, Q, the stress is zero beneath the load, and takes maximum compressive value at the surface ahead of the load and maximum tensile value on the surface but 'behind' it. The state of stress in polar coordinates is very simple, but it is not a great deal of practical use. The transformation to a Cartesian set and a shift of origin provide something of very much more utility, as we can then find the state of stress arising beneath distributed surface tractions $(p(x), q(x))$ representing the contact pressure and shear traction distribution, respectively, Fig. 1.4.

The state of stress is given by

$$\sigma_{xx}(x, y) = \frac{-2}{\pi y}\int_{\text{contact}} \frac{p(\xi)(x-\xi)^2 y^2 - q(\xi)(x-\xi)^3 y}{((x-\xi)^2 + y^2)^2}\mathrm{d}\xi \tag{1.6}$$

$$\sigma_{yy}(x, y) = \frac{-2}{\pi y}\int_{\text{contact}} \frac{p(\xi)y^4 - q(\xi)(x-\xi)y^3}{((x-\xi)^2 + y^2)^2}\mathrm{d}\xi \tag{1.7}$$

$$\sigma_{xy}(x, y) = \frac{-2}{\pi y}\int_{\text{contact}} \frac{p(\xi)(x-\xi)y^3 - q(\xi)(x-\xi)^2 y^2}{((x-\xi)^2 + y^2)^2}\mathrm{d}\xi. \tag{1.8}$$

In practice, these integrals may be quite hard to evaluate but, in principle, they may be used to find the internal stress beneath a contact. If the contact is sliding so that $q(x) = \pm f p(x)$ along the whole surface, the symmetry of the results within Eqs. (1.6), (1.7), and (1.8) possesses useful properties.

But, to solve the contact problem itself, we need the displacement field produced by the line forces, so to find this we first find the strain field under conditions of plane strain and then integrate the strains to find the displacements. We specialise the results to the surface with the outcome

$$u(x) = - P\left[\frac{(1-2\nu)(1+\nu)}{2E}\right]\mathrm{sgn}(x) + + Q\left[\frac{2(1-\nu)(1+\nu)}{\pi E}\right]\ln|x| + \frac{c_1(1+\nu)}{E} \tag{1.9}$$

$$v(x) = - P\left[\frac{2(1-\nu)(1+\nu)}{\pi E}\right]\ln|x| + + Q\left[\frac{(1-2\nu)(1+\nu)}{2E}\right]\mathrm{sgn}(x) + \frac{c_2(1+\nu)}{E}, \tag{1.10}$$

where $v(x)$ is the surface normal displacement, $u(x)$ is the surface tangential displacement, E is Young's modulus, and ν is Poisson's ratio. These results merit comment; first, observe that, as we might intuitively expect, the normal force, P, causes a logarithmically varying surface normal depression, Fig. 1.5, with the depth of the depression becoming unbounded as the point of application of the load is approached. Perhaps rather less obvious is that the normal force also causes surface particles to be drawn in, parallel with the surface, towards the loaded point, and with the magnitude of the displacement independent of the position of the particle being considered. Similarly, the shear force causes surface particles to be pushed/drawn along with a magnitude varying logarithmically with position. But note, also, that ahead of the shear force material is pushed down by a constant amount whilst behind it it is raised, forming a step, Fig. 1.5. Lastly, note that each equation incorporates an arbitrary constant so that some datum depth must be chosen at which the displacements are arbitrarily set to zero.

Partly to eliminate the need for an arbitrary datum depth, we develop a contact formulation in terms of the gradient. So, it is straightforward to find the surface slope, $\mathrm{d}v/\mathrm{d}x$, at point x due to an element of pressure, $p(\xi)\mathrm{d}\xi$ at point ξ, but the effect of the shear force needs a little more thought. If the shear force, Q, is smeared over a small length $\mathrm{d}x$ (and $Q = q(x)\mathrm{d}x$), we see that the surface slope is affected only by the surface shear traction *at that point*, and not elsewhere; by contrast the contact pressure at any point influences the slope at all points. Therefore,

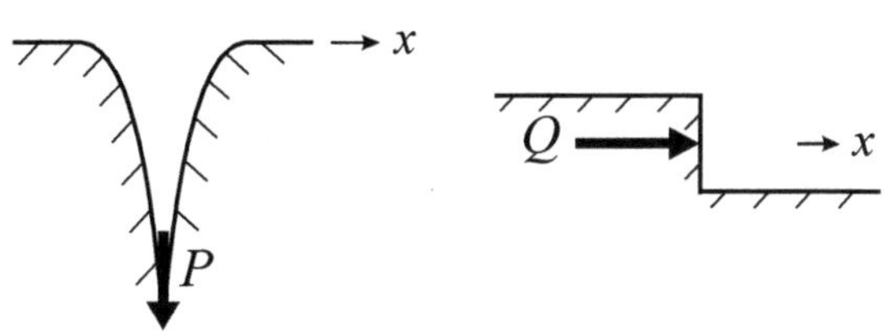

Fig. 1.5 Surface displacements due to surface forces

$$\frac{\mathrm{d}v}{\mathrm{d}x} = \frac{2(1-\nu)(1+\nu)}{\pi E} \int_{\text{contact}} \frac{p(\xi)\mathrm{d}\xi}{x-\xi} - \frac{(1-2\nu)(1+\nu)}{E} q(x). \tag{1.11}$$

Similar results follow immediately for the surface tangential displacement gradient which, we note, is the surface strain

$$\varepsilon_{xx} = \frac{\mathrm{d}u}{\mathrm{d}x} = \frac{(1-2\nu)(1+\nu)}{E} p(x) + \frac{2(1-\nu)(1+\nu)}{\pi E} \int_{\text{contact}} \frac{q(\xi)\mathrm{d}\xi}{x-\xi}. \tag{1.12}$$

The next step is to look at the relative surface normal displacements and relative surface tangential strains when two elastic half-planes, not necessarily having the same elastic properties, are brought into contact. The contact pressure and interfacial shear traction are obviously mutual but we must work in a common coordinate set so that some signs are reversed and we can write down the relative change in surface slope as

$$\frac{1}{A} \frac{\mathrm{d}(v_1 - v_2)}{\mathrm{d}x} = \frac{1}{\pi} \int_{\text{contact}} \frac{p(\xi)\mathrm{d}\xi}{x-\xi} - \beta q(x), \tag{1.13}$$

where A is the composite plane strain compliance (the sum of the reciprocal plane strain moduli) and is explicitly given by

$$A = 2\left\{\frac{1-\nu_1^2}{E_1} + \frac{1-\nu_2^2}{E_2}\right\} = \frac{4}{E^*}, \tag{1.14}$$

where the last term applies when the contacting bodies are elastically similar, and this defines the plane strain elastic modulus of a material, E^*.

The subscripts refer to bodies 1 and 2 as shown in the figure, and β is one of the two Dundurs' constants.[1] The last term in Eq. (1.13) is zero when both bodies have the same elastic constants, E^*, or when

$$\frac{A\beta}{2} = \frac{(1+\nu_1)(1-2\nu_1)}{E_1} - \frac{(1+\nu_2)(1-2\nu_2)}{E_2} \tag{1.15}$$

vanishes. Similarly, the difference in surface strains is given by

$$\frac{1}{A} \frac{\mathrm{d}(u_1 - u_2)}{\mathrm{d}x} = \frac{1}{\pi} \int_{\text{contact}} \frac{q(\xi)\mathrm{d}\xi}{x-\xi} + \beta p(x). \tag{1.16}$$

If either (a) the two bodies are made from the same material (or certain other special combinations such that β vanishes) or (b) the contact is perfectly lubricated so that shear tractions cannot be supported, the two equations are uncoupled and take on these simpler forms

[1]Dundurs (1969) showed that plane elastic problems composed of two bodies depend not on the obvious three dimensionless properties ν_1, ν_2, E_1/E_2 but, in fact, on only two quantities (α, β). Half-plane problems have a further reduced dependence on only the solitary quantity β.

$$\frac{\mathrm{d}(v_1 - v_2)}{\mathrm{d}x} = \frac{A}{\pi}\int_{\text{contact}} \frac{p(\xi)\mathrm{d}\xi}{x - \xi} \tag{1.17}$$

$$\frac{\mathrm{d}(u_1 - u_2)}{\mathrm{d}x} = \frac{A}{\pi}\int_{\text{contact}} \frac{q(\xi)\mathrm{d}\xi}{x - \xi}. \tag{1.18}$$

An example in the use of these equations is to establish the contact pressure distribution when one convex body is pressed into another, using Eq. (1.17). If the form of the contacting profile is symmetric, and we think of the relative separation when the bodies are just touching but no external load is applied as $h(x)$, contact will be made over a symmetrical interval, and we shall denote the ends of the contact by $[-a \quad a]$. Then,

$$\frac{\mathrm{d}(v_1 - v_2)}{\mathrm{d}x} = \frac{\mathrm{d}h}{\mathrm{d}x} = \frac{A}{\pi}\int_{-a}^{a} \frac{p(\xi)\mathrm{d}\xi}{x - \xi} \qquad -a \leq x \leq a. \tag{1.19}$$

It is important to note that the interval of integration and the interval over which the LHS is set to the integral is matched; this is necessary for the integral equation to be properly posed. In setting the gradient of the gap to the shape formed by the contact pressure we are, effectively, ensuring that there is intimate contact within the proposed contact region. And, we are also ensuring that external to the putative region of contact there is no contact pressure. These constitute two of the requirements of contact embodied in the so-called Signorini conditions. The other two have to be verified *a posteriori* and are (a) that the contact pressure is compressive throughout the contact and (b) that external to the contact there is no interpenetration of the surfaces so that, for example, $\mathrm{d}v/\mathrm{d}x > \mathrm{d}h/\mathrm{d}x,\ x > a$.

Inversion of Integral Equations

It is not possible, here, to develop, in full, the inversion procedures for the Cauchy integral equations of the type developed in the last section. They are said to have Cauchy kernels because, when the observation (or collocation) point, x, approaches the integration point, ξ, a '$1/x$' type singularity arises. The inversion of the equations is, in principle, achievable through the evaluation of an integral, but first it is necessary to decide what kind of end-point behaviour is expected. So, the extent of contact is unknown *a priori*, and we seek a 'bounded both ends' solution. If we press a rigid punch having a straight front face width of width $2a$ into an elastic half-plane (which we would also require to be incompressible—$\nu = 1/2$—in order to ensure that Dundurs' second constant vanishes, the size of the contact would be known and we would require a 'singular both ends' solution. We will not consider mixed ends conditions in this brief introductory text, for which see (Hills et al. 1993).

If the integral equation has the generic form

$$\frac{1}{\pi}\int_{-a}^{a}\frac{f(s)\mathrm{d}s}{s-x}=g(x) \qquad -a\leq x\leq a, \tag{1.20}$$

its solution is given by

$$f(x)=-\frac{w(x)}{\pi}\int_{-a}^{a}\frac{g(s)\mathrm{d}s}{w(s)(s-x)}+Cw(x), \tag{1.21}$$

where, if we seek a 'singular both ends' solution

$$w(x)=\frac{1}{\sqrt{a^2-x^2}} \qquad C\neq 0, \tag{1.22}$$

whereas, if we seek a 'bounded both ends' solution

$$w(x)=\sqrt{a^2-x^2} \qquad C=0, \tag{1.23}$$

and, in addition, we must enforce the consistency condition

$$\int_{-a}^{a}\frac{g(s)\mathrm{d}s}{w(s)}=0. \tag{1.24}$$

When these integrals are analytically intractable numerical integration procedures to solve these equations very efficiently are available, and the key reference is an extended paper by Erdogan et al. (1973).

Hertz' Problem

As an example application consider the problem of pressing together two elastically similar circular cylinders, of radii R_1, R_2, having their axes parallel, as shown in Fig. 1.6. The circular profiles may be approximated by parabolae, as may be seen by applying Pythagoras' theorem and taking the first term in a binomial expansion, so that

$$2Rh(x)\simeq x^2, \tag{1.25}$$

where

$$\frac{1}{R}=\frac{1}{R_1}+\frac{1}{R_2}, \tag{1.26}$$

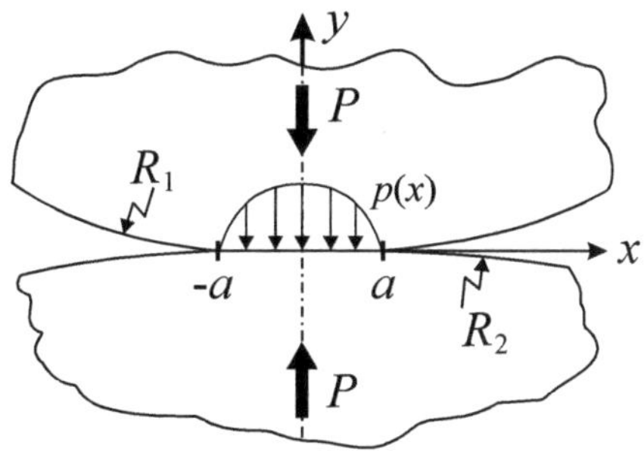

Fig. 1.6 The Hertz contact problem

and hence

$$\frac{\mathrm{d}h}{\mathrm{d}x} \simeq \frac{x}{R}. \tag{1.27}$$

Substitution of this geometric result into Eq. (1.20) gives

$$\frac{x}{R} = \frac{A}{\pi} \int_{-a}^{a} \frac{p(\xi)\mathrm{d}\xi}{x - \xi} \qquad -a \leq x \leq a, \tag{1.28}$$

and the inversion procedure gives

$$p(x) = -\frac{1}{\pi AR}\sqrt{a^2 - x^2} \int_{-a}^{a} \frac{r\mathrm{d}r}{\sqrt{a^2 - r^2}(r - s)} = -\frac{\sqrt{a^2 - x^2}}{AR}. \tag{1.29}$$

At the moment this is in an unfamiliar form. But, if we write down the normal equilibrium condition, viz.

$$P = \int_{-a}^{a} p(x)\mathrm{d}x, \tag{1.30}$$

and substitute in this result, we find that

$$a^2 = \frac{2APR}{\pi}, \tag{1.31}$$

so that

$$p(x) = -\frac{2P}{\pi a}\sqrt{1 - (x/a)^2}. \tag{1.32}$$

Further Examples and Summary

In the case of a rigid, square-ended punch pressed onto an incompressible half-plane we seek a 'singular both ends' solution, and, as the front face is flat, $\mathrm{d}h/\mathrm{d}x = 0$. Therefore,

$$\frac{A}{\pi}\int_{-a}^{a}\frac{p(r)\mathrm{d}r}{s-r} = 0 \qquad -a \leq x \leq a, \tag{1.33}$$

so that

$$p(x) = \frac{C}{\sqrt{a^2-x^2}}, \tag{1.34}$$

and the imposition of normal equilibrium gives the value of the constant, C, so that

$$p(x) = \frac{P}{\pi\sqrt{a^2-x^2}}. \tag{1.35}$$

Figure 1.7 provides a comparison of the profiles of common indenters against the corresponding resulting pressure distribution. The values for cylinders and a flat punch, already analysed, are included, together with a shallow wedge. Care must be taken when evaluating the integral within the interval $(-a \leq x \leq a)$ when it is Cauchy, and external to that interval, when it is regular.

Lastly, caution is urged when one of the bodies has a profile where other features are present over a length scale not big compared with the contact half-width.

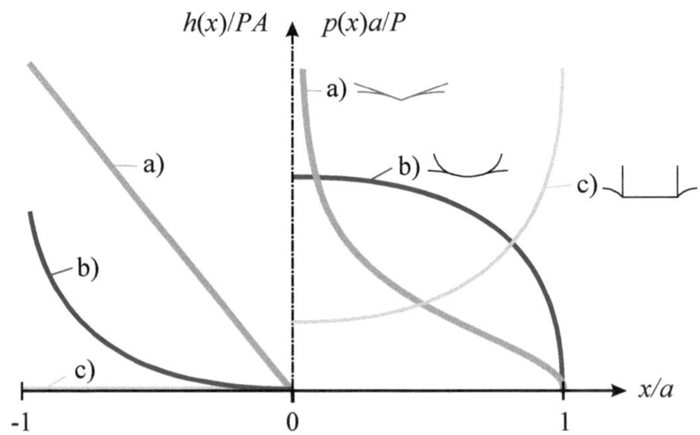

Fig. 1.7 Comparison of different indenter profiles and resulting pressure distributions for a) a wedge, b) Hertzian contact, and c) a complete contact

Examples include internal corners at, for example, firtree root corners and external corners, for example, in the case of a railway line head where the line head width is only 2–3 times the contact size.

Complete Contacts

Because it is not possible to form simple closed-form elasticity solutions to shapes such as rectangles, except by series representation, it is equally impossible to produce simple recipes for the state of stress for complete contacts generally, even those where the bodies have simple shapes, such as that shown in Fig. 1.8. It is inevitable, therefore, that numerical methods, such as the finite element method, will have to be used. But, we can obtain at least some useful information in the neighbourhood of the corners of these contacts, and it is these regions which are often of the most practical interest, as it is at or near them that there is the greatest possibility of fretting, or where cracks may start. If we 'zoom in' with a microscope so that the field of view includes just the surfaces near the contact edge, Fig. 1.8, we see something which is simply two wedges—one a half-plane and the other, in this case, a quarter plane. The bodies may be locally separated, or in contact but slipping, or adhered, depending on the loading on the bodies overall, but we shall start off with the simplest assumption—that they are, at the edge, in intimate contact and adhered, so that they may, together, be thought of as a monolithic three-quarter plane. This is simply a special case of a wedge of included angle 2α with free surfaces, and was studied in detail by Williams in a very celebrated solution which we now present.

Williams' Solution

Figure 1.9 shows a semi-infinite wedge of internal angle 2α, and with a polar axis set centred on its apex, and our aim is to state what we can about the stress field locally (before the features outside our field of view become important). We will assume a

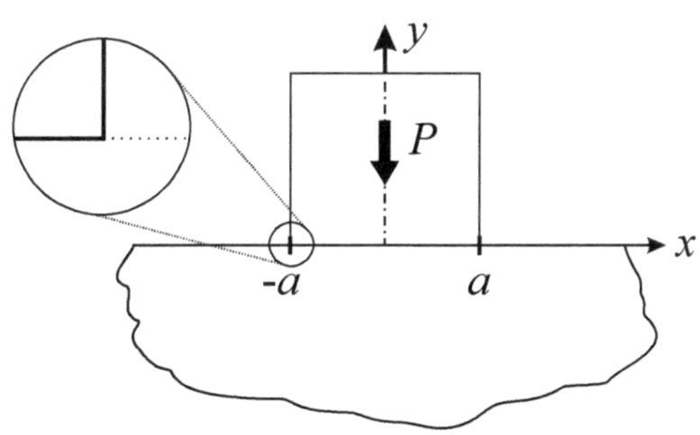

Fig. 1.8 Square block on half-space forming a complete contact

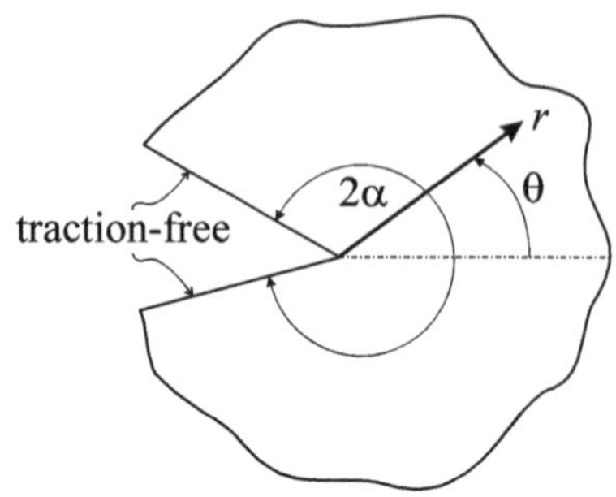

Fig. 1.9 Axis set used in elastic wedge solution

variable separable type solution, and we will initially assume that the state of stress varies with r in a power order manner. For consistency with most solutions published in the literature, we will actually assume that $\sigma_{ij} \sim r^{\lambda-1}$, and as the corresponding Airy stress function in polar form (see section "Half-Plane Problems") is seen to be two orders of r higher, we assume a solution of the form

$$\phi = r^{\lambda+1} f(\theta). \tag{1.36}$$

When this is substituted into the biharmonic equation it is seen that the variation of the solution with r assumed is acceptable and it turns the partial differential equation into the following ordinary differential equation

$$\left(\frac{d^2}{d\theta^2} + (\lambda+1)^2\right)\left(\frac{d^2}{d\theta^2} + (\lambda-1)^2\right) f = 0. \tag{1.37}$$

For $\lambda \neq \pm 1$, the four possible solutions to this equation correspond to the following form for the Airy stress function

$$\begin{aligned}\phi = & r^{\lambda+1} \left(A_1 \cos(\lambda+1)\theta + A_2 \sin(\lambda+1)\theta + \right.\\ & \left. + A_3 \cos(\lambda-1)\theta + A_4 \sin(\lambda-1)\theta\right).\end{aligned} \tag{1.38}$$

When we evaluate the derivatives needed we see that the corresponding state of stress is given by

$$\begin{aligned}\sigma_{rr} = & \lambda r^{\lambda-1} \left[-A_1 (\lambda+1) \cos(\lambda+1)\theta - A_2 (\lambda+1) \sin(\lambda+1)\theta - \right.\\ & \left. - A_3 (\lambda-3) \cos(\lambda-3)\theta - A_4 (\lambda-3) \sin(\lambda-3)\theta\right]\end{aligned} \tag{1.39}$$

$$\begin{aligned}\sigma_{\theta\theta} = & \lambda r^{\lambda-1} \left[A_1 (\lambda+1) \cos(\lambda+1)\theta + A_2 (\lambda+1) \sin(\lambda+1)\theta + \right.\\ & \left. + A (\lambda+1) \cos(\lambda-1)\theta - + A (\lambda+1) \sin(\lambda-1)\theta\right]\end{aligned} \tag{1.40}$$

$$\sigma_{r\theta} = \lambda r^{\lambda-1}\left[A_1(\lambda+1)\sin(\lambda+1)\theta - A_2(\lambda+1)\cos(\lambda+1)\theta + \right.$$
$$\left. + A_3(\lambda-1)\sin(\lambda-1)\theta - A_4(\lambda-1)\cos(\lambda-1)\theta\right]. \tag{1.41}$$

The requirements that the free boundaries of the wedge be devoid of tractions means that

$$\sigma_{\theta\theta} = \sigma_{r\theta} = 0 \qquad \theta = \pm\alpha. \tag{1.42}$$

When this is applied to the second and third equations above, and then sums and differences are taken to tidy up the appearance of the equations, we find that

$$A_1(\lambda+1)\sin(\lambda+1)\alpha + A_3(\lambda-1)\sin(\lambda-1)\alpha = 0 \tag{1.43}$$

$$A_1(\lambda+1)\cos(\lambda+1)\alpha + A_3(\lambda-1)\cos(\lambda-1)\alpha = 0 \tag{1.44}$$

$$A_2(\lambda+1)\cos(\lambda+1)\alpha + A_4(\lambda-1)\cos(\lambda-1)\alpha = 0 \tag{1.45}$$

$$A_2(\lambda+1)\sin(\lambda+1)\alpha + A_4(\lambda-1)\sin(\lambda-1)\alpha = 0, \tag{1.46}$$

or, we may write these in matrix form, where the first pair of homogeneous equations corresponds to a symmetric solution

$$\begin{bmatrix} (\lambda+1)\sin(\lambda+1)\alpha & (\lambda-1)\sin(\lambda-1)\alpha \\ (\lambda+1)\cos(\lambda+1)\alpha & (\lambda-1)\cos(\lambda-1)\alpha \end{bmatrix} \begin{Bmatrix} A_1 \\ A_3 \end{Bmatrix} = \begin{Bmatrix} 0 \\ 0 \end{Bmatrix}, \tag{1.47}$$

and the second pair to an antisymmetric solution

$$\begin{bmatrix} (\lambda+1)\cos(\lambda+1)\alpha & (\lambda-1)\cos(\lambda-1)\alpha \\ (\lambda+1)\sin(\lambda+1)\alpha & (\lambda-1)\sin(\lambda-1)\alpha \end{bmatrix} \begin{Bmatrix} A_2 \\ A_4 \end{Bmatrix} = \begin{Bmatrix} 0 \\ 0 \end{Bmatrix}. \tag{1.48}$$

Each pair of equations will possess a solution if, and only if, the determinant of the matrix vanishes, and these characteristic equations, when expanded out, are

$$\lambda \sin 2\alpha \pm \sin 2\lambda\alpha = 0, \tag{1.49}$$

where the $+$ sign is adopted for the symmetric solution and the $-$ sign for the antisymmetric solution. Solutions to these equations may be found by numerical methods. For either sign, there are many solutions, and each corresponds to a valid eigensolution of the wedge problem posed. We should not be surprised by this because we have specified the solution to the problem displayed in Fig. 1.9 only by requiring the free surfaces to be just that—traction free.

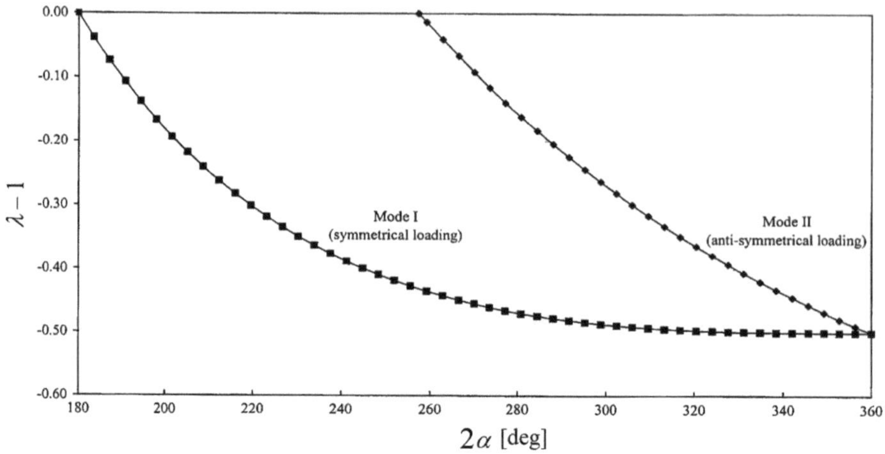

Fig. 1.10 Williams' solution eigenvalues $\lambda - 1$

We have made no assertions at all about how the region might be loaded. Thus, we can expect to find a series of physically acceptable answers to this very idealised problem. What we wish to do is to look at the dominant solution, that is, the one which would dwarf all others as the observation point approached the wedge apex. Therefore, we wish to find the value of λ which is 'smallest' (furthest to the left on a directed number scale), and there will be two such values of interest, one for symmetric loading and one for antisymmetric loading. We recall that the state of stress corresponding to a particular value of λ will vary as $r^{\lambda-1}$, and therefore plot, in Fig. 1.10, the relevant values of $\lambda - 1$ found. For symmetric loading, we see that there is no singular eigensolution for all external corners ($2\alpha < 180°$), and that as the value of α is increased the value of the smallest eigenvalue becomes smaller still, until when $2\alpha = 360°$, and the notch is folded rounded to form a semi-infinite crack, λ ($= \lambda_I$ say) $= 1/2$, and the state of stress is square root singular. Turning to the antisymmetric solution, we see that the dominant term remains bounded (non-singular) not only for all external corners but also for modest internal corners, until 2α approaches 257.4°. Further increase in α cause the value of λ_{II} to become smaller (but always greater than λ_I), until a crack results when this solution, too, becomes square root singular in terms of stress state.

Thus, for any value of α, we can see how quickly the dominant symmetric and antisymmetric solutions decay with radius as the observation point moves away from the wedge apex. But we can go rather further. If we back substitute the value of λ_I into Eq. (1.47) and multiply out either equation, we can effectively determine the ratio A_1/A_3 so, for mode I loading, for example, we can rewrite the state of stress in the following form:

$$\left\{\begin{array}{c}\sigma_{r\theta}\\ \sigma_{\theta\theta}\\ \sigma_{rr}\end{array}\right\} = \frac{A_3\lambda_I(\lambda_I+1)}{r^{1-\lambda_I}}\left\{\begin{array}{c}\frac{A_1}{A_3}\sin(\lambda_I+1)\theta+\sin(\lambda_I-1)\theta\\ \frac{A_1}{A_3}\cos(\lambda_I+1)\theta+\cos(\lambda_I-1)\theta\\ -\frac{A_1}{A_3}\cos(\lambda_I+1)\theta-\cos(\lambda_I-1)\theta\end{array}\right\}. \quad (1.50)$$

This makes it clear that the spatial variation of stress, for each value of λ, is fully specified. The variation with both r and θ is fully specified. For any value of λ, there is only one degree of freedom in the solution, and that is the overall multiplier (here A_3) which must be found, and its value is determined by collocating this semi-infinite solution into whatever finite problem is being studied. Note that this symmetrical solution induces no shear traction, $\sigma_{r\theta}$, on the bisector ($\theta = 0$) line, and similarly the antisymmetric solution induces no direct stress (neither σ_{rr} nor $\sigma_{\theta\theta}$) on that line, so that they effectively uncouple along it. The multipliers on the solution are known as generalised stress intensity factors (generalised, because the nomenclature is more particularly used for the special case of a crack tip), and are usually given the symbols K_I, K_{II}, and we may therefore write the state of stress corresponding to the dominant values of λ in the form

$$\sigma_{ij}(r,\theta) = K_I r^{\lambda_I - 1} g_{ij}^I(\theta) + K_{II} r^{\lambda_{II}-1} g_{ij}^{II}(\theta), \quad (1.51)$$

where the functions $g_{ij}^k(\theta)$ are the eigenvectors, as described above, scaled so that, along the bisector line, $g_{\theta\theta}^I(0) = 1$ and $g_{r\theta}^{II}(0) = 1$. With these definitions, the multipliers for any given problem may be found by looking at the stress state along $\theta = 0$ developed by the external loads on the component, and finding[2]

$$K_I = Lt_{r\to 0} r^{1-\lambda_I}\sigma_{\theta\theta}(r) \quad (1.52)$$

and

$$K_{II} = Lt_{r\to 0} r^{1-\lambda_{II}}\sigma_{r\theta}(r). \quad (1.53)$$

A further class of solutions may be developed when the wedge is subject to antisymmetric loading but this will not be covered here. It, too, becomes singular when 2α exceeds π, and, in terms of order of singularity, always lies between the λ_I, λ_{II} curves, $\lambda_{III} - 1$ becoming 1/2 when 2α reaches 2π radians.

Contact Edges

Figure 1.11 shows the problem of a square elastic block, of side $2a$, pressed onto an elastically similar half-plane by a normal force, P, together with a shear force, Q

[2]Note that, in defining crack tip stress intensity factors, for historical reasons the stress intensity factors have an extra factor of $\sqrt{2\pi}$ included in both Eq. (1.51) in the denominator, and in the right hand side of the definitions of stress intensity factor, Eqs. (1.52) and (1.53).

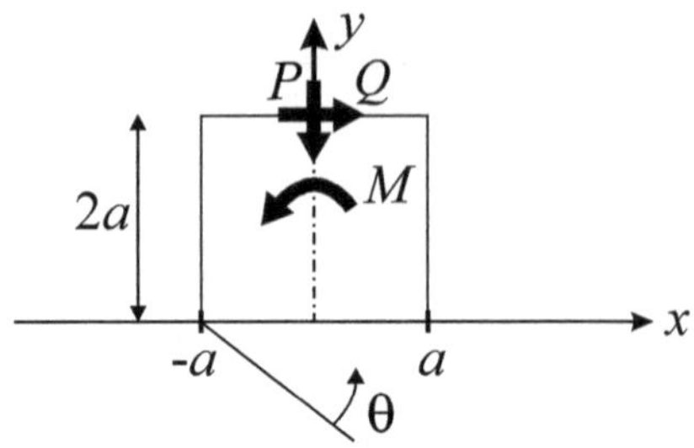

Fig. 1.11 Square block on half-plane subject to a normal and shear force

which we shall actually apply along the upper surface but with an applied moment, $M = 2aQ$ so as to make it statically equivalent to application through the plane of contact.

We do this so that local effects associated with the point of application of the force do not abruptly affect the state of stress along the interface, but have a 'smooth' influence there. What we wish to do is to find out the conditions along the interface for various ratios of Q/P—to see if intimate contact is maintained, and to see if there are any regions of slip. Intuition tells us that, if the shear force acts to the right, the likely first point of slip and separation might be the left hand corner, and so it is this which we examine in some detail. As stated at the beginning of this section, we assume that it is both in intimate contact and adhered, so that locally the corner behaves like a wedge of include angle $3\pi/2$, and we set up a local polar axis set there with θ measured from the 'notch' bisector. Using the results found in the previous section, and taking $2\alpha = 3\pi/2$, we see that we can write down the local state of stress in the form

$$\sigma_{\theta\theta}(r, \pi/4) = K_I r^{\lambda_I - 1} g^I_{\theta\theta}(\pi/4) + K_{II} r^{\lambda_{II} - 1} g^{II}_{\theta\theta}(\pi/4), \tag{1.54}$$

$$\sigma_{r\theta}(r, \pi/4) = K_I r^{\lambda_I - 1} g^I_{r\theta}(\pi/4) + K_{II} r^{\lambda_{II} - 1} g^{II}_{r\theta}(\pi/4), \tag{1.55}$$

and hence the ratio of the shear to the direct traction is simply

$$\frac{\sigma_{r\theta}(r, \pi/4)}{\sigma_{\theta\theta}(r, \pi/4)} = \frac{K_I r^{\lambda_I - 1} g^I_{\theta\theta}(\pi/4) + K_{II} r^{\lambda_{II} - 1} g^{II}_{\theta\theta}(\pi/4)}{K_I r^{\lambda_I - 1} g^I_{r\theta}(\pi/4) + K_{II} r^{\lambda_{II} - 1} g^{II}_{r\theta}(\pi/4)}. \tag{1.56}$$

For a three-quarter plane, the eigenvalues are $\lambda_I = 0.5445$ and $\lambda_{II} = 0.9085$, so that the symmetric term is only just less than square root singular, whereas the antisymmetric term is only quite weakly singular, and therefore regardless of the values of K_I, K_{II} it will always be possible to choose observation points which are sufficiently close to the contact edge for the symmetric terms to dominate and we can see that

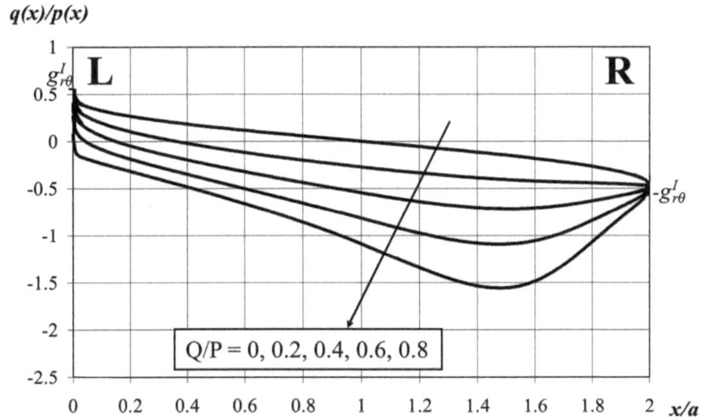

Fig. 1.12 Traction ratio along the interface for various ratios Q/P

$$\frac{\sigma_{r\theta}(r, \pi/4)}{\sigma_{\theta\theta}(r, \pi/4)} \to \frac{g^I_{\theta\theta}(\pi/4)}{g^I_{r\theta}(\pi/4)} = 0.543. \tag{1.57}$$

Thus, providing that the coefficient of friction is at least this big, and the sign of $K_I g^I_{\theta\theta}(\pi/4) < 0$ so that intimate contact is maintained, the corners of the contact will stick. The sign of this ratio implies that the square block will tend to 'spread' as it is pushed into contact. Note that we have found this result without any need to calibrate the asymptote—no values have been ascribed to K_I, K_{II}. Also, it should be appreciated that the individual traction components become infinitely big as the contact corner is approached ($\sigma_{ij} \sim r^{-0.45}$), but their ratio is finite. A finite element study of the problem shown in Fig. 1.11 was carried out, and the two components were assumed to be bonded together along the interface. The implied traction ratio along the interface, for various ratios Q/P is shown in Fig. 1.12. Convergence of the model is imperfect at the edges but convincingly converges at the ratio derived above. For low values of Q/P, the magnitude of the traction ratio at interior points is less than at the ends, so that a coefficient of friction of 0.543 guarantees adhesion along the whole interface. Above $Q/P \simeq 0.3$, the traction ratio at internal points exceeds that at the edges, so that slip from an interior point will start first, unless the coefficient of friction is high enough. For example, if $f \approx 0.7$ and $Q/P \approx 0.4$ slip will start from an interior point (roughly a quarter point), and then further increase in the shear force will cause the asymptotic solution to be eroded from an interior point, towards the apex. The solution and the arguments above hinge on there being a hinterland of adhered material.

The point of first separation is the trailing edge of contact. The value of the shear force which causes this may be determined directly by looking at the output from the finite element analysis but will lack precision. A better way is to use the finite element results to calibrate the generalised 'notch' intensity factors, with the result

$$\left\{ \begin{array}{l} K_I \, a^{\lambda_I - 1} \\ K_{II} \, a^{\lambda_{II} - 1} \end{array} \right\} = \begin{bmatrix} -0.157 & +0.179 \\ -0.130 & -0.274 \end{bmatrix} \left\{ \begin{array}{l} P/2a \\ Q/2a \end{array} \right\}. \tag{1.58}$$

Note that, because the local solution is referred to the 'notch' bisector, both elements of loading (P, Q) excite both eigensolutions (K_I, K_{II}) and, in turn, each eigensolution contributes to both traction components ($\sigma_{\theta\theta}, \sigma_{r\theta}$) arising along the contact interface. So, the condition for edge closure ($K_I < 0$) is found by expanding the upper row and closure is maintained if $Q/P < 0.877$. This calculation assumes, of course, that there is no slip, and so is only precisely accurate when the slip zone is both small and remote from the edge, but in practice it is an extremely accurate predictor, much more precise than a direct interpretation of the finite element output, without collocation of asymptotic forms, would allow.

We have established the condition for adhesion at edges of the contact when normal contact is first made. The question arises 'What happens when this inequality is not satisfied, i.e $f < 0.543$?' In order to answer this question convincingly, we need to conduct a new eigenvalue analysis—one where, instead of the half-plane and the wedge being bonded together to form a new wedge, in this case of internal angle 270°, we instead look at what happens when a quarter plane (or, more generally, a semi-infinite wedge) slides with friction over an elastically similar half-plane. There is insufficient space in this brief introductory text to look at this in detail, and for a full explanation the original papers by Comninou (1976) should be looked at, together with papers in which their results are applied to a range of practical problems, including this one (Churchman and Hills 2006a). In essence, two series representations of the kind described above for the Williams solution are put together, one for each body and then, in addition to ensuring that the faces are traction free, conditions are written down for the interface which ensure continuity of displacement and direct traction in the $\theta-$direction, together with a requirement that the shear traction be a constant (the coefficient of friction) multiplied by the normal traction, at all points. It should be noted that, in connection with this, we give a sign to the coefficient of friction, f. It is positive at a trailing edge (where the outwards normal faces the opposite direction to the velocity of the contact defining body (the quarter plane), and is negative at the leading edge, where the outward normal to the contact defining body has the same sense as the body's velocity).

When the contact is formed both edges tend to slip outwards, and they are both, therefore, like leading edges. If a shear force is applied which acts to the right, the RH edges suffer a change in shear traction which is in the same sense as when the contact was formed—it continues to be a leading edge—and therefore the slip zone simply grows. It is at the trailing edge where changes arise, because what was formerly a leading edge now becomes trailing, and so the change in magnitude of the shear traction (a reduction) is now in the opposite sense to the slip displacement giving rise to instantaneous stick. What happens next can be found from the asymptotics. We state, without proof, that at the leading edge of frictional sliding contact between a quarter plane and an elastically similar half-plane, the tractions are given by an eigensolution of the form

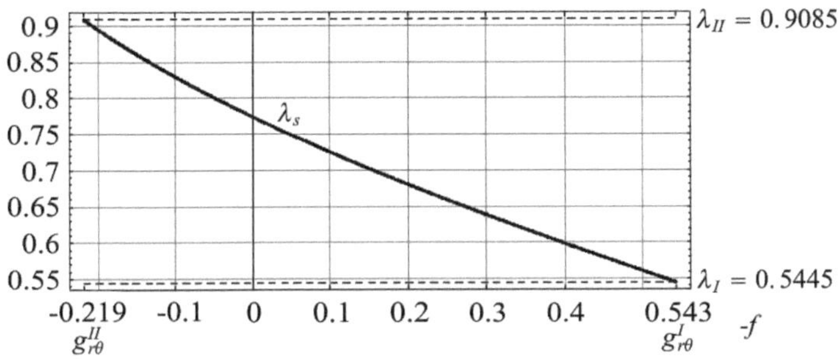

Fig. 1.13 Eigenvalues λ for a sensible range of values of the coefficient of friction

$$\frac{q(s)}{f} = p(s) = K_S s^{\lambda_s - 1}, \tag{1.59}$$

where λ_s is shown in Fig. 1.13 for a sensible range of values of the coefficient of friction. This calculation is relevant only where the coefficient of friction is less than the naturally occurring traction ratio at the adhered contact corner, and it will be noted that, in these cases, $\lambda_S > \lambda_I$. It follows that the slipping corner shows a singular behaviour which is more weakly singular than an adhered corner. Therefore, when we now exert a finite shear force the change in solution corresponds to the adhered eigensolution, but, because this is more strongly singular than the slipping one, there will be local separation accompanied by a reversal of the slip direction adjacent to the point of separation, i.e. we will now have

$$p(s) = K_S s^{\lambda_S - 1} + \Delta K_I g_{\theta\theta}^{I}(\pi/4) s^{\lambda_I - 1} + \Delta K_{II} g_{\theta\theta}^{II}(\pi/4) s^{\lambda_{II} - 1} \tag{1.60}$$

$$q(s) = f K_S s^{\lambda_S - 1} + \Delta K_I g_{r\theta}^{I}(\pi/4) s^{\lambda_I - 1} + \Delta K_{II} g_{r\theta}^{II}(\pi/4) s^{\lambda_{II} - 1}, \tag{1.61}$$

where K_S is the multiplier on the slipping eigensolution. The order of strength of the terms is—weakest K_{II}, middle K_S and strongest K_I. The actual length of separation is tiny in comparison with the contact half-width (Churchman and Hills 2006b).

A summary of the results found in this family of calculations is provided in Fig. 1.14. It should be noted that, for practical purposes, all contacts of this class become fully stuck as, even, when the coefficient of friction is too small to ensure adhesion to the contact edge in the first cycle of loading, some shaking down to a practically stuck state (save for a very, very small region adjacent to the contact edge) will occur. The exception is when the friction is sufficiently high to inhibit all edge slip but the application of a shear load cause slip to start from a point well within the contact. In such cases, though, the island of slip is surrounded by adhered material, limiting the slip displacement.

We turn, now, briefly, to a consideration of what happens when we have a square-ended elastic body sliding over an elastically similar half-plane, in the presence of

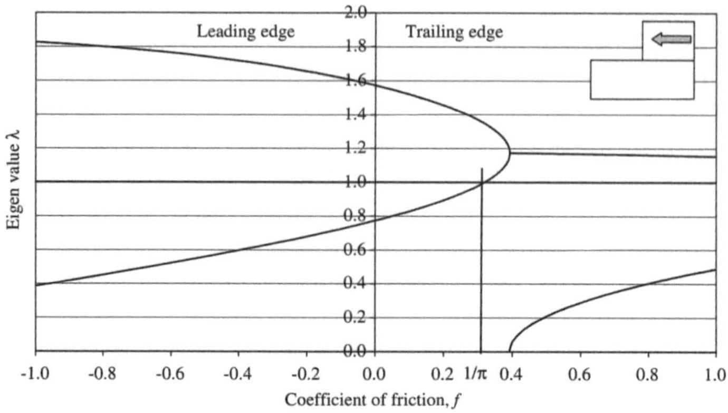

Fig. 1.14 Dominant eigenvalue for a rectangular block sliding on an elastically similar half-plane. Note at RH edge eigenvalue becomes complex and real and imaginary parts are displayed

friction. We have mentioned that an asymptotic solution exists for a quarter plane sliding over an elastically similar half-plane, and now show the dominant eigenvalue for the problem as a function of the coefficient of friction, Fig. 1.14.

This shows that, as the coefficient of friction is increased, the order of the singular behaviour increases at the leading edge. At the trailing edge, however, increasing the coefficient of friction reduces the order of the singularity, and, when f exceeds $1/\pi$ (about 0.318) λ_S exceeds unity, so that the state of stress becomes bounded. A further increase in the coefficient of friction, to value just a little less than 0.4, (see Fig. 1.14) causes the value to become complex, and this implies separation of the surfaces—very locally this becomes an incomplete contact so that the contact pressure becomes square root bounded. The results of a finite element analysis of the whole square-ended contact, and further results for shapes other than rectangular pads are given in the papers by Karuppanan (2008).

Lastly, we ask what the transition in behaviour is between these two classes of problem—a stationary contact and a sliding one—if we first apply a normal load alone, and then gradually increase it until the body slides. As might be expected from our consideration of a stationary contact there are two kinds of behaviour, depending on whether the coefficient of friction is sufficient to maintain adhesion to the contact edges upon application of a normal load alone, and the results are summarised in Fig. 1.15. For the case when the friction is high enough to maintain full stick everywhere, we recall that the asymptotes for corner behaviour can only be eroded from the hinterland—an interior point—and extend outwards until finally the corners slip, and this is exactly what we observe. With a coefficient of friction of 0.8 the whole contact remains stuck until the shear force is almost 60% of the value needed to causes sliding, and then the quarter point nearer to the leading edge starts to slip, Fig. 1.15. Slip breaks through to the leading edge when the shear force is about three-quarters of that needed to cause sliding, and then the very last point to

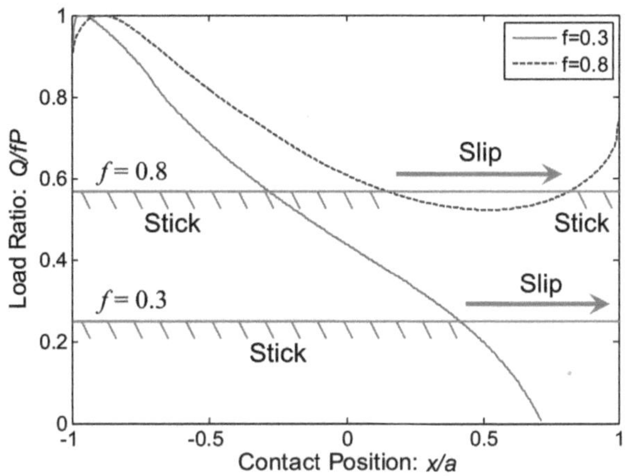

Fig. 1.15 The evolution of contact behaviour during transition from normal load to sliding for two coefficients of friction

slip is at the trailing edge. The problem is quite hard to maintain numerically stable when sliding is approached.

On the other hand, when the coefficient of friction is 0.3, slip zones of opposite sign, of approximate extent one-quarter of the half-width, are established on formation of the contact. An infinitesimal shear force causes the trailing edge to stick, and there will then be a minute separation region, too small to resolve in the figure. But the leading edge slip zone increases smoothly in size as the shear force is increased, exceeding half of the width of the contact when the shear force is about 40% of the sliding value and, as with the higher coefficient of friction, the last point to slip is at the trailing edge.

Incomplete Contacts in Partial Slip

Contacts may slide, or they may roll, or they may be stationary, meaning that there is no rigid-body movement between the contacting bodies. But it does not follow that, for the last class, there are no regions of slip. In this section, we will introduce the partial slip solution for contacts which are capable of being represented using half-plane theory, where slip is resisted by pointwise Coulomb friction, and where the bodies are made from the same material. The most general kind of problem of this kind we can have is shown in Fig. 1.16, where the contact is subject to a normal load, P, moment, M, shear force, Q, and tensions parallel with the free surface whose difference is σ—we shall consider the last to be present only in body B, for simplicity. In the most general case, all of these effects may be present, and they may all vary in some general way with time. The first two affect only the normal load problem, and induce no shear tractions while the last two induce only shear tractions and induce no change in the contact pressure, as has been shown (section "Half-Plane Problems").

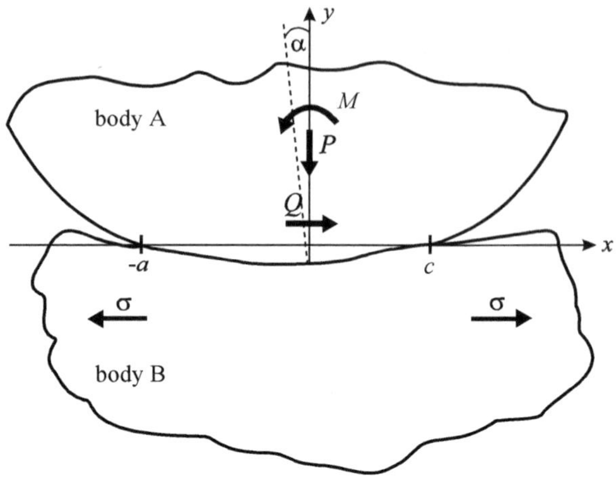

Fig. 1.16 Generic half-plane contact subject to P, M, Q, and σ

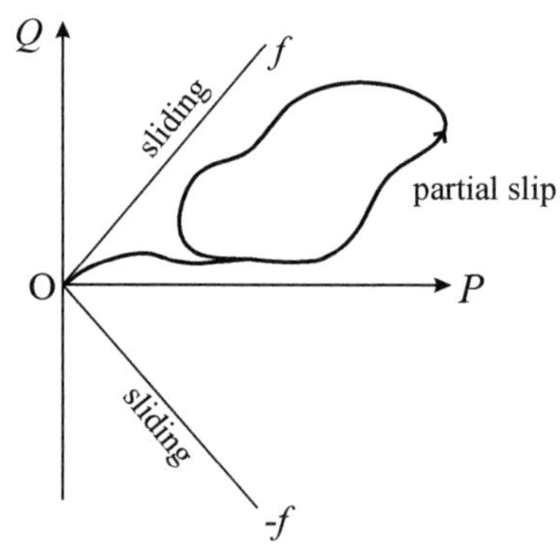

Fig. 1.17 Load history displayed in P-Q space

To start with, take the simpler problem where only the normal load and shear force are present. Note that the behaviour of this kind of problem depends not only where we are in $P - Q$ space, Fig. 1.17, but how we got there, and first we will find the condition for the contact being in a fully stuck state. In section "Further Examples and Summary", it was shown that a constant normal displacement over a patch $[-a \quad a]$ is produced by a pressure distribution of the form

$$p(x) = \frac{P}{\pi\sqrt{a^2 - x^2}} \tag{1.62}$$

and, as the integral equations relating normal surface displacement to contact pressure and tangential surface displacement to shear tractions are the same, it follows immediately that a shear tractions distribution

$$q(x) = \frac{Q}{\pi\sqrt{a^2 - x^2}} \tag{1.63}$$

corresponds to a constant surface tangential displacement. Suppose that we have already formed a contact whose extent is currently $[-a \quad a]$ and we impose a small increase in normal load ΔP and a small change in shear force ΔQ. If we assume that this element of loading leads to a wholly stuck contact there will be no gradient in the tangential displacement within the contact, so that we may expect the changes in tractions to be given by the equations above, with P replaced by ΔP and Q replaced by ΔQ. The changes in traction distributions are therefore similar and, regardless of whether, before this change was made, the contact was in a sliding, partial slip, or adhered state, and it will now be stick everywhere provided that (Barber et al. 2011)

$$\frac{\Delta |Q|}{\Delta P} < f. \tag{1.64}$$

Upon unloading slip will always result unless the unloading trajectory follows the loading trajectory.

A consequence of the result just found is that, if the normal load is applied and held constant, a change in the shear force will produce some partial slip, and we shall now look at this case in more detail. This kind of problem was originally studied first by Cattaneo (1938) and subsequently by Mindlin (1949) for a Hertzian contact, but Ciavarella (1998) and Jäger (1998) subsequently showed that the ideas apply equally to any geometry of contact. Consider, first, forming the normal contact. It was shown in section "Half-Plane Problems" that

$$E^* \frac{\mathrm{d}h}{\mathrm{d}x} = \frac{4}{\pi} \int_{-a}^{a} \frac{p(\xi)\mathrm{d}\xi}{x - \xi} \qquad -a \le x \le a, \tag{1.65}$$

where $E^*$is the plane strain elastic modulus and $\mathrm{d}h/\mathrm{d}x$ is the 'overlap' (or gap) function denoting the front-face relative profile of the contacting bodies, and a is the contact half-width. This would normally be inverted to discover the contact pressure distribution, $p(x)$, and then the contact law found by exerting overall normal equilibrium by setting

$$P = \int_{-a}^{a} p(x)\mathrm{d}x. \tag{1.66}$$

When the contact is formed, strains develop in the surfaces of both bodies parallel with the free surface—but, as they are elastically similar and the pressure experienced mutual, these strains are equal so that no shear tractions develop. When an increasing shear force is applied, this will be resisted by the development of shear tractions along the interface. Intuitively, we expect the central portion, where the contact pressure is highest, to remain stuck, and so in this region, we expect the relative normal strains to be preserved, at zero. Towards the edges of the contact, as the pressure falls smoothly to zero, we expect slip zones to be present and limiting friction to be attained. To represent this state, assume that the central stick patch occupies the region $[-b \quad b]$, and write the shear traction down as the sliding distribution ($q(x) = fp(x)$) over the whole contact, together with an, as yet unknown, corrective distribution within the stick interval. The relative surface direct strain, ε_{xx},we write down in terms of the following integral equation, which is analogous to the normal integral equation:

$$E^* \frac{\mathrm{d}\varepsilon_{xx}}{\mathrm{d}x} = \frac{4}{\pi} \int_{\text{contact}} \frac{q(\xi)\mathrm{d}\xi}{x-\xi} = \frac{4f}{\pi} \int_{-a}^{a} \frac{p(\xi)\mathrm{d}\xi}{x-\xi} + \frac{4}{\pi} \int_{-b}^{b} \frac{q'(\xi)\mathrm{d}\xi}{x-\xi}. \tag{1.67}$$

But we know that $\varepsilon_{xx} = 0$ when $-b \le x \le b$, so that

$$0 = \frac{4f}{\pi} \int_{-a}^{a} \frac{p(\xi)\mathrm{d}\xi}{x-\xi} + \frac{4}{\pi} \int_{-b}^{b} \frac{q'(\xi)\mathrm{d}\xi}{x-\xi} \qquad -b \le x \le b, \tag{1.68}$$

and by comparison with the first integral, and also bearing in mind that we wish to look at values of x in the range $-a < b < x < b < a$, we see that

$$-E^* \frac{\mathrm{d}h}{\mathrm{d}x} = \frac{4}{\pi} \int_{-b}^{b} \frac{q'(\xi)\mathrm{d}\xi}{x-\xi} \qquad -b \le x \le b. \tag{1.69}$$

It follows from a further comparison of the normal and tangential integral equations that the corrective shear traction distribution, $q'(x)$, must just be a scaled form of the contact pressure, corresponding to the pressure which would be present at a lower load, and the value of the corrective shear force, $Q' = \int_{-b}^{b} q'(x)\mathrm{d}x$, is given by

$$Q = fP - Q'. \tag{1.70}$$

This apparently fairly obvious way of looking at things is a very important result. It is exceptionally powerful when considering, for example, a rough contact or one with multiple contact areas (Dini and Hills 2009), because without it the partial slip aspect would be very intractable to conduct as a separate calculation. We have shown the results, here, for a two-dimensional problem and the technique applies only approximately to three-dimensional problems, unless the bodies are made from material displaying no Poisson effect, but the full solution of, for example, Hertz' problem for spheres shows only a small discrepancy (Munisamy et al. 1994).

There is insufficient room, here, for a full description of all types of solution possible. The chapter by Barber describes briefly more complex loading trajectories where the normal and shear forces form closed loops in P-Q space, but we will conclude with a brief description of the steady-state solution between two points 1, 2 in P-Q-σ space where the trajectory is in the form of a straight line. This is of some practical significance, because there are many cases of fretting contacts where a steady force develops a particular value of these quantities and then vibration, for example, introduces small changes in them all, so that

$$P_1 = P_{\text{mean}} - \Delta P/2 \qquad P_2 = P_{\text{mean}} + \Delta P/2, \tag{1.71}$$

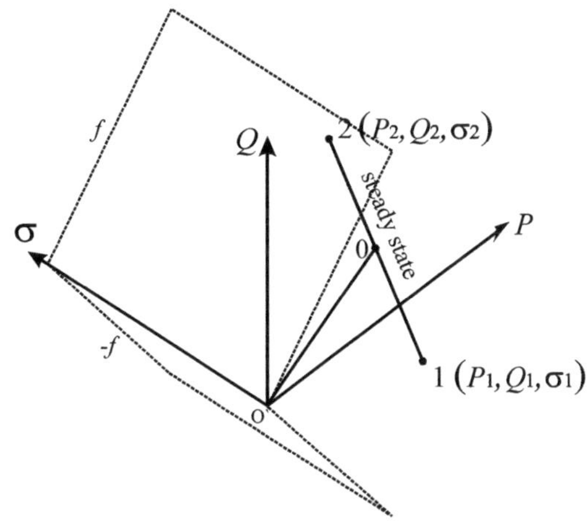

Fig. 1.18 Load history displayed in P-Q-σ space

$$Q_1 = Q_{\text{mean}} - \Delta Q/2 \qquad Q_2 = Q_{\text{mean}} + \Delta Q/2, \tag{1.72}$$

$$\sigma_1 = P_{\text{mean}} - \Delta\sigma/2 \qquad \sigma_2 = \sigma_{\text{mean}} + \Delta\sigma/2. \tag{1.73}$$

It is easiest to think of the loading trajectory as a straight line between the end points, Fig. 1.18, but this might be relaxed to include some hysteresis provided that the slip zones advance monotonically during load changes in each direction. Because the normal load changes so will the size of the contact, and we also know that the slip zones will increase in size during each load change, and that as load reversals are experienced the entire contact will stick. Provided that the bulk tension effects are small, the shear traction in the slip regions will be the same at each end of the contact, and the maximum size of the slip zones will always be reached immediately before the reversal of load. Lastly, we know that the size and position of the stick zones at these extremes—otherwise the 'permanent stick zone'—must be same so that material flows both in and out at each end of the contact and material is preserved. Thus, the problem is solved when once we know $a_1 = a(P_1), a_2 = a(P_2)$ together with the extent of the permanent stick zone $[-m \quad n]$ which might, in principle, be a function of several quantities in Eqs. (1.72) and (1.73).

Refer to Fig. 1.18 which is a sketch of the loading problem in P-Q-σ space, and Fig. 1.19 which shows the expected layout of the stick and slip zones as the end points 1, 2 are approached. When the load reverses the contact becomes fully stuck, and the slip zones grow monotonically as loading heads towards the other end of the trajectory, so that Fig. 1.19 shows the maximum extent of the slip zones and the minimum extent of the stick zone, i.e. the permanent stick zone. An important feature

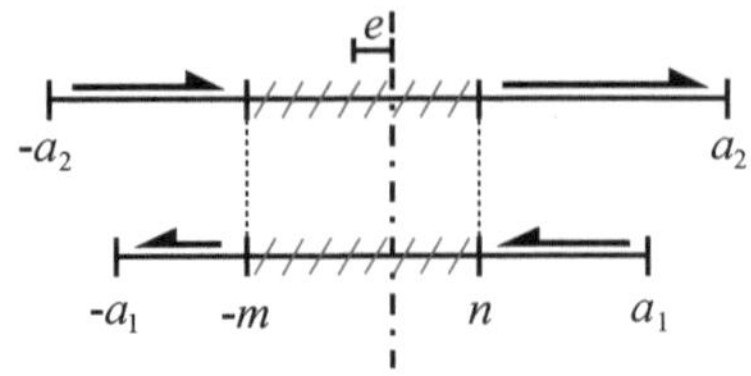

Fig. 1.19 Expected layout of the stick and slip zones as point 1, 2 is approached

of the sketch is that the permanent stick zone is of the same extent and at the same location at each end of the loading regime. This follows directly from a consideration of conservation of material; slip displacements in one direction at any point in the contact at some point in the cycle must be precisely matched by a return of those slipping points to their original locations at another point in the cycle, bearing in mind that this is a steady state. If this principle were violated it would follow that slip displacement would increase in a ratcheting fashion. Also, because it never slips, the permanent stick zone includes a surface strain difference $\Delta\varepsilon_{xx}(x)$ which remains constant in time and, in particular, is locked in strains difference at points 1, 2.

From the integral equations developed in section "Half-Plane Problems", and the Ciavarella-Jäger method described at the beginning of this section, we may write down the differential surface strain at point 1, as usual using the concept of a corrective shear traction over the permanent stick interval $[-m \quad n]$ superimposed on a sliding distribution as

$$\Delta\varepsilon_{xx1}(x)\frac{E^*}{4} = -\frac{1}{\pi}\int_{-a_1}^{a_1}\frac{fp_1(\xi)\,\mathrm{d}\xi}{\xi - x} + \frac{1}{\pi}\int_{-m}^{n}\frac{q_1^*(\xi)\mathrm{d}\xi}{\xi - x} + \frac{\sigma_1}{4}, \tag{1.74}$$

and, at point 2 there is an analogous state with the direction of the sliding shear reversed, i.e.

$$\Delta\varepsilon_{xx2}(x)\frac{E^*}{4} = \frac{1}{\pi}\int_{-a_2}^{a_2}\frac{fp_2(\xi)\,\mathrm{d}\xi}{\xi - x} + \frac{1}{\pi}\int_{-m}^{n}\frac{q_2^*(\xi)\mathrm{d}\xi}{\xi - x} + \frac{\sigma_2}{4}, \tag{1.75}$$

and, as these two quantities are matched over the stated interval, and making use of the solution for normal contact, as before

$$2fE^*\frac{\mathrm{d}h}{\mathrm{d}x} + \frac{\Delta\sigma}{4} = -\frac{1}{\pi}\int_{-m}^{n}\frac{(q_2^*(\xi) - q_1^*(\xi))\mathrm{d}\xi}{\xi - x} \qquad -m < x < n, \tag{1.76}$$

where $\Delta\sigma = \sigma_1 - \sigma_2$. Also, the range of shear force experienced, ΔQ, is given by

$$\Delta Q = f\,(P_2 + P_1) + \int_{-m}^{n}\left(q_2^*\,(\xi) - q_1^*\,(\xi)\right)\mathrm{d}\xi. \tag{1.77}$$

Details of the inversion of integral equation (1.76) are given in the original paper (Andresen et al. 2018), but a key point is that the LHS includes only two terms; the profile, implying a corrective shear traction which is a scaled form of the contact pressure, and a constant which does not appear in the solution, save in the consistency condition. For the particular case of a Hertzian contact, the inversion is straightforward and leads to the result that the offset of the stick zone

$$\frac{n-m}{2} = -\frac{\Delta\sigma R}{4fE^*} \tag{1.78}$$

is independent of the magnitude of the shear force range, and the extent of the stick zone

$$m+n = 2\sqrt{\frac{R}{\pi E^*}\left[2P_{\text{mean}} - \frac{\Delta Q}{f}\right]} \tag{1.79}$$

is independent of the magnitude of the range of bulk tension applied. A further important general result (not just for the Hertzian geometry) is that the size of the permanent stick zone depends on just three of the six possible quantities given above, viz. ΔQ, $\Delta\sigma$, P_{mean}. Another possibly unexpected feature of the solution, apart from its relative simplicity, is that we may solve for the size of the permanent stick zone independently of a consideration of the normal contact law. The size and position of the permanent stick zone depend only on the contacting profile within the stick zone, and the overall size of the contact does not appear in the solution.

Further Types of Contact

Far fewer general results are available for the remaining two types of contact, and we provide, here, some example results.

Receding Contacts

Before any external forces are applied, receding contacts conform, in the same way that complete contacts do. In some cases the initial extent of the contact is defined by the limits of conformality, such as the thin rectangular strip resting on the elastically similar half-plane shown in Fig. 1.20. Indeed, if the contact pressure is applied over a significant fraction of the upper surface, the contact may well extend along the whole of the lower length of the rectangle and not recede, but be complete. If, on the other hand, pressure on the upper surface stops well short of the ends they will tend to lift off, and so the contact will recede. Receding contacts of this kind have the property that the application of an infinitesimal load causes the contact to 'snap' to a reduced length, and an increase in the applied pressure does not cause a change in the contact

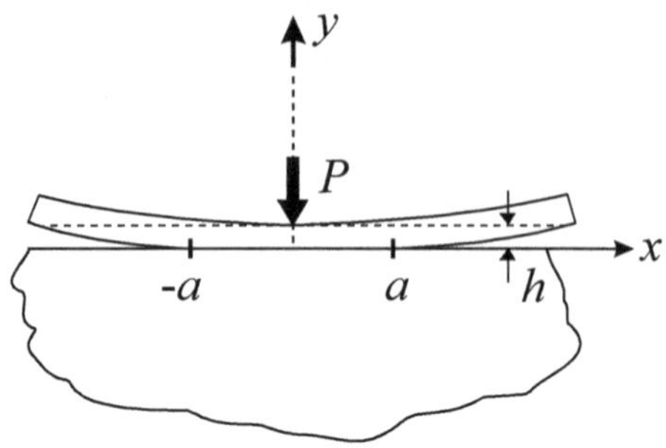

Fig. 1.20 Receding contact

length, so that the problem becomes linear, in the sense that the magnitude of the stresses and displacements (including the lift-off angle) are simply proportional to the applied load. If contacts of this class are to be modelled by the finite element method, it is probably best to represent the contacting pair as a monolithic entity, and then to look for the development of interfacial tension near the edges. If tension arises (and this will depend, also, on the coefficient of friction present) then the contact is clearly going to recede and a good tip is to model the layer with a very small gap between it and the half-plane in the majority of the length over which tension was observed, and to permit further separation. With this modification, when the contact lifts (at a very small load) the number of node pairs where separation occurs is much smaller than if the gap was not present. Finite element programmes do not cope well with the sudden separation of a large length of the interface and this helps considerably.

The tendency for a strip to lift off is very strongly controlled by the presence of friction along the interface. If we have a layer of thickness c resting on an elastically similar half-plane, we can think of the pair, together as a half-plane of additional 'depth' h, and, if a line load is applied, Fig. 1.20, use the Flamant solution to discover what happens. We can convert the solution derived in section "Half-Plane Problems" into Cartesian coordinates, and this then gives

$$p(x) \equiv \sigma_{yy}(x, h) = \frac{-2Ph^3}{\pi \left(x^2 + h^2\right)^2}, \tag{1.80}$$

$$q(x) \equiv \sigma_{xy}(x, h) = \frac{-2Pxh^2}{\pi \left(x^2 + h^2\right)^2}, \tag{1.81}$$

so that, *prima facie*, it would seem that intimate contact will be maintained between layer and substrate. But notice that, if we take the traction ratio

$$\frac{\sigma_{xy}}{\sigma_{yy}} = \frac{x}{h} \tag{1.82}$$

keep h constant and increase x we can see that the traction ratio becomes infinitely large, and it would therefore require an infinitely large coefficient of friction to inhibit all slip. For a finite coefficient of friction, there will be a region of slip, and consequent separation. In order to solve problems of this class, one possibility is to set up an integral equation using dislocations as the kernel, rather than surface forces as was done for half-plane problems. These are very effective at permitting discrete regions of slip and separation whilst continuity of material is maintained elsewhere.

The solution is known for the state of stress induced by an edge dislocation in a half-plane, in closed form. For the purposes of the present problem, it is best if the origin lies on the surface of the true half-plane, on top of which lies a layer of thickness c, so that the 'augmented half-plane' (the actual half-plane and layer together) occupies the region $y < c$, Fig. 1.20. The tractions arising on the putative interface line, $y = 0$, generated by a dislocation, also on the line, installed at point ξ and having Burgers vector component (b_x, b_y) are given by

$$\begin{Bmatrix} \sigma_{xy}(x,0) \\ \sigma_{yy}(x,0) \end{Bmatrix} = \frac{E}{4\pi\left(1-\nu^2\right)} \begin{bmatrix} \frac{1}{x-\xi} + G_{xxy}(x,\xi) & G_{yxy}(x,\xi) \\ G_{xyy}(x,\xi) & \frac{1}{x-\xi} + G_{yyy}(x,\xi) \end{bmatrix} \begin{Bmatrix} b_x(\xi,0) \\ b_y(\xi,0) \end{Bmatrix} \tag{1.83}$$

in plane strain, where the functions G_{ijk} account for the presence of the free surface and are given explicitly in (Chaise et al. 2014). Note that there are Cauchy singular terms—the glide dislocation generates a singular behaviour in the shear traction and the climb dislocation develops a singular behaviour in the direct traction, but these singularities will integrate out when we have distributions of dislocations. We expect the solution to be inherently symmetrical about the $x = 0$ line, and for regions of separation to develop when $|x| > b$, and regions of slip will be present just 'inboard' of the separation points, $a < |x| < b$. The total direct and shear tractions arising along the line $y = 0$ are therefore given by

$$\begin{aligned} N(x) = p(x) + \frac{E}{4\pi\left(1-\nu^2\right)} \Bigg[& \int_{-\infty}^{-b} \left[\frac{1}{x-\xi} + G_{yyy}(x,\xi)\right] B_y(\xi)\,\mathrm{d}\xi + \\ & + \int_{b}^{\infty} \left[\frac{1}{x-\xi} + G_{yyy}(x,\xi)\right] B_y(\xi)\,\mathrm{d}\xi + \int_{-\infty}^{-a} G_{xyy}(x,\xi)\, B_x(\xi)\,\mathrm{d}\xi + \\ & + \int_{a}^{\infty} G_{xyy}(x,\xi)\, B_x(\xi)\,\mathrm{d}\xi \Bigg] \end{aligned} \tag{1.84}$$

$$
\begin{aligned}
S(x) = & q(x) + \frac{E}{4\pi\left(1-\nu^2\right)}\left[\int_{-\infty}^{-b} G_{yxy}(x,\xi)\, B_y(\xi)\, \mathrm{d}\xi + \right. \\
& + \int_b^{\infty} G_{yxy}(x,\xi)\, B_y(\xi)\, \mathrm{d}\xi + \int_{-\infty}^{-a}\left[\frac{1}{x-\xi} + G_{xxy}(x,\xi)\right] B_x(\xi)\mathrm{d}\xi + \\
& \left. + \int_a^{\infty}\left[\frac{1}{x-\xi} + G_{xxy}(x,\xi)\right] B_x(\xi)\mathrm{d}\xi\right], \qquad (1.85)
\end{aligned}
$$

respectively, where $B_i(x) = \mathrm{d}b_i/\mathrm{d}x, i = x, y$, and represent the primary unknowns in the problem, the dislocation densities. We now exploit the inherent symmetry in the problem and observe that the glide dislocations are antisymmetrically disposed while the climb dislocations are symmetrically arranged, so that the dislocation *densities* have the properties

$$
B_x(x) = B_x(-x) \qquad B_y(x) = -B_y(-x). \qquad (1.86)
$$

Further, the kernels themselves have symmetry properties which we can exploit, and write

$$
F_{yyy}(x,\xi) = -\frac{1}{x+\xi} + G_{yyy}(x,\xi) - G_{yyy}(x,-\xi) \qquad (1.87)
$$

$$
F_{xxy}(x,\xi) = \frac{1}{x+\xi} + G_{xxy}(x,\xi) + G_{xxy}(x,-\xi) \qquad (1.88)
$$

$$
F_{xyy}(x,\xi) = G_{xyy}(x,\xi) + G_{xyy}(x,-\xi) \qquad (1.89)
$$

$$
F_{yxy}(x,\xi) = G_{yxy}(x,\xi) - G_{yxy}(x,-\xi). \qquad (1.90)
$$

With these substitutions, the integral representations of the dislocation densities may be written down compactly in the conventional form as

$$
\begin{aligned}
N(x) = & p(x) + \frac{E}{4\pi\left(1-\nu^2\right)}\left[\int_b^{\infty}\left[\frac{1}{x-\xi} + F_{yyy}(x,\xi)\right] B_y(\xi)\, \mathrm{d}\xi + \right. \\
& \left. + \int_a^{\infty} F_{xyy}(x,\xi)\, B_x(\xi)\, \mathrm{d}\xi\right] \qquad (1.91)
\end{aligned}
$$

$$
\begin{aligned}
S(x) = & q(x) + \frac{E}{4\pi\left(1-\nu^2\right)}\left[\int_b^{\infty} F_{yxy}(x,\xi)\, B_y(\xi)\, \mathrm{d}\xi + \right. \\
& \left. + \int_a^{\infty}\left[\frac{1}{x-\xi} + F_{xxy}(x,\xi)\right] B_x(\xi)\mathrm{d}\xi\right]. \qquad (1.92)
\end{aligned}
$$

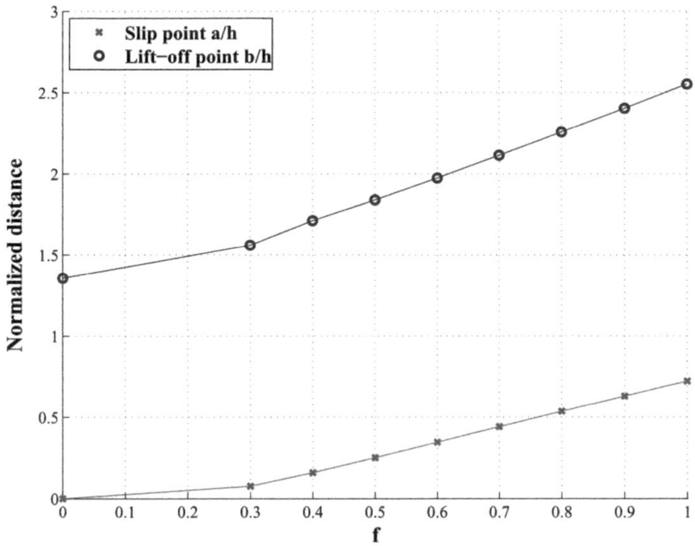

Fig. 1.21 Slip and lift-off points as a function of the coefficient of friction

Lastly, we impose the requirements that both components of traction vanish in the open region and that the shear traction be proportional to the direct traction in the slipping region with the constant of proportionality being the coefficient of friction. This gives rise to the integral equations

$$N(x) = 0 \qquad b \leq x < \infty \tag{1.93}$$

$$S(x) = -fN(x) \qquad a \leq x < \infty. \tag{1.94}$$

These two simultaneous integral equations with generalised Cauchy kernels must be solved by a numerical method and details of this may be found in the original paper (Chaise et al. 2014). It is a Gaussian formulation which correctly allows both for the Cauchy character of the kernel and the end-point behaviour. Note (a) that we require the dislocation density to go smoothly to zero in a square root bounded fashion as the observation point approaches the closure point or stick-slip transition point, (b) that the primary unknowns in the problem, apart from the dislocation density, are the closure and stick-slip transition points, and their location is dependent on the coefficient of friction but does not depend on the applied load. Figure 1.21 summarises the results found.

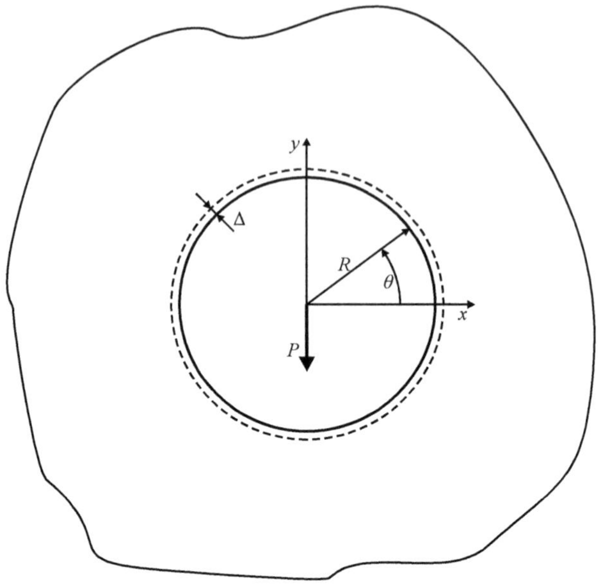

Fig. 1.22 Shrink-fitted disk in an infinite plane subject to a point load

Other Kinds of Recession

Not all receding contacts exhibit the 'snapping' phenomenon. Consider the problem of an oversized circular disk, of radius $R + \Delta$ ($\Delta \ll R$) fitted into an infinite plane containing a circular hole of radius R, so that there is interference between the two bodies, and the interfacial contact pressure is easily found from Lame's thin-walled cylinder solution, under conditions of plane strain, to be

$$\sigma_{rr}(R, \theta) = \frac{E^* \Delta}{R}, \tag{1.95}$$

and now suppose that a radial force, P, is applied to the centre of the disk, as shown in Fig. 1.22. The state of stress induced within an infinite plane is given by Barber (2010)

$$\sigma_{rr}(R, \theta) = \frac{3 + \nu}{4\pi} \frac{P \sin\theta}{R} \tag{1.96}$$

$$\sigma_{r\theta}(R, \theta) = \frac{1 - \nu}{4\pi} \frac{P \cos\theta}{R} \tag{1.97}$$

$$\sigma_{\theta\theta}(R, \theta) = = \frac{1 - \nu}{4\pi} \frac{P \sin\theta}{R} \tag{1.98}$$

and the point where the radial stress is most quickly diminished is $\theta = \pi/2$. Intimate contact is maintained there providing that

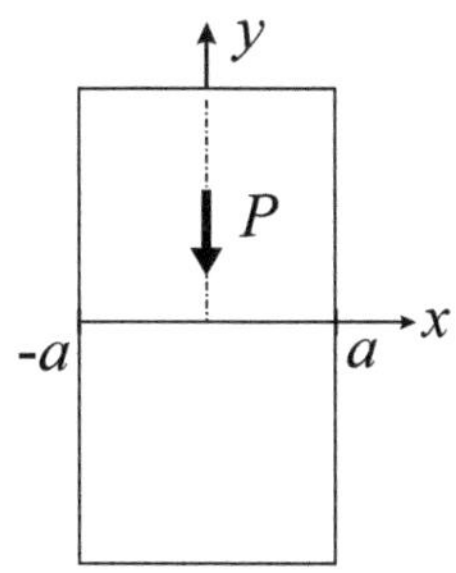

Fig. 1.23 Common edge contact

$$\sigma_{rr,\max} \leq 0, r = R \quad \rightarrow \quad \frac{P}{\sigma_0 R} \leq \frac{4\pi}{3+\nu} \tag{1.99}$$

but if this inequality is exceeded a region of tension will develop. The arc of separation maybe found by, for example, deploying edge dislocations around the interface again, as described in section "Other Kinds of Recession". As the load is steadily increased, the arc of contact between disk and hole will steadily and smoothly recede.

Note that if, in contrast, there is no interference between disk and hole, but the two have an exactly matched radius, the contact becomes one of the 'snapping' kind, and the angle to which the contact jumps is precisely the same limit found above. On the other hand, if the disk is very slightly smaller than the hole in which it sits, the contact becomes an advancing or incomplete one, and if the difference in size between disk and hole is very small, the contact angle to which the contact advances for sufficiently large loads is also this same angle.

Common Edge Contacts

When two bodies are brought into contact and where their edges align, Fig. 1.23, the local contact pressure is expected to take a finite value, in contrast to other forms of contact. As has been recommended elsewhere, a good way of studying contacts of this class is to think of the two contacting bodies as being in intimate contact in the presence of full stick, and then to look for violations of the Signorini inequalities, and the friction law. With the simple straight edge shown in Fig. 1.23, it is clear that, at the edge ($x \rightarrow 0$), the only non-zero component of stress must be that which is not a traction, viz. $\sigma_{yy} = -p < 0$. Therefore, an infinitesimal coefficient of friction would be sufficient to prevent all slip and, providing that intimate contact was maintained, one might expect this very simple asymptotic principle to persist, and therefore be eroded only from interior points, so that the edge will be the last point to slip. This is easy to illustrate by looking at the classical solution for a beam with a shear force gradually exerted, Fig. 1.24.

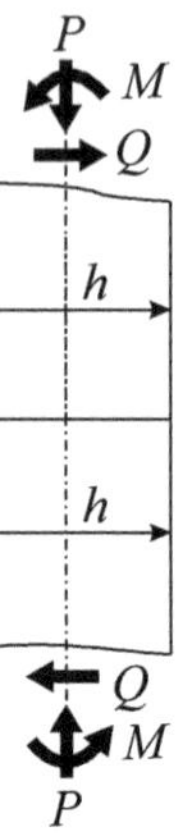

Fig. 1.24 Beam with a shear force gradually exerted

If the beam is of depth $2h$, and is subject to an axial force, P (per unit depth), there is uniform compressive stress across the interface of magnitude $\sigma_{yy} = -P/2h$. Let the beam also be suffering a linearly varying bending moment of gradient $\mathrm{d}M/\mathrm{d}x$, and let the interface between the two separate parts of the beam be located, for simplicity, at precisely the point where the moment vanishes. A shear force, $Q(= \mathrm{d}M/\mathrm{d}x)$, is also present and let us suppose that the bending moment gradient and hence the shear force are increasing in magnitude. Elementary results for the shear stress distribution associated with a shear force (Barber 2001) show that it is of the parabolic form

$$\sigma_{xy}(x, y) = \frac{3Q}{4h}\left(1 - \frac{x^2}{h^2}\right). \tag{1.100}$$

So, the first point to slip will be a pair of particles on the centreline of the beam, and slip will start when

$$\frac{Q}{fP} = \frac{2}{3}. \tag{1.101}$$

At higher values of shear force a region of slip will develop from the centreline, and this may be modelled using either the solution for an edge dislocation in a strip as a kernel, or by employing a series solution (Kartal et al. 2010). The last points to slip are, indeed, the outer edges of the interface when, of course, $\frac{Q}{fP} \to 1$.

If the interface meets the edges not at a right angle but at some other angle, it is very easy to employ Mohr's circle (or to use a different transform method) to find the tractions present on the plane and it is easy to see that, if the angle the interface makes with the surface normal is θ, the minimum coefficient of friction to ensure edge adhesion is $f > \tan\theta$. It is now not absolutely certain that the contact edge will be the last point to slip in any particular problem but, for practically arising values of θ, this still seems quite likely.

One class of problem where it is quite likely that the edge will slip *first* is when there is anti-plane loading present. Again elementary strength of materials knowledge is all that is needed to illustrate this point. If we have two cylinders of radius R, made of the same material pressed together by a force, P, the pressure is uniform over the interface and of magnitude $\sigma_{zz} = -p = \frac{P}{\pi R^2}$. We now apply an increasing torque, T, to the assembly, which will give rise to a shear traction on the interface of magnitude

$$\sigma_{rz}(r, \theta) = \frac{2Tr}{\pi R^4}, \tag{1.102}$$

so that slip will start, at the outer edge of the contact, when

$$\frac{T}{PR} = \frac{f}{2}. \tag{1.103}$$

The limit state condition (free spinning) will be attained when

$$T = \frac{fP}{\pi R^2} \int_0^R 2\pi r^2 \mathrm{d}r \qquad i.e. \frac{T}{PR} = \frac{2f}{3}. \tag{1.104}$$

Summary

Incomplete contacts are best modelled, wherever possible, using half-plane (or space) theory, and with the solution for surface line (point) force as the starting point. This works well even in partial slip problem where we have to establish a stick condition over part of the contact. Incomplete contacts have the property that the contacting surfaces do not conform and contacts advance with increasing load. All other kinds of contact (complete, receding, and common edge) have the property that the contacting surfaces, at least initially, do conform and there is therefore the possibility that the contact remains intimate and does not slip. As a consequence, a 'good bet' to start off any analysis is to assume that the bodies are, indeed, in intimate contact and are also stuck everywhere so that the two bodies are formed into a monolith with the interface merely a trace line in the combined solid. This makes analysis easier, in the first instance, and then, if there is implied slip, glide dislocations' may be distributed where there is slip, and climb dislocations where there is tension (and therefore separation). Of course for this to be a viable approach, the solution is needed for a dislocation in the domain of the contact (contacting pair treated as a single body). Note that these ideas apply also to anti-plane loading, where the kernel needed for representing anti-plane slip is a screw dislocation. These ideas have been used successfully in a range of problems and recently even to incomplete contacts (Moore and Hills 2018).

References

Andresen, H. N., Hills, D. A., Barber, J. R., & Vázquez, J. (2019) Frictional half-plane contact problems subject to alternating normal and shear loads and tension. *International Journal of Solids and Structures, 168*, 166–171.

Barber, J. R. (2001). *Intermediate mechanics of materials*. McGraw-Hill.

Barber, J. R. (2010). *Elasticity* (3rd ed.). Springer.

Barber, J. R., Davies, M., & Hills, D. A. (2011). Frictional elastic contact with periodic loading. *International Journal of Solids and Structures, 48*, 2041–2047.

Cattaneo, C. (1938). Sul contato di due corpo elastici. *Atti della Accademia Nazionale dei Lincei. Classe di Scienze Fisiche, Matematiche e Naturali. Rendiconti Lincei, 27*, 342–348, 434–436, 474–478.

Chaise, T., Paynter, R. J. H., & Hills, D. A. (2014). Contact analysis of a semi-infinite strip pressed onto a half plane by a line force. *International Journal of Mechanical Sciences, 81*, 60–64.

Churchman, C. M., & Hills, D. A. (2006a). General results for complete contacts subject to oscillatory shear. *Journal of the Mechanics and Physics of Solids, 54*, 1186–1205.

Churchman, C. M., & Hills, D. A. (2006b). Slip zone length at the edge of a complete contact. *International Journal of Solids and Structures, 43*, 2037–2049.

Ciavarella, M. (1998). The generalised cattaneo partial slip plane contact problem, part i theory, part ii examples. *International Journal of Solids and Structures, 35*, 2349–2378.

Ciavarella, M., Baldini, A., Barber, J. R., & Strozzi, A. (2006). Reduced dependence on loading parameters in almost conforming contacts. *International Journal of Mechanical Sciences, 48*, 917–925.

Comninou, M. (1976). Stress singularity at a sharp edge in contact problems with friction. *Journal of Applied Mathematics and Physics, 27*, 493–499.

Dini, D., & Hills, D. A. (2009). Frictional energy dissipation in a rough hertzian contact. *ASME Journal of Tribology, 131*, 021401-1–8.

Dundurs, J. (1969). Discussion of edge-bonded dissimilar orthogonal wedges under normal and shear loading (by D. Bogey). *Journal of Applied Mechanics, 36*, 342–348, 650–652.

Erdogan, F., Gupta, G. D., & Cook, T. S. (1973). Numerical solution of singular integral equations. In G.C. Sih (Ed.), *Methods of analysis and solutions of crack problems*. Noordhoff.

Hertz, H. (1881). über die berührung fester elastischer körper. *Journal für die reine und angewandte Mathematik, 92*, 156–171.

Hills, D. A., Nowell, D., & Sackfield, A. (1993). *Mechanics of elastic contacts*. Butterworth-Heinemann.

Jäger, J. (1998). A new principle in contact mechanics. *Journal of Tribology, 120*, 677–684.

Kartal, M. E., Hills, D. A., Barber, J. R., & Nowell, D. (2010). Torsional contact between elastically similar flat-ended cylinders. *International Journal of Solids and Structures, 47*(10), 1375–1380.

Karuppanan, S., & Hills, D. A. (2008). Frictional complete contacts between elastically similar bodies subject to normal and shear load. *International Journal of Solids and Structures, 17*, 4662–4675.

Mindlin, R. D. (1949). Compliance of elastic bodies in contact. *Journal of Applied Mechanics, Transactions ASME, 16*, 259–268.

Moore, M. R., & Hills, D. A. (2018). Solution of half-plane contact problems by distributing climb dislocations. *International Journal of Solids and Structures, 147*, 61–66.

Moore, M. R., Ramesh, R., Hills, D. A., & Barber, J. R. (2018). Half-plane partial slip contact problems with a constant normal load subject to a shear force and a differential bulk tension. *Journal of the Mechanics and Physics of Solids*, *118*, 245–253.

Munisamy, R. L., Hills, D. A., & Nowell, D. (1994). Static axi-symmetric hertzian contacts subject to sharing forces. *ASME Journal of Applied Mechanics*, *61*(2), 278–283.

Persson, A. (1964). *On the stress distribution of cylindrical bodies in contact*. Doktosavhandlingar vid Chalmers Tekniska Hogskola.

Truman, C. E., Sackfield, A., & Hills, D. A. (1995). Contact mechanics of wedge and cone indenters. *International Journal of Mechanical Sciences*, *37*, 261–275.

Chapter 2
Contact Problems Involving Friction

J. R. Barber

Abstract The Coulomb friction law is simple to apply in the formulation of elastic contact problems, but it is also a rich source of unexpected physical phenomena, including ranges of unstable dynamic response, history-dependence, 'wedging' and mathematical problems of existence and uniqueness of solution. We first explore the implications of the law in the context of simple discrete systems and demonstrate the importance of interaction [coupling] between the normal and tangential contact problems, particularly in problems of periodic loading. The discussion is then extended to problems of the elastic continuum, and to cases where elastodynamic effects must be included [for example, the interaction of a seismic disturbance with a frictional interface]. It is shown that finite element formulations of elastodynamic problems with Coulomb friction are inherently ill-posed and alternative friction laws that avoid this difficulty are discussed.

Frictional Forces and Tractions

If two bodies are pressed together, both normal and tangential contact forces and tractions [force per unit area] may be developed. The normal tractions result from interatomic force potentials and [when compressive] resist the interpenetration of material. However, the mechanism responsible for tangential or *frictional* tractions is much less clear and indeed has been a subject of scientific speculation for hundreds of years. Frictional tractions play a crucial rôle in a wide range of engineering and natural systems, including for example frictional damping in nominally static bolted joints, fretting fatigue (e.g. in turbine blade roots), the settlement of soils or masonry structures, tectonic plate movement during earthquakes, frictional 'wedging' during automatic assembly processes and frictional slip in belt drives.

J. R. Barber (✉)
Department of Mechanical Engineering, University of Michigan, Ann Arbor, MI, USA
e-mail: jbarber@umich.edu

M. Paggi and D. Hills (eds.), *Modeling and Simulation of Tribological Problems in Technology*, CISM International Centre for Mechanical Sciences 593,
https://doi.org/10.1007/978-3-030-20377-1_2

Microslip and Gross Slip [Sliding]

In the contact of rigid bodies, there is a sharp dichotomy between *sliding*, where there is a non-zero relative tangential velocity $\boldsymbol{V}$, and *stick*, where $|\boldsymbol{V}| = 0$. However, if the bodies are deformable, it is possible for the parts of the nominal contact area to be in a state of stick, whilst the rest is slipping. The extent of slip in these cases is limited by the deformation of the materials. It is therefore generally very small and is referred to as *microslip*. By contrast, once the entire contact area slips, substantial slip displacements can accumulate and generally, we need to define the deformations of the two bodies in separate coordinate systems.

The Coulomb Friction Law

In engineering applications, it is often assumed that the magnitude of the friction force $\boldsymbol{Q}$ during sliding is proportional to the normal force P—i.e. $|\boldsymbol{Q}| = fP$, where f is an experimentally determined constant known as the *coefficient of friction*. This is known variously as *Amontons'* or *Coulomb's law of friction*.

The Coulomb law is at best only an approximation to the frictional behaviour of actual solids. The ratio $|\boldsymbol{Q}|/P$ has been shown to vary with normal force P and sliding speed $|\boldsymbol{V}|$ as well as other material and geometric features, and at small length scales such as in the atomic force microscope [AFM], significant deviations from the law are observed. Also, experiments are mostly restricted to the measurement of force resultants under sliding conditions, and it does not necessarily follow that the same proportionality exists between the tractions in a pointwise sense—i.e. $|\boldsymbol{q}(x, y)| = fp(x, y)$ for all points (x, y) in the contact area that are experiencing microslip. However, the law has the virtue of simplicity and deviations from it at the macroscale are often not much larger than the inevitable variance of experimental measurements. We shall therefore use the Coulomb law in most of this discussion, though some alternative treatments will be introduced towards the end.

Physical Explanations of Coulomb's Law

Most authors attribute the approximate constancy of the friction coefficient to the inevitable roughness of the contacting surfaces. For example, Bowden and Tabor (1950) argued that contact would be restricted to a relatively small region near the peaks of the rough surfaces and that the resulting actual contact pressures would be sufficiently high to cause plastic deformation. By likening this process to that in the hardness test, they argued that the total area of actual contact would be $A = P/H$, where H is the hardness of the softer material. Bowden and Tabor further argued that the materials would bond together in this area and that a limiting value S of shear traction would be needed to break these bonds. We therefore obtain $|\boldsymbol{Q}| = SA = SP/H$ and hence $|\boldsymbol{Q}| = fP$ with $f = S/H$.

An alternative explanation depends on conceiving the rough surfaces as comprising a statistically distributed ensemble of 'asperities' which act independently of each other. The force resultants can then be expressed as a function of this distribution and of the normal approach of the surfaces. Many theories of this kind were proposed in the 1950s, but a significant discovery due to Greenwood and Williamson (1966) is that with practical [e.g. Gaussian] height distributions of asperities, the relation between normal force and total actual contact area is almost linear, regardless of the assumed micromechanics at the individual asperity. A similar argument then predicts approximate linearity between normal and tangential force resultants during sliding.

A weakness of these theories is that real surfaces exhibit multiscale features and the predictions depend on the scale at which asperities are defined. However, other models of rough surface contact (Archard 1957; Persson 2001; Barber 2013) reach similar conclusions based on the multiscale character of rough surfaces without relying on a description in terms of asperities.

Mathematical Definition of Coulomb's Law

If Coulomb's law is assumed to apply in a pointwise sense, each point in the contact area must be in a state either of stick $\boldsymbol{V} = \boldsymbol{0}$, or slip $|\boldsymbol{V}| > 0$. In slip regions, the local tangential traction must oppose the relative velocity, so

$$\boldsymbol{q}(x, y) = -fp(x, y)\boldsymbol{e}_V; \quad |\boldsymbol{V}| > 0, \tag{2.1}$$

where

$$\boldsymbol{e}_V = \frac{\boldsymbol{V}}{|\boldsymbol{V}|} \tag{2.2}$$

is a unit vector defining the direction of slip.

In stick regions, we have

$$|\boldsymbol{q}(x, y)| \le fp(x, y); \quad \boldsymbol{V} = \boldsymbol{0}. \tag{2.3}$$

In other words, the magnitude of the tangential traction is not larger than that during slip. This assumes that the 'coefficient of static friction' is the same as that for 'dynamic' or 'kinetic' friction. The case where this assumption is relaxed will be discussed later.

Two-dimensional problems. In two-dimensional problems, the sliding speed V and the tangential traction q are scalars and slip, when it occurs, can only be either to the right or to the left. The friction law is therefore simplified to

$$q(x, y) = -fp(x, y)\,\mathrm{sgn}(V); \quad V \neq 0$$
$$-fp(x, y) < q(x, y) < fp(x, y); \quad V = 0,$$

where sgn(·) is the signum function defined such that $\mathrm{sgn}(z) = 1$ if $z > 0$ and $\mathrm{sgn}(z) = -1$ if $z < 0$.

We can then distinguish three possible contact states:

$$\begin{array}{lll} \text{forward slip} & V > 0; & q(x, y) = -fp(x, y) \\ \text{backward slip} & V < 0; & q(x, y) = fp(x, y) \\ \text{stick} & V = 0; & -fp(x, y) < q(x, y) < fp(x, y). \end{array} \tag{2.4}$$

If the state is known for all points (x, y) in the contact area, the governing equations for the elastic contact problem are then linear.

History-Dependence

The Coulomb law introduces history-dependence into the contact problem, even if all the contacting bodies are linear elastic. This is demonstrated by the simple example of Fig. 2.1. A rigid block is pressed against a frictional rigid plane by a constant normal force P and then loaded by a time-varying tangential force $F(t)$. Tangential motion [displacement] u is resisted by a linear spring of stiffness k.

If the spring is initially undeformed and the force $F(t)$ increases monotonically from zero, the body will move to the right and the motion will be opposed by a friction force fP and a spring force $ku(t)$. If the loading rate is slow enough for inertial effects to be neglected [the *quasi-static* assumption], we therefore obtain

$$F(t) = fP + ku(t) \quad \text{or} \quad u(t) = \frac{F(t) - fP}{k}. \tag{2.5}$$

However, if $F(t)$ increases to a maximum value $F_{\max}$ and then falls monotonically, the body will stick at the point

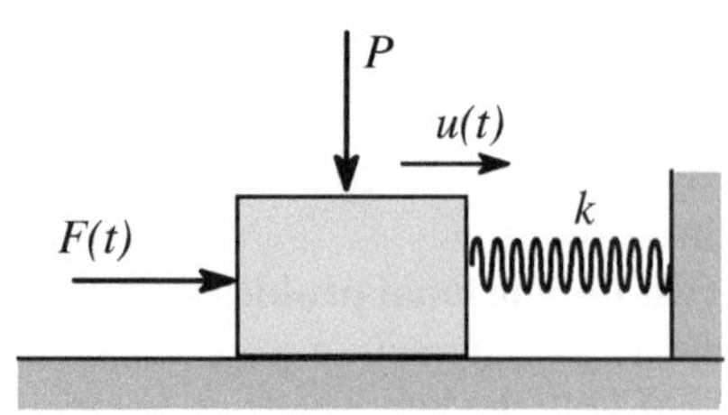

Fig. 2.1 A simple frictional problem exhibiting load history-dependence

$$u_{\max} = \frac{F_{\max} - fP}{k} \tag{2.6}$$

until the condition for backward slip [$\dot{u} < 0$] is reached when

$$ku_{\max} - F(t) = fP \quad \text{or} \quad F(t) = F_{\max} - 2fP. \tag{2.7}$$

Further reduction in $F(t)$ will then lead to a displacement

$$u(t) = \frac{F(t) + fP}{k}. \tag{2.8}$$

The stick location. Suppose after some initial period, the force $F(t)$ is maintained constant and equal to F_1. Under quasi-static conditions, the displacement $u(t) = u_1$ must also be constant, so we have stick and $Q = F_1 - ku_1$. The stick inequality $-fP \le Q \le fP$ then tells us that

$$-fP \le F_1 - ku_1 \le fP \quad \text{or} \quad \frac{F_1 - fP}{k} \le u_1 \le \frac{F_1 + fP}{k}, \tag{2.9}$$

but within this range, u_1 can take any value. This uncertainty can be resolved only by tracking the displacement in time from some known initial condition, using the friction law to determine the *changes* in displacement due to changes in applied forces.

The 'Rate' and 'Static' Problems

The problem of determining the evolution of the frictional state due to a known loading history $\boldsymbol{F}(t)$ can be stated in differential form, such that we determine the time derivative $\dot{\boldsymbol{u}}(t)$ of the contact displacements as a function of the loading rate $\dot{\boldsymbol{F}}(t)$ and parameters such as $\boldsymbol{u}(t)$ defining the instantaneous state. This incremental problem is known as the *rate problem*. In principle, the solution of the rate problem could then be integrated in time to define the solution of the evolution problem.

In the special case of proportional loading, where $\boldsymbol{F}(t) = g(t)\boldsymbol{F}_0$ and $g(t)$ increases monotonically from zero, the velocities in Eqs. (2.1)–(2.4) can be replaced by instantaneous tangential displacements $\boldsymbol{u}$, defining what is usually known as the *static problem*. Here, we shall be mainly concerned with more complex non-monotonic loading, to which the static formulation does not apply.

Unfortunately, we encounter a difficulty here. Mathematicians have been unable to prove that the rate problem for a general elastic system is well-posed in either the discrete or the continuum formulation. This is a particular problem for finite element codes, since these are usually designed to be user-friendly and need to be 'robust', meaning that meaningful results are returned when the problem is physically well-defined.

More precisely, existence and uniqueness theorems have been proved for frictional problems under the stipulation that *the friction coefficient f be sufficiently small* (Haslinger and Nedlec 1983; Kikuchi and Oden 1988). In mathematical language, this means that for a given elastic system, there exists some critical coefficient of friction f_{cr} such that if $f < f_{cr}$, the rate problem has a unique solution for all loading scenarios $\boldsymbol{F}(t)$. But unfortunately, f_{cr} is system dependent and usually cannot be easily determined. And even if we can determine f_{cr}, how will a frictional system evolve with a given $\boldsymbol{F}(t)$ if $f > f_{cr}$? To explore these issues, we shall first consider a very simple frictional system due to Klarbring (1990).

The Klarbring Model

Figure 2.2a shows a point mass M supported by a generalized two-dimensional massless spring $\boldsymbol{K}$ and making frictional contact with a rigid plane surface, with friction coefficient f. The mass is loaded by a time-varying force $\boldsymbol{F} = \{F_1, F_2\}$ and as a result experiences displacements $\boldsymbol{u} = \{v, w\}$ in directions $\{x_1, x_2\}$ respectively.

Figure 2.2b shows a free-body diagram of the mass, including the spring force $-\boldsymbol{K}\boldsymbol{u}$ and the normal and tangential contact reactions p, q, respectively. We assume that the loading rate is sufficiently slow for quasi-static conditions to apply, so the mass is always in equilibrium, leading to the equations

$$\begin{aligned} F_1 + q - k_{11}v - k_{12}w &= 0 \\ F_2 + p - k_{21}v - k_{22}w &= 0. \end{aligned} \tag{2.10}$$

Fig. 2.2 **a** The Klarbring model, **b** free-body diagram of the mass M

At any given time, the system must be in one of the following four states:

$$\begin{array}{llll} \dot{v}=0 & w=0 & |q|\le fp & \text{stick} \\ q=-fp & w=0 & \dot{v}\;>0 & \text{forward slip} \\ q=fp & w=0 & \dot{v}\;<0 & \text{backward slip} \\ q=0 & p=0 & w\;>0 & \text{separation.} \end{array} \tag{2.11}$$

We also require that the normal contact force be non-tensile—i.e. $p \ge 0$.

Monotonic Proportional Loading

Suppose the force components F_1, F_2 are increased monotonically and in proportion. For example, $\boldsymbol{F}(t) = \boldsymbol{C}t$, where $\boldsymbol{C} = \{C_1, C_2\}$ is a time-independent vector. This is a case where the static and rate problems are equivalent, and the system can remain in the same state for all t.

Stick. If that state is stick, we have $v = w = 0$. so $q = -C_1 t$, $p = -C_2 t$ and stick is possible if and only if

$$fC_2 < C_1 < -fC_2. \tag{2.12}$$

Separation. If the state is separation, we have $p = q = 0$, and solving for w, we obtain

$$w = \frac{k_{11}C_2 - k_{21}C_1}{(k_{11}k_{22} - k_{12}k_{21})}. \tag{2.13}$$

The denominator must be positive since $\boldsymbol{K}$ is positive definite, so separation is possible if and only if

$$k_{11}C_2 - k_{21}C_1 > 0 \quad \text{or} \quad C_2 > \frac{k_{21}}{k_{11}}C_1. \tag{2.14}$$

Similar calculations can be performed on the equations and inequalities for forward slip and backward slip. The resulting inequalities define the admissible domains for each state in F_1F_2-space. Figure 2.3 shows the resulting diagram for the case where $k_{21} > 0$ and $f < k_{11}/k_{21}$. Notice that one and only one state is possible for any given set of values of $\{C_1, C_2\}$. In other words, at least under proportional loading, a solution exists and is unique.

More General Loading Scenarios

Suppose now that we define a loading scenario that starts with a period in which $|F_1| < fF_2$, but which later crosses the stick-backward slip boundary in Fig. 2.3.

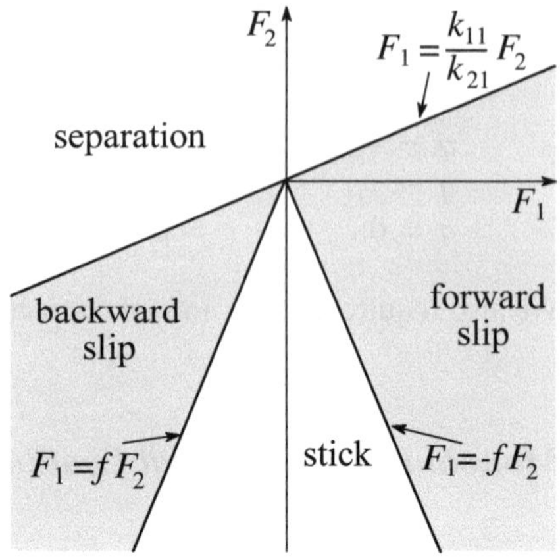

Fig. 2.3 State diagram for proportional loading for a case where $k_{21} > 0$ and $f < k_{11}/k_{21}$

The system will therefore start with a period of stick and then transition to backward slip. In other words, the mass will move to the left by a distance just sufficient to maintain the condition $q = fp$. This modifies the state diagram to that shown in Fig. 2.4a. In other words, the vertex of the stick sector moves along the separation boundary.

When contact transitions to separation or vice versa, this always occurs through the vertex of the stick sector and in a transition to contact, the initial state [backward slip, stick, or forward slip] depends on the local slope of the force trajectory. Figure 2.4b illustrates a case of a transition from separation to forward slip.

In summary, the system has a well-defined and unique response to all possible loading scenarios. Similar results are obtained for all cases for which

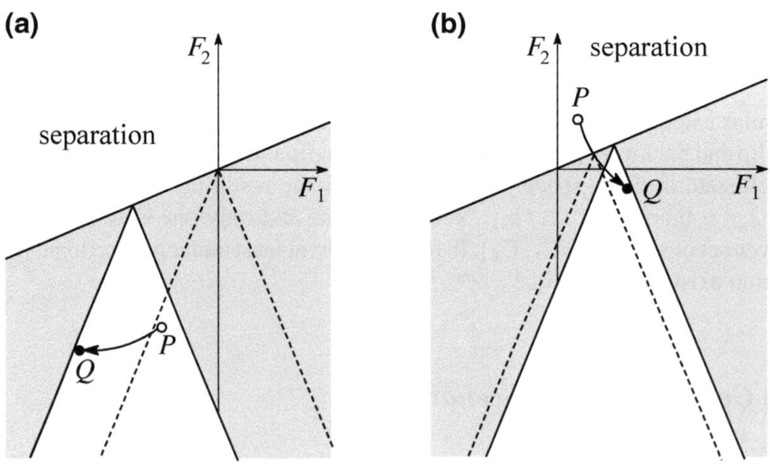

Fig. 2.4 **a** Response to a period of backward slip, **b** transition from separation to forward slip

$$f < f_{cr} \quad \text{where} \quad f_{cr} = \frac{|k_{21}|}{k_{11}} \tag{2.15}$$

defines the critical coefficient of friction.

The Case $f > f_{cr}$

A similar analysis for the case where $f > f_{cr}$ shows that with proportional loading there exists a sector in $F_1 F_2$-space where stick, slip and separation are *all* possible. Figure 2.5 illustrates this for a case where $k_{21} > 0$, in which case the slip direction in the multiple-solution sector is backward.

The non-uniqueness occurs strictly only at the origin. For example, if the contact remains stuck as we move a finite distance into the multiple-solution sector, it must then remain stuck, since a finite disturbance would be needed to move to separation [say]. The same applies to the other states in this sector, except that backward slip is dynamically unstable. If the inertia terms are reintroduced into Eq. (2.10), it can be shown that a state of backward slip at constant speed satisfies the equations of motion identically, but an infinitesimal perturbation on this state would grow exponentially until a transition occurred to either stick or separation (Cho and Barber 1998).

Loading scenarios requiring a jump. Figure 2.6 shows a non-proportional loading scenario that starts in the stick sector, but which then passes through the multiple solution sector into that where separation is the only option.

In this case, stick must occur as far as A and at this point the inequality $|q| < fp$ is satisfied in the strict sense, so no slip can occur. In fact, stick must continue until we reach the point B, after which there must be a transition to separation, but this implies an instantaneous change in both displacements $\{v, w\}$. In other words, under the quasi-static assumption, the system must experience a discontinuous jump in displacements. It is no coincidence that existence theorems for frictional contact

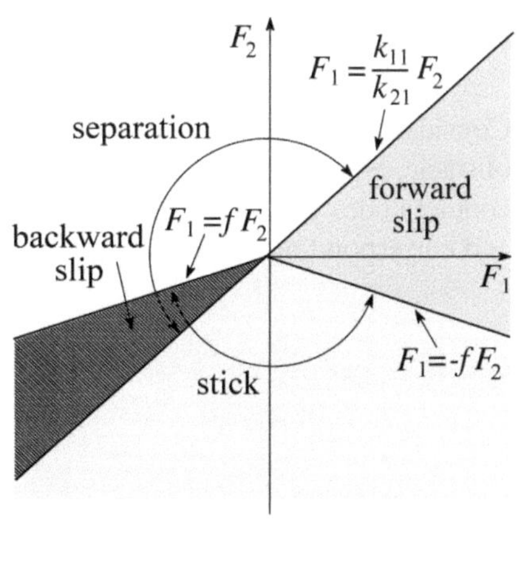

Fig. 2.5 State diagram for proportional loading for a case where $k_{21} > 0$ and $f > f_{cr}$

Fig. 2.6 A loading scenario passing through the multiple solution sector

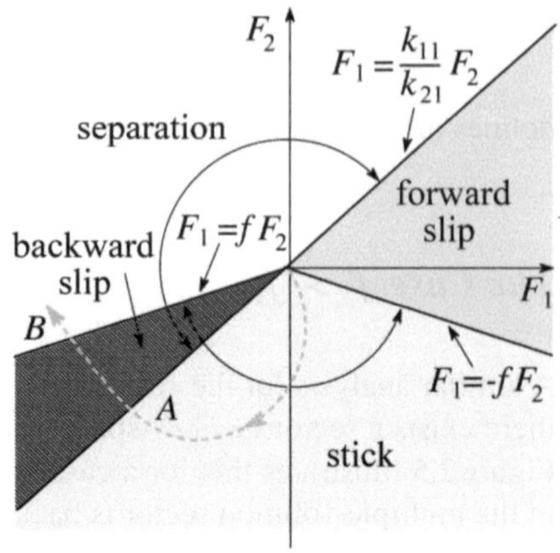

problems can be proved if the condition that displacement be a continuous function of load be relaxed (Martins et al. 1992).

The actual behaviour at such a jump will necessarily be elastodynamic and will occur on the time scale of the natural frequencies of the system. If the system is critically damped, it will tend monotonically to the new [separation] state, but if it is sub-critically damped, it may bounce off the support a few times before settling into a decaying oscillation about the new state.

If the loading rate is 'slow', this dynamic event is fast relative to the loading rate, and in the limit, we can say that these transitions occur instantaneously. The rate problem is strictly ill-posed, but we can still define a quasi-static solution to the evolution problem, provided we accept that infinitesimal changes in load sometimes require finite changes of displacement. Notice however that the non-uniqueness at the origin in Fig. 2.5 still remains.

Finite Element Discretization of a Two-Dimensional Frictional Contact Problem

Consider the case where a two-dimensional elastic body is in contact with a rigid obstacle, and the solution is approximated using the finite element method. At the contact nodes $j = (1, N)$, we anticipate normal and tangential nodal forces p_j, q_j and corresponding displacements w_j, v_j as shown in Fig. 2.7.

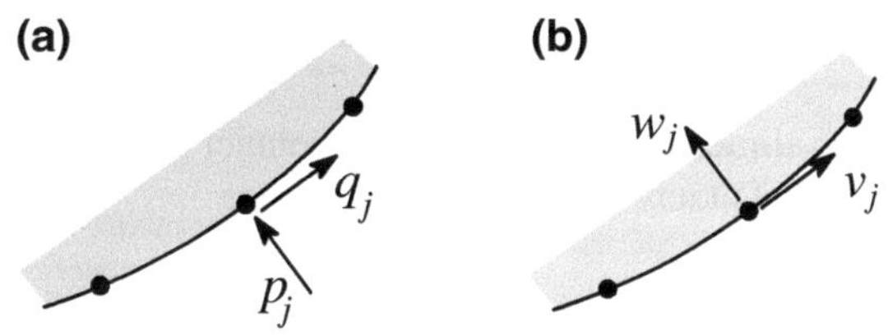

Fig. 2.7 **a** Nodal forces and **b** nodal displacements in a two-dimensional finite element discretization

Static Reduction

The entire elastic system will be discretized using M nodes $[M > N]$, but we can partition the stiffness matrix such that the nodal forces

$$\boldsymbol{F} \equiv \begin{Bmatrix} \boldsymbol{F}^{\mathrm{I}} \\ \boldsymbol{F}^{\mathrm{E}} \\ \boldsymbol{F}^{\mathrm{C}} \end{Bmatrix} = \boldsymbol{K}\boldsymbol{u} \equiv \begin{bmatrix} \boldsymbol{K}^{\mathrm{II}} & \boldsymbol{K}^{\mathrm{IE}} & \boldsymbol{K}^{\mathrm{IC}} \\ \boldsymbol{K}^{\mathrm{EI}} & \boldsymbol{K}^{\mathrm{EE}} & \boldsymbol{K}^{\mathrm{EC}} \\ \boldsymbol{K}^{\mathrm{CI}} & \boldsymbol{K}^{\mathrm{CE}} & \boldsymbol{K}^{\mathrm{CC}} \end{bmatrix} \begin{Bmatrix} \boldsymbol{u}^{\mathrm{I}} \\ \boldsymbol{u}^{\mathrm{E}} \\ \boldsymbol{u}^{\mathrm{C}} \end{Bmatrix}, \tag{2.16}$$

where $\boldsymbol{u}^{\mathrm{I}}$ is a vector comprising the displacements at internal [unloaded] nodes, and $\boldsymbol{u}^{\mathrm{E}}$ comprises externally loaded non-contact nodes.

We then use the equation $\boldsymbol{F}^{\mathrm{I}} = \boldsymbol{0}$ and the known values of the external loads $\boldsymbol{F}^{\mathrm{E}}(t)$ to eliminate $\boldsymbol{u}^{\mathrm{I}}$, $\boldsymbol{u}^{\mathrm{E}}$ (Thaitirarot et al. 2014), obtaining an equation of the form

$$\begin{Bmatrix} \boldsymbol{q} \\ \boldsymbol{p} \end{Bmatrix} = \begin{Bmatrix} \boldsymbol{q}^{w}(t) \\ \boldsymbol{p}^{w}(t) \end{Bmatrix} + \begin{bmatrix} \boldsymbol{A} & \boldsymbol{B}^{T} \\ \boldsymbol{B} & \boldsymbol{C} \end{bmatrix} \begin{Bmatrix} \boldsymbol{v} \\ \boldsymbol{w} \end{Bmatrix}, \tag{2.17}$$

where $p_i^w(t), q_i^w(t)$ are the contact (nodal) forces that would be produced by the external loads $\boldsymbol{F}^{\mathrm{E}}(t)$ if the nodes were all welded in contact at $\boldsymbol{v} = \boldsymbol{w} = \boldsymbol{0}$.

The reduced stiffness matrix

$$\boldsymbol{K} = \begin{bmatrix} \boldsymbol{A} & \boldsymbol{B}^{T} \\ \boldsymbol{B} & \boldsymbol{C} \end{bmatrix} \tag{2.18}$$

is symmetric and positive definite, so $\boldsymbol{A}$ and $\boldsymbol{C}$ are also symmetric and positive definite and of dimension $N \times N$. The matrix $\boldsymbol{B}$ defines the coupling between normal contact forces and tangential displacements and vice versa. We shall find that $\boldsymbol{B}$ has a critical effect on the history-dependence of the frictional contact problem. It is *not* necessarily symmetric or positive definite.

Notice that we have developed this description from a finite element discretization of a continuum problem, but the resulting equations are mathematically equivalent to those from a generalization of the Klarbring model to N nodes—i.e. to a system comprising N rigid bodies connected by a generalized spring $\boldsymbol{K}$ and making frictional contact with N obstacles.

Frictional Contact Conditions

For a two-dimensional system, each contact node i must be in one of the four states of Eq. (2.11) at any given time t and hence

$$\begin{array}{llll}
\dot{v}_i = 0 & w_i = 0 & |q_i| \le f p_i & \text{stick} \\
q_i = -f p_i & w_i = 0 & \dot{v}_i \ > 0 & \text{forward slip} \\
q_i = f p_i & w_i = 0 & \dot{v}_i \ < 0 & \text{backward slip} \\
q_i = 0 & p_i = 0 & w_i \ > 0 & \text{separation.}
\end{array} \tag{2.19}$$

Solution Algorithms

Various algorithms are available for the solution of problems defined by given functions $\boldsymbol{p}^w(t), \boldsymbol{q}^w(t)$ and conditions (2.19).

Iteration on the set of states. One of the simplest involves the use of a 'sufficiently small' constant time step defining a set of instants y_k. Then, assuming that all the nodal values are known at time t_k, we:

(i) Assume the state at each node remains the same at t_{k+1}.
(ii) Use the *equations* from (2.19) to calculate the values of p_i, q_i, v_i, w_i at time t_{k+1} for all i.
(iii) Check the state *inequalities* from (2.19).
(iv) If an inequality is violated at node j, make an appropriate change of state at that node and re-solve the equations.
(v) Repeat until the state assumptions for this time step satisfy the appropriate inequalities at all N nodes.
(vi) Advance by one time step.

Gauss–Seidel iteration. For larger values of N, it is more efficient to combine this algorithm with a Gauss–Seidel approach (Ahn and Barber 2008):

(i) We solve the equations for p_j, q_j, v_j, w_j for node j under the assumption that the corresponding values for nodes $i \ne j$ remain at their most recent estimated values.
(ii) The state at node j [only] is also changed if necessary to satisfy the inequalities at that node.
(iii) We then track through the nodes repeatedly until a suitable convergence criterion is satisfied.

LCP solution. An alternative approach is to use an LCP [*Linear Complementarity Problem*] algorithm:

(i) Assuming the states remain the same and that the loads change linearly with time, solve the resulting linear equations for the time t_{k+1} at which each of the inequalities will first be violated.

(ii) Choose the smallest of these times—i.e. the time at which an inequality is first violated.
(iii) Use the nature of the violation to predict what will happen next—e.g. if p_j is predicted to become negative starting at $t = \tau$, set $t_{k+1} = \tau$ and make a transition [pivot] at node j to separation for the next time step.

Evolution in Displacement Space

The instantaneous configuration of the system is completely defined by the displacement components v_i, w_i, so the evolution of the system due to varying loads can be described by the motion of a point in a space defined by these coordinates. We consider a simpler case where all nodes are in contact [$w_i = 0$ for all i] and we track the evolution in v_i-space.

The two-node case. The process is most easily visualized in a system with only two nodes. The contact forces are then given by

$$\begin{aligned} p_1 &= p_1^w + B_{11}v_1 + B_{12}v_2 \\ q_1 &= q_1^w + A_{11}v_1 + A_{12}v_2 \\ p_2 &= p_2^w + B_{21}v_1 + B_{22}v_2 \\ q_2 &= q_2^w + A_{21}v_1 + A_{22}v_2. \end{aligned} \tag{2.20}$$

Stick at node 1 [for example] requires that $-fp_1 \le q_1 \le fp_1$ and hence

$$\begin{aligned} (A_{11} - fB_{11})\,v_1 + (A_{12} - fB_{12})\,v_2 &\le fp_1^w - q_1^w && \text{I} \\ (A_{11} + fB_{11})\,v_1 + (A_{12} + fB_{12})\,v_2 &\ge -fp_1^w - q_1^w && \text{II} \quad . \end{aligned} \tag{2.21}$$

Each of these two inequalities excludes the domain on one side the straight line defined by the corresponding equality. There are two more similar inequalities for node 2, so the stick domain is defined by the four straight lines in Fig. 2.8 (Ahn et al. 2008).

During periods of slip, the corresponding inequality is satisfied as an equality, and the point $P(v_1, v_2)$ must lie *on* the corresponding line. The corresponding direction of slip is indicated by arrows on Fig. 2.8. If both nodes are slipping, P must lie at the intersection of the corresponding two lines. Also, if the excluded regions overlap so that there is no admissible region, then at least one of the two nodes must be in a state of separation.

Loading scenario. The slope of the line

$$(A_{11} - fB_{11})\,v_1 + (A_{12} - fB_{12})\,v_2 = fp_1^w - q_1^w$$

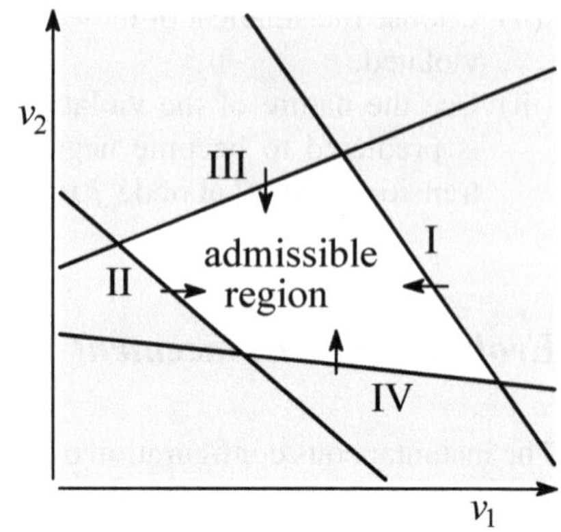

Fig. 2.8 The inequalities (2.21) exclude the region to the right of line I and to the left of line II respectively. During forward slip at node 1 [$\dot{v}_1 > 0$], the instantaneous point must lie *on* the line II

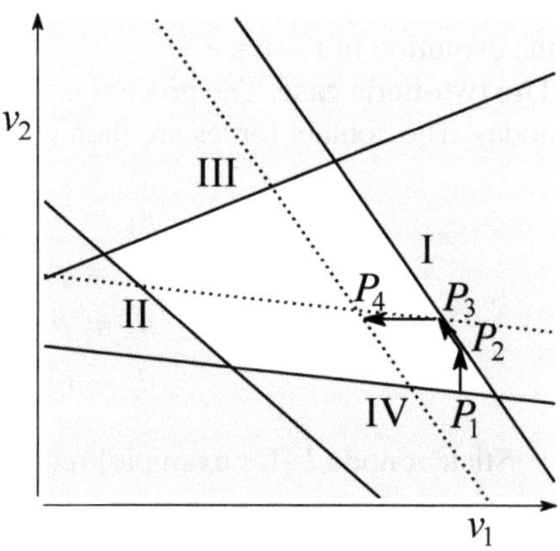

Fig. 2.9 Motion of the point $P(v_1, v_2)$ due to changes in the external loads causing constraints IV and I to move in succession

[for example] depends only on the matrices $\boldsymbol{A}$, $\boldsymbol{B}$ and the coefficient of friction f. So as p_i^w, q_i^w change, the lines in Fig. 2.8 move [excluding more or less space], but they retain the same slope. In effect, they 'push' the point $P(v_1, v_2)$ around the space.

Figure 2.9 shows a case where the constraint IV advances to the dotted line, after which constraint I advances. If the initial condition was defined by the point P_1, the advance of IV initially causes only node 2 to slip until we reach P_2 where node 1 also starts to slip. When I advances, only node 1 slips and P moves to P_4.

The critical coefficient of friction. It is clear that for this kind of evolution to be possible, slip at any node must move P away from the excluded region, rather than into it. In the case of constraint I in Eq. (2.21), this requires that

$$A_{11} - fB_{11} > 0 \quad \text{and hence} \quad f < \frac{B_{11}}{A_{11}}. \tag{2.22}$$

This is clearly similar to the criterion $f < k_{21}/k_{11}$ for the single node Klarbring model [Eq. (2.15)], and similar criteria can be written for the other constraints.

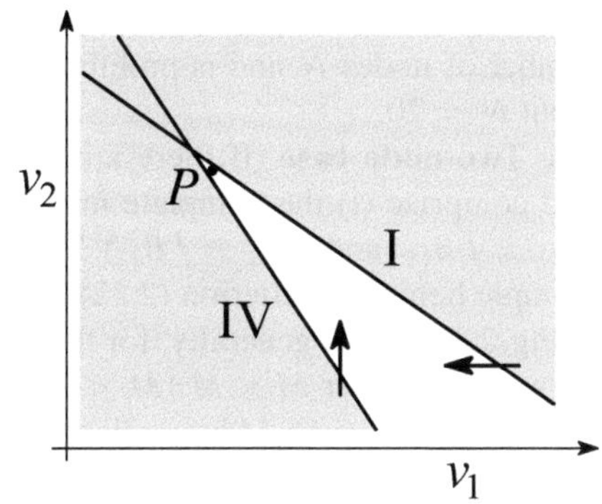

Fig. 2.10 If the constraints I, IV move so as to increase the shaded area, P would need to move downwards and to the right, but these directions are not compatible with the slip directions for these constraints

However, Fig. 2.10 shows a situation in which both constraints I and IV satisfy conditions of the form (2.22)—i.e. separately they permit slip at their respective nodes as they advance—but in which in combination the two constraints can 'trap' P in such a way that the permitted directions of slip are impossible.

With larger numbers of nodes $[N > 2]$, v-space becomes an N-dimensional hyperspace and each constraint excludes the hypervolume on one side of a hyperplane of dimension $(N - 1)$. The instantaneous admissible stick space is then a polyhedron in this hyperspace, whose facets are defined by segments of these hyperplanes. Trapping of the instantaneous point $P(v_1, v_2, \ldots, v_N)$ can involve the interaction of multiple constraints. For example, P can be trapped in a vertex or along an edge.

Klarbring's P-Matrix Condition

Klarbring (1999) has shown that the discrete frictional rate problem is mathematically well-posed if and only if every matrix of the form

$$\boldsymbol{A} + f \boldsymbol{\Lambda} \boldsymbol{B}$$

is a P-matrix, where $\boldsymbol{A}$, $\boldsymbol{B}$ are defined in Eq. (2.18) and $\boldsymbol{\Lambda}$ is any diagonal matrix each of whose diagonal elements is either $+1$ or -1. A P-matrix is a positive definite matrix, all of whose principal minors are also positive definite.

The matrix $\boldsymbol{A}$ is a P-matrix, so Klarbring's criterion is always satisfied for $f = 0$. However, if $\boldsymbol{B} \neq \boldsymbol{0}$, there exists some critical coefficient $f = f_{\text{cr}}$ above which the criterion is not satisfied.

We can find f_{cr} by

(i) Evaluating the determinant of all principal minors of $(\boldsymbol{A} + f \boldsymbol{\Lambda} \boldsymbol{B})$ for all possible matrices $\boldsymbol{\Lambda}$.
(ii) Equating each such determinant to zero and solving for f.
(iii) Setting f_{cr} equal to the minimum of these values.

However, the number of calculations involved increases combinatorially with the number of nodes N and is prohibitively computer-intensive for values bigger than about $N = 20$.

The Two-node case. If there are only two nodes, the determinants to be evaluated comprise (i) the complete matrix $(\boldsymbol{A} + f\boldsymbol{\Lambda}\boldsymbol{B})$ and (ii) the diagonal elements $(A_{11} \pm fB_{11})$ and $(A_{22} \pm fB_{22})$. The latter is associated with a single constraint, an example being the criterion (2.22), whereas the former precludes the two-node trap of Fig. 2.10. More generally, for the N-node system, a criterion involving a principal minor of order $M \times M$ $[M < N]$ involves the point P becoming trapped at the intersection between M hyperplanes.

Ahn (2010) developed a dynamic solution to the two-node system and showed that in scenarios involving a two-node trap, an initial instability causes the system to evolve dynamically to a unique state involving a displacement discontinuity at both nodes. It seems likely that similar dynamic responses would be predicted for N-node systems at higher order traps, but the present author is unaware of any proof, and one can only speculate as to whether the final state reached would be unique.

Wedging

We define a frictional elastic system as 'wedged' if it exists in a non-trivial state of stick [i.e. the internal forces are non-zero] even though the imposed external loads and/or displacements are zero. It is analogous to a state of residual stress in an elastic–plastic system that has been loaded past the yield state and then unloaded.

Wedging is important in automated assembly processes, since components may 'jam' in an incorrect configuration during assembly. But in other contexts [e.g. screwed fasteners], wedging is *necessary* for the satisfactory performance of the device.

It is clear that some coupling [e.g. $\boldsymbol{B} \neq \boldsymbol{0}$] is needed for a wedged state, since if there is no coupling, the solution for the normal forces is trivial [they will all be zero] and hence none of the contacts could support a frictional force.

Figure 2.11 shows a simple mechanical system that can become wedged when the force F is removed if the coefficient of friction is sufficiently high, or the angle of the wedge sufficiently low.

Ideally, we would like to be able to determine a critical coefficient of friction f_{w} below which wedging is impossible.

Wedging for the Klarbring model. For the Klarbring model of Fig. 2.2 to be wedged, we must have $w = 0$ and $\boldsymbol{F} = \boldsymbol{0}$, so the equilibrium equation (2.10) reduce to

$$q = k_{11}v; \quad p = k_{21}v. \tag{2.23}$$

We can always choose the sign of v to satisfy the inequality $p > 0$, and wedging is possible if and only if $|q| < fp$ and hence

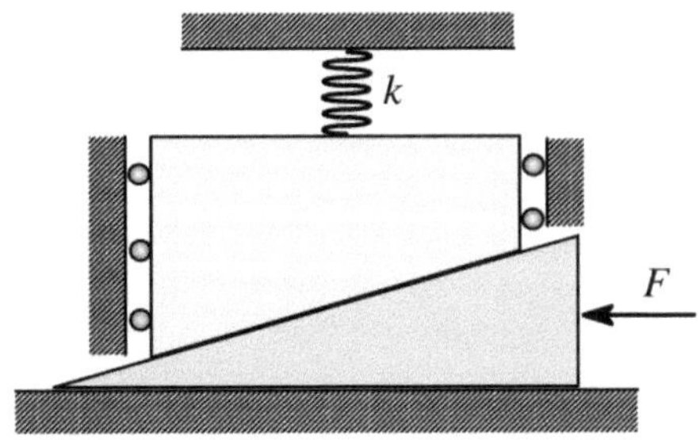

Fig. 2.11 A frictional elastic system susceptible to wedging

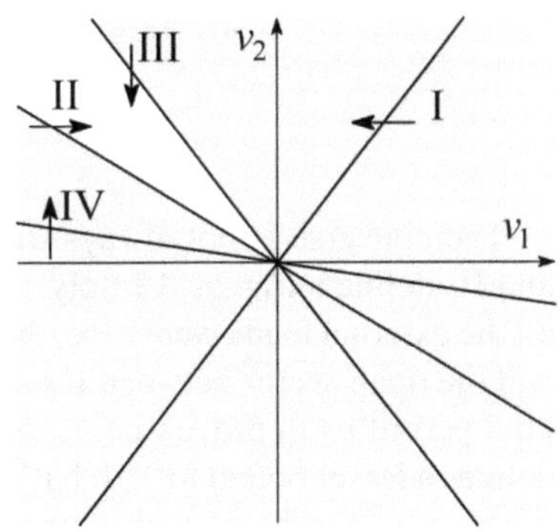

Fig. 2.12 The unshaded sector defines the region in which wedging is possible

$$f > f_{\mathrm{w}} = \frac{k_{11}}{|k_{21}|}. \tag{2.24}$$

In this case, the critical coefficient of friction is identical to the value of f_{cr} given by Eq. (2.15), which is also a special case of Klarbring's Condition of section "Klarbring's P-Matrix Condition". However, the two problems are not identical for the more general N-node case, as we shall see.

The N-node case. If there are no external loads $[\boldsymbol{p}_w = \boldsymbol{q}_w = \boldsymbol{0}]$, the right-hand sides of all the constraints analogous to (2.21) are zero. It follows that if a non-trivial wedged state $\boldsymbol{v}_0$ satisfies all of the frictional constraints [i.e. lies strictly within the admissible region for all nodes], then the state $\lambda\boldsymbol{v}_0$ will also satisfy these constraints for all $\lambda > 0$.

In v-space, all the constraints for this homogeneous case define hyperplanes that pass through the origin and the region where wedging is possible comprises a sector radiating from the origin, bounded by hyperplanes and open to infinity. This is illustrated for the two-node case in Fig. 2.12.

In this particular case, the two constraints defining the wedging sector also comprise a two-node trap in the evolution problem, but this is not always the case. For example, if we interchange the lines defining II and IV, the interaction of constraints II and III would still define a trap, but the wedged region would be defined by the smaller sector between III and IV. More generally, it can be shown that if there exists an open sector allowing wedging, then at least one trap exists, implying that $f > f_{\mathrm{cr}}$.

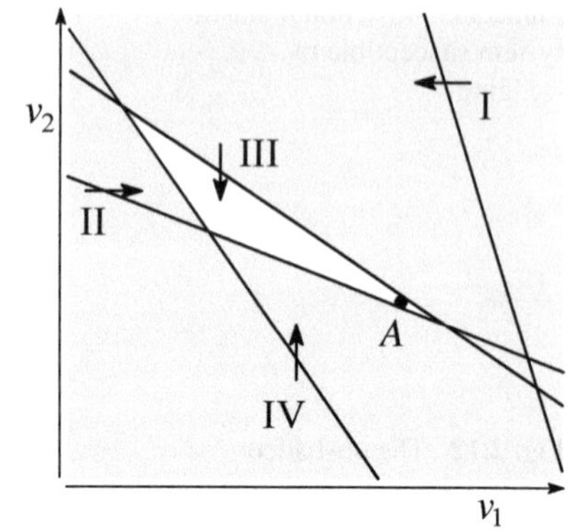

Fig. 2.13 A two-node case where the rate problem is ill-posed, but wedging is not possible

The converse is not always true. Figure 2.13 shows a case where constraints II and III define a trap at A implying $f > f_{\mathrm{cr}}$, but wedging is not possible because if all the external loads were zero, the constraint lines would all pass through the origin and the open sector between II and III would be excluded by IV. Arguments of this kind permit us to prove that $f_{\mathrm{w}} \geq f_{\mathrm{cr}}$ for the two-node system [in other words, f_{cr} defines a lower bound to f_{w}], but to the present author's knowledge, no such general proof exists for the N-node case.

Hild's Eigenvalue Problem

Hild and coworkers (Hassani et al. 2003; Hild 2004) have developed a method for determining an upper bound for f_{w} by considering the condition under which there are no applied loads and all points in the contact area are in a state of incipient slip—i.e. $|q| = fp$. If a non-trivial state can be identified for some particular value f_k, then this state will also define a wedged state for any coefficient of friction $f > f_k$ and hence f_k constitutes an upper bound to f_{w}.

For the discrete system defined by Eq. (2.17), if there are no external loads [$\boldsymbol{p}^w = \boldsymbol{q}^w = \boldsymbol{0}$], and if all the nodes are in contact [$\boldsymbol{w} = \boldsymbol{0}$], we obtain

$$\boldsymbol{q} = \boldsymbol{A}\boldsymbol{v}; \quad \boldsymbol{p} = \boldsymbol{B}\boldsymbol{v}. \tag{2.25}$$

If the limiting condition $|q_i| = fp_i$ is satisfied at all nodes, we then obtain the generalized eigenvalue equation

$$\boldsymbol{A}\boldsymbol{v} = f\boldsymbol{\Lambda}\boldsymbol{B}\boldsymbol{v}, \tag{2.26}$$

where once again $\boldsymbol{\Lambda}$ is a diagonal matrix each of whose diagonal components is either $+1$ or -1.

For any given eigenvalue f_k of (2.26), the determinant

$$|\boldsymbol{A} - f_k \boldsymbol{\Lambda} \boldsymbol{B}| = 0 \tag{2.27}$$

and this is one of the conditions from "Klarbring's P-Matrix criterion" defining f_{cr}. It follows that $f_k \geq f_{cr}$, where the inequality arises because there are other conditions in Klarbring's condition involving principal minors of this matrix that do not correspond to states involving incipient slip at all nodes.

More significantly, an eigenvalue of (2.26) defines an incipient slip state if and only if the corresponding eigenfunction defines values of p_i that all have the same sign [which can then be chosen so as to make $p_i > 0$ for all i]. In the special case where $N = 1$, there are no principal minors and there is only one term, so the two criteria necessarily coincide.

A Relaxation Procedure

Suppose that we impose arbitrary values of v_i whilst keeping all nodes in contact [$\boldsymbol{w} = \boldsymbol{0}$]. This will usually imply that some of the resulting p_i will be negative. Suppose we identify one such node [say j] and 'release' it, meaning that we set $p_j = q_j = 0$ whilst keeping all other nodal displacements unchanged. Repeating this process we can find a state where (i) all the nodes that remain in contact satisfy the condition $p_i > 0$ and (ii) at these nodes, there are non-zero tangential displacements v_i and hence also tangential forces q_i. This will define a wedged state if

$$f > \max\left(\frac{|q_i|}{p_i}\right) \equiv f_1, \tag{2.28}$$

where only the contacting nodes are considered. Suppose that the maximum value [which therefore defines f_1] occurs at node k.

Imagine now that we had some physical procedure that would allow us to reduce the coefficient of friction gradually at all nodes. When f falls slightly below f_1, node k will start to slip, but the other nodes will remain stuck. This phase of the process can be modelled by assuming slip at node k and stick at the remaining contact nodes until the coefficient of friction becomes low enough for a second node to reach incipient slip. This process can be repeated, also allowing nodal separation when required, until we reach a state where only one node remains stuck. The coefficient of friction at which this last node reaches incipient slip then defines a tighter upper bound to f_w than does f_1.

Periodic Loading

Engineering systems involving nominally static loading frequently also experience a periodic loading component. Examples include bolted joints subject to vibration, aero engine blade roots, periodic machine operating cycles, thermal stresses associated with diurnal temperature variation and many others. In these situations, periodic microslip can occur, often leading eventually to the initiation and propagation of fretting fatigue cracks.

Systems of this kind generally accumulate a large number of cycles during their lifetime, so we are principally interested in the steady state, rather than in the transient behaviour during the first few cycles. This is often overlooked in the theoretical and numerical literature, where there are numerous treatments of systems loaded monotonically from an unloaded state. Notice that the steady state may be reached after one or two cycles, or it may be approached asymptotically. The latter case implies a larger number of time steps in a numerical solution, but often this can be reduced by identifying the proportional change in residual tractions and displacements during a single cycle and approximating the limit by extrapolation.

Shakedown

In some circumstances, the steady state involves stick throughout the contact interface. In other words, some microslip occurs during the first few cycles, but the resulting residual stresses inhibit slip in subsequent cycles. This phenomenon is known as *shakedown*.

A similar effect is observed in the cyclic loading of elastic–plastic bodies, where residual stresses are associated with inelastic strains. For this case, *Melan's theorem* (Melan 1936) states essentially that 'the system will shake down if it can'—in other words, if a distribution of inelastic strain can be identified such that the resulting stress state lies within the yield surface at all points throughout the cycle.

For many years it was believed that a similar theorem could be applied to the microslip problem. In other words [in the discrete formulation]:

If there exists a set of time-invariant nodal slips such that the resulting nodal forces would satisfy the frictional stick condition at all nodes throughout the cycle, then the system will eventually shake down, though not necessarily to the state so identified.

However, we now know that this theorem can be proved if and only if the system is uncoupled (Klarbring et al. 2007). Counter-examples [i.e. loading scenarios for which both cyclic slip and shakedown are possible depending on the initial conditions] can be found for all systems where this condition is not satisfied.

In the two-dimensional discrete problem, this condition requires that the coupling matrix $\boldsymbol{B} = \boldsymbol{0}$ in Eq. (2.16), so that the contact tractions are defined by

$$\begin{Bmatrix} \boldsymbol{q} \\ \boldsymbol{p} \end{Bmatrix} = \begin{Bmatrix} \boldsymbol{q}^w(t) \\ \boldsymbol{p}^w(t) \end{Bmatrix} + \begin{bmatrix} \boldsymbol{A} & \boldsymbol{0} \\ \boldsymbol{0} & \boldsymbol{C} \end{bmatrix} \begin{Bmatrix} \boldsymbol{v} \\ \boldsymbol{w} \end{Bmatrix}, \tag{2.29}$$

or

$$\begin{aligned} \boldsymbol{q} &= \boldsymbol{q}^w(t) + \boldsymbol{A}\boldsymbol{v} \\ \boldsymbol{p} &= \boldsymbol{p}^w(t) + \boldsymbol{C}\boldsymbol{w}. \end{aligned} \tag{2.30}$$

However, the theorem also applies to uncoupled three-dimensional systems and to problems of the continuum.

The most common example of an uncoupled system is one in which contact occurs on a plane which is also a plane of symmetry. Also, if the contact problem is one which can be modelled by representing the bodies as elastic half spaces, the uncoupled assumption will be satisfied if Dundurs' bimaterial constant $\beta = 0$, where

$$\beta = \left[\frac{(1-2\nu_1)}{G_1} - \frac{(1-2\nu_2)}{G_2} \right] \Big/ \left[\frac{(1-\nu_1)}{G_1} + \frac{(1-\nu_2)}{G_2} \right], \tag{2.31}$$

and G_i, ν_i are the shear modulus and Poisson's ratio respectively for the contacting materials $i = 1, 2$ (Dundurs 1969). Situations in which $\beta = 0$ [or at least $\beta \ll 1$] include:

- One material is relatively rigid [$G_1 \gg G_2$] and the other is incompressible [$\nu_2 = 0.5$].
- Both materials are incompressible [$\nu_1 = \nu_2 = 0.5$].

This is important in many modern applications involving polymers and biomaterials, for which ν might be quite close to 0.5.

Ponter's Theorem

More generally, it can be shown (Andersson et al. 2014) that if an uncoupled two or three-dimensional discrete frictional problem is subjected to periodic loading, the following quantities are independent of initial conditions in the steady state:

- the set $\mathcal{T}$ of nodes that never slip (the *permanent stick zone*).
- the status of any given node i (stick, slip or separation) at any time t.
- the frictional tractions $\boldsymbol{q}_i(t)$ for all nodes $i \notin \mathcal{T}$.
- the nodal velocities $\dot{\boldsymbol{v}}_i(t)$.
- the frictional energy dissipation per cycle.

The tractions $\boldsymbol{q}_i(t)$ for $i \in \mathcal{T}$ will generally depend on initial conditions, but since these nodes never slip, this does not affect the frictional dissipation. The frictional Melan's theorem can be regarded as the special case of Ponter's theorem in which all nodes are in $\mathcal{T}$ and the frictional dissipation is zero.

Proof **The normal contact problem**. Since the system is uncoupled, the normal contact tractions $p_i(t)$ and normal displacements (gaps) $w_i(t)$ are uniquely determined by the instantaneous normal loads. They are therefore independent of initial conditions.

The tangential contact problem. The evolution of the periodic state is described by the time-dependence of the tangential displacements $\boldsymbol{v}(t)$. The corresponding tangential tractions are given by

$$\boldsymbol{q}(t) = \boldsymbol{A}\boldsymbol{v}(t). \tag{2.32}$$

The energy norm Suppose there exist two distinct transient solutions *(orbits)* $\boldsymbol{v}_1(t), \boldsymbol{v}_2(t)$ which correspond to the same external loading $\boldsymbol{p}^w(t), \boldsymbol{q}^w(t)$ but different initial conditions $\boldsymbol{v}_1(0), \boldsymbol{v}_2(0)$.

We define an *energy norm*

$$\mathcal{E} = \frac{1}{2}\left(\boldsymbol{v}_1(t) - \boldsymbol{v}_2(t)\right)^T \cdot \boldsymbol{A}\left(\boldsymbol{v}_1(t) - \boldsymbol{v}_2(t)\right), \tag{2.33}$$

which can be regarded as a scalar measure of the *difference* between the two orbits.

The time derivative of $\mathcal{E}$ can then be written

$$\begin{aligned}\dot{\mathcal{E}} &= \left(\dot{\boldsymbol{v}}_1(t) - \dot{\boldsymbol{v}}_2(t)\right)^T \cdot \boldsymbol{A}\left(\boldsymbol{v}_1(t) - \boldsymbol{v}_2(t)\right)\\ &= \left(\dot{\boldsymbol{v}}_1(t) - \dot{\boldsymbol{v}}_2(t)\right)^T \cdot \left(\boldsymbol{q}_1(t) - \boldsymbol{q}_2(t)\right).\end{aligned} \tag{2.34}$$

Non-zero contributions to $\dot{\mathcal{E}}$ can arise from node i at time t *if and only if*

$$\dot{\boldsymbol{v}}_1^{(i)}(t) - \dot{\boldsymbol{v}}_2^{(i)}(t) \neq 0 \quad \text{and} \quad \dot{\boldsymbol{q}}_1^{(i)}(t) - \dot{\boldsymbol{q}}_2^{(i)}(t) \neq 0. \tag{2.35}$$

This implies that node i is in contact and slipping at time t in one or both of orbits 1 and 2.

Slip in orbit 1 and stick in orbit 2. Consider the case where node i is slipping in orbit 1 but stuck in orbit 2. We then have

$$\boldsymbol{q}_i^1 = -\frac{f p_i(t)\boldsymbol{v}_i^1(t)}{|\boldsymbol{v}_i^1(t)|} \quad \text{and} \quad \boldsymbol{v}_i^2 = \boldsymbol{0}; \quad |\boldsymbol{q}_i^2(t)| < f p_i(t). \tag{2.36}$$

Figure 2.14 shows a cross section through the friction cone at node i and time t. Since the node is slipping in orbit 1, $\boldsymbol{q}_i^1$ must lie on the dashed circle $|\boldsymbol{q}_i| = f p_i$ and must oppose the slip velocity $\dot{\boldsymbol{v}}_i^1$ as shown. In orbit 2 where node i is stuck, $\boldsymbol{q}_i^2$ must lie strictly within the circle as shown.

It is clear from the figure that all states satisfying these conditions will correspond to cases where the scalar product $\dot{\boldsymbol{v}}_i^1(t) \cdot \left(\boldsymbol{q}_i^1(t) - \boldsymbol{q}_i^2(t)\right) < 0$, and since $\dot{\boldsymbol{v}}_i^2 = \boldsymbol{0}$, this means that the contribution of this node to $\dot{\mathcal{E}}$ is negative.

Slip in both orbits. We next consider the case where node i is slipping at time t in both orbits, but in different directions. The cross section through the friction cone is

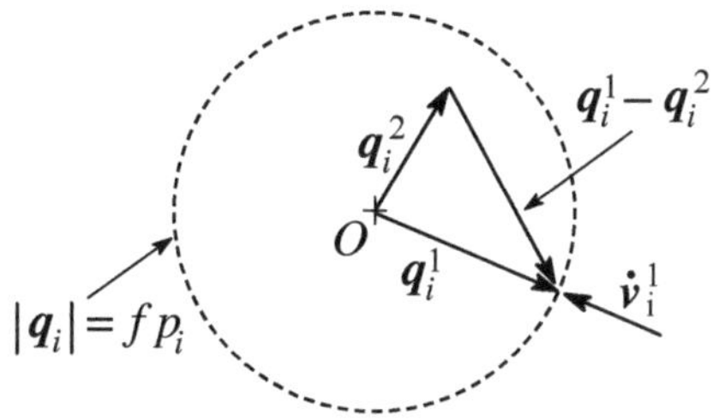

Fig. 2.14 Cross section through the friction cone at node i and time t

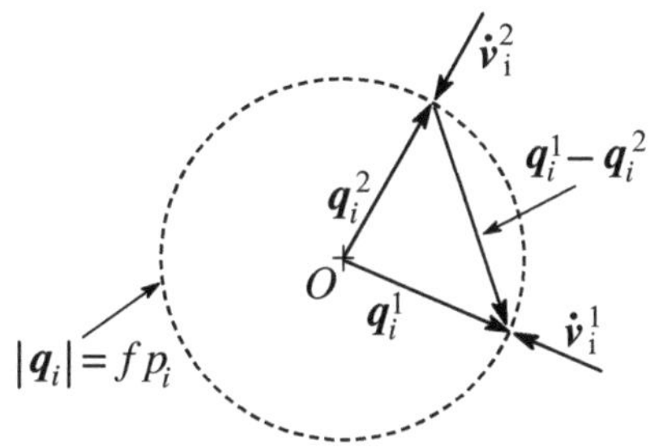

Fig. 2.15 Cross section through the friction cone when both nodes are slipping, but in different directions

shown in Fig. 2.15. In particular, we note that the traction $\boldsymbol{q}_i$ must oppose the slip velocity $\dot{\boldsymbol{v}}_i$ in both orbits.

It is clear from the figure that the scalar products

$$(\boldsymbol{q}_i^1 - \boldsymbol{q}_i^2) \cdot \dot{\boldsymbol{v}}_i^2 > 0 \quad \text{and} \quad (\boldsymbol{q}_i^1 - \boldsymbol{q}_i^2) \cdot \dot{\boldsymbol{v}}_i^1 < 0, \tag{2.37}$$

so $(\boldsymbol{q}_i^1 - \boldsymbol{q}_i^2) \cdot (\dot{\boldsymbol{v}}^1 - \dot{\boldsymbol{v}}^2) < 0$ and once again the contribution of this node to $\dot{\mathcal{E}}$ is negative.

Summary. Similar proofs show that the contribution of any node to $\dot{\mathcal{E}}$ is strictly negative whenever the state is different in orbits 1 and 2, or if both are slipping, whenever the slip directions are different. In all other cases the contribution to $\dot{\mathcal{E}}$ is zero. Thus, $\mathcal{E}$ decreases monotonically and must eventually converge [possibly asymptotically] on a positive value or zero.

The norm $\mathcal{E}$ is non-negative and non-increasing and hence, when both orbits have reached their steady states, it must be independent of time. We conclude that this condition is satisfied if and only if

1. The state of any given node i at any time t is the same in the two orbits.
2. When this unique state is one of slip, the slip direction is the same in the two orbits.

From (1), it follows that *The set $\mathcal{T}$ of nodes that never slip in the steady state is the same for both orbits.*

The homogeneous 'difference' problem. If the time-varying states of the nodes and the slip directions are assumed given, the remaining governing equations for the problem are linear and permit linear superposition. The difference $\boldsymbol{v}(t) = \boldsymbol{v}^1(t) - \boldsymbol{v}^2(t)$

corresponds to the solution of these equations with zero external force. During stick periods at node i, the slip velocity $\dot{\boldsymbol{v}}_i = \mathbf{0}$, whereas during slip periods, the tractions in the two orbits are the same, giving $\boldsymbol{q}_i(t) = \boldsymbol{q}_i^1(t) - \boldsymbol{q}_i^2(t) = \mathbf{0}$. The solution of this *homogeneous* problem is $\dot{\boldsymbol{v}}_i = \mathbf{0}$, so distinct orbits can differ by at most a time-invariant set of nodal displacements.

At nodes that are slipping, the slip velocities are the same in the two orbits, both in magnitude and direction. Since the normal forces are also the same, the frictional forces and hence the frictional energy dissipation are the same in the two orbits. We would expect fretting fatigue damage to correlate with frictional energy dissipation. It follows that for uncoupled systems, fretting damage should be independent of initial conditions [such as the assembly protocol for a bolted joint].

The permanent stick zone. When $\mathcal{E} \neq 0$ in the steady state, the difference between the two orbits can lie only in a time-invariant difference between $\boldsymbol{v}_1(t)$ and $\boldsymbol{v}_2(t)$. This difference can exist if and only if there exists a non-null permanent stick zone $\mathcal{T}$ comprising those nodes that never slip during the steady state. The 'system memory' resides in the slip displacements in these nodes.

Counter-examples. If in any given system there is coupling, we can always devise a loading scenario that defines a counter-example to the theorem. Suppose we apply large time-invariant normal loads to all the nodes except node i, sufficient to ensure that nodes $j \neq i$ never slip in the steady state.

If the initial conditions involve tangential displacements $\boldsymbol{v}_0$ of these stuck nodes, and if there are appropriate off-diagonal terms in $\boldsymbol{B}$, the normal load at the 'slipping' node i will depend on $\boldsymbol{v}_0$. So, the frictional force and the energy dissipation will also depend on $\boldsymbol{v}_0$.

Coupled Systems

If the system is coupled [e.g. if $\boldsymbol{B} \neq \mathbf{0}$], the steady-state response to periodic loading will generally depend on the initial conditions or the loading history. This implies that the system must in some sense possess memory. This is characteristic of hysteretic systems [those where the response depends on the direction but not the rate of deformation]. Based on the previous discussion, we anticipate that the memory must reside in the slip displacements $\boldsymbol{v}_i$ at nodes that are instantaneously stuck.

Cyclic load factor. Suppose the external loading takes the form

$$\begin{Bmatrix} \boldsymbol{q}^w(t) \\ \boldsymbol{p}^w(t) \end{Bmatrix} = \begin{Bmatrix} \boldsymbol{q}_0^w \\ \boldsymbol{p}_0^w \end{Bmatrix} + \lambda \begin{Bmatrix} \boldsymbol{q}_1^w(t) \\ \boldsymbol{p}_1^w(t) \end{Bmatrix}, \tag{2.38}$$

where $\boldsymbol{q}_0^w$, $\boldsymbol{p}_0^w$ are constant mean loads, $\boldsymbol{q}_1^w(t)$, $\boldsymbol{p}_1^w(t)$ are normalized periodic loads, and λ is a scalar load factor.

We anticipate the existence of critical values of λ, such that the steady state comprises:

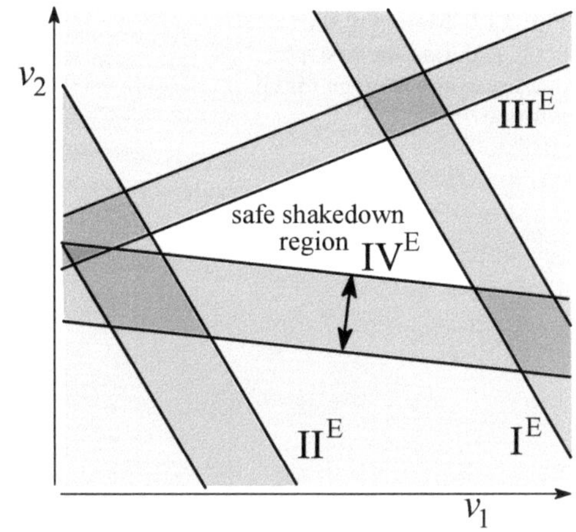

Fig. 2.16 Range of variation of the constraints I, II, III, IV during a periodic loading cycle

$0 < \lambda < \lambda_1$	Shakedown for all initial conditions.
$\lambda_1 < \lambda < \lambda_2$	Shakedown or cyclic slip depending on initial conditions.
$\lambda_2 < \lambda < \lambda_3$	Cyclic slip, but dissipation depends on initial conditions. At least some nodes must be permanently stuck.
$\lambda_3 < \lambda < \lambda_4$	Unique cyclic slip (or one of a few such states) approached asymptotically. The permanent stick zone $\mathcal{T} = \emptyset$.
$\lambda > \lambda_4$	Unique cyclic slip reached after a few cycles. There is some time during each cycle when all nodes slip.

We shall illustrate this sequence in the context of the two-dimensional two-node system.

In the corresponding $v_1 v_2$-diagram, each of the four constraints I, II, III, IV moves during the loading cycle between two extreme positions, For example, constraint IV moves in the range indicated by arrows in Fig. 2.16. We denote the extreme positions—i.e. those which exclude the largest area of the diagram—by I^E, IIE,IIIE, IVE as shown.

For shakedown to be possible, there must exist a *safe shakedown region* that is never excluded by any of the constraints. This region is defined by the extreme constraints, as shown in Fig. 2.16. However, whether this region is reached during a sequence of periodic loading cycles can depend on the initial condition $\boldsymbol{v}(0)$.

Figure 2.17 shows a loading scenario in which constraint IV [which controls slip at node 2 with $\dot{v}_2 > 0$] advances to IVE and then recedes, after which constraint I [which controls slip at node 1 with $\dot{v}_1 < 0$] advances to I^E. If the initial condition is defined by the point P_1, this loading sequence will 'push' P to P_4. Repeating this sequence will result in P approaching the bottom left corner of the safe shakedown region [labelled SD in this figure] asymptotically, through geometrically decreasing alternating vertical and horizontal segments.

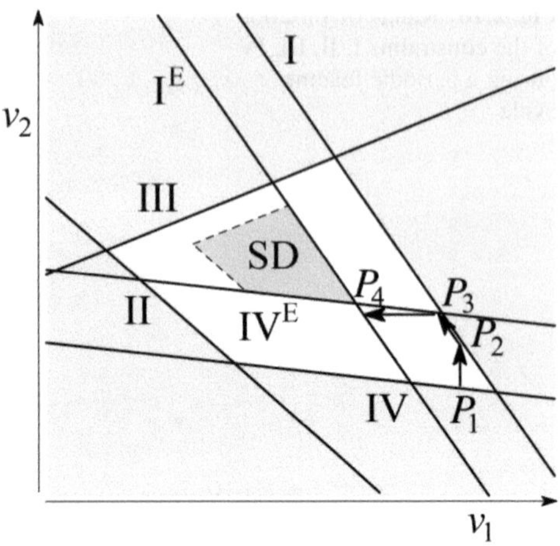

Fig. 2.17 Motion of the point P during the first loading cycle, from an initial position P_1

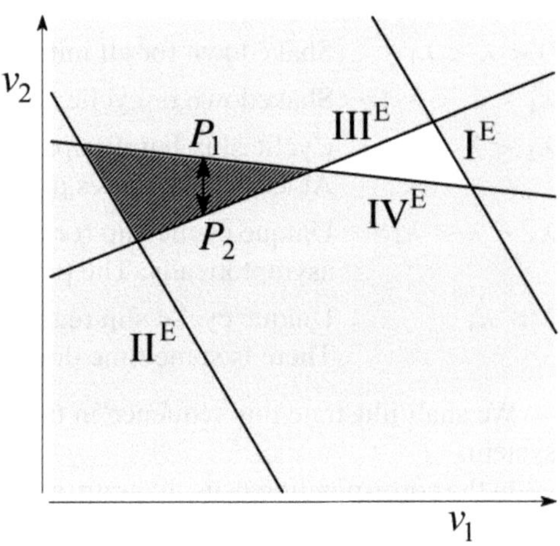

Fig. 2.18 A scenario in which the steady state can comprise cyclic slip [e.g. between points P_1 and P_2] or shakedown [in the unshaded triangle] depending on initial conditions

Ahn et al. (2008) have shown that P always reaches the safe shakedown space if this is a quadrilateral, but if it is triangular, shakedown depends on the initial condition. For example, Fig. 2.18 shows a case where the steady state can comprise cyclic slip between points P_1 and P_2, since the constraint II never moves far enough to the right to push P into the safe region. In this case, node 1 comprises the permanent stick zone $\mathcal{T}$. However, if the initial position lies to the right of the dark shaded triangle, P will always reach the safe [unshaded] region and the system will shake down.

Calculation of λ_1, λ_2. If λ is gradually increased from zero, the safe shakedown space will be a quadrilateral for $\lambda < \lambda_1$ and a triangle for $\lambda_2 > \lambda > \lambda_1$. Figure 2.19 illustrates this case for (a) $\lambda = \lambda_1^-$ and (b) $\lambda = \lambda_1^+$. Clearly the limiting case $\lambda = \lambda_1$ corresponds to the situation where the three constraints IIE, IIIE, IVE intersect at a common point. More generally, for the two-node system, we consider all possible

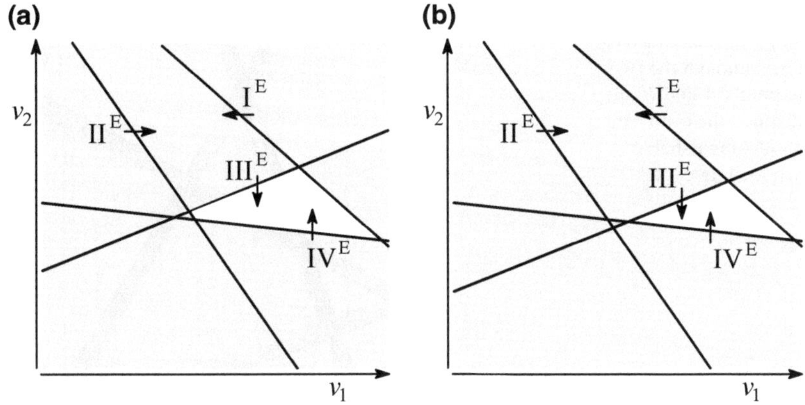

Fig. 2.19 The safe shakedown space [unshaded] for **a** $\lambda = \lambda_1^-$, **b** $\lambda = \lambda_1^+$

combinations of three extreme constraints. Treating these as equalities, we obtain three equations for the three unknowns v_1, v_2, λ. The lowest value of λ so obtained defines λ_1.

If λ is further increased, the triangular safe shakedown space will get smaller until eventually it shrinks to a point when $\lambda = \lambda_2$. This value can be determined by requiring that the three constraint lines defining the triangle should intersect in a common point. In Fig. 2.19, the three equations are those associated with I^E, IIIE, and IVE. More generally, there are four combinations of three constraint equations from the set of four, and λ_1, λ_2 are two of the four values of λ obtained from these combinations. The lowest positive value defines λ_1, but λ_2 is not always the second lowest.

The N-node system. For an N-node system, the safe shakedown space is an N-dimensional polyhedron in v_i-hyperspace and arguments similar to those above show that it will always be reached if all the $2N$ constraint hyperplanes are active in defining it (Ahn et al. 2008). The load factor λ_1 is then the lowest value at which one facet of this hypervolume shrinks to a point.

To determine λ_1, we choose both extreme constraints from one node [for which there are N choices] and one from each pair of extreme constraints at each of the other $N - 1$ nodes [2^{N-1} choices]. The corresponding equations are then solved for v_i, $i = (1, N)$ and λ. This procedure defines a set of $2^{N-1}N$ candidate values for λ, the lowest of which is λ_1. Like Klarbring's P-matrix criterion, the procedure becomes computationally impractical for N much larger than 20. Alternatively, optimization methods can be used to approach the correct limit more rapidly.

Asymptotic approach to a unique steady state. Consider a loading scenario where all the nodes slip at least once during each cycle, so the permanent stick zone $\mathcal{T}$ is null. At any given time, the system memory must reside in those nodes that are instantaneously stuck. This implies that the memory is 'exchanged' between nodes

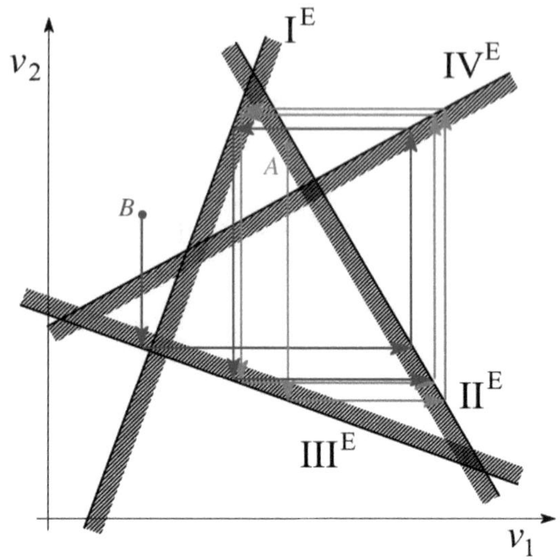

Fig. 2.20 A unique rectangular orbit acting as an attractor. Although the two starting points A and B are quite distinct, the orbits are almost indistinguishable after two loading cycles

during each cycle. We should expect some 'degradation' of memory during each interchange, so the system should tend asymptotically to a unique steady state.

Figure 2.20 shows the evolution of a two-node in which the extreme positions of the constraints are reached in the sequence I $\to$ II $\to$ III $\to$ IV $\to$ I $\to$ .... The response is shown for two initial positions A and B, but the results converge rapidly on a unique rectangular orbit. In other words, this orbit serves as an attractor in the iterative procedure.

However, if the coupling is stronger, implying a bigger difference between the slopes of constraints I and II, and between III and IV, a steady-state orbit can act as a repeller. Klarbring's P-matrix condition limits the degree of coupling if the rate problem is to be well-posed, but there exist well-posed problems leading to divergent orbits (Andersson et al. 2013).

A typical example is shown in Fig. 2.21, where the dashed rectangle represents the steady-state solution which is unique. In this case, the divergent response alternates from one side to the other of the steady state, so qualitatively similar orbits are obtained for all starting positions. The predicted deviation from the steady state increases without limit. However, with finite loading parameters, the position of the constraints must lie in the finite domain at all times and hence eventually we must encounter a situation when both nodes slip simultaneously.

For example, suppose the constraints in Fig. 2.21 advance to their extreme positions and then retract to a 'rest' position before the next constraint in the sequence moves. Figure 2.22 shows possible rest positions, with the regions that are excluded at all times indicated by solid colours.

In diverging from the unstable steady state, the transient orbit must eventually reach one of these boundaries at which point there will be a period during which both nodes slip. At this instant, since there are no stuck nodes, the system retains no

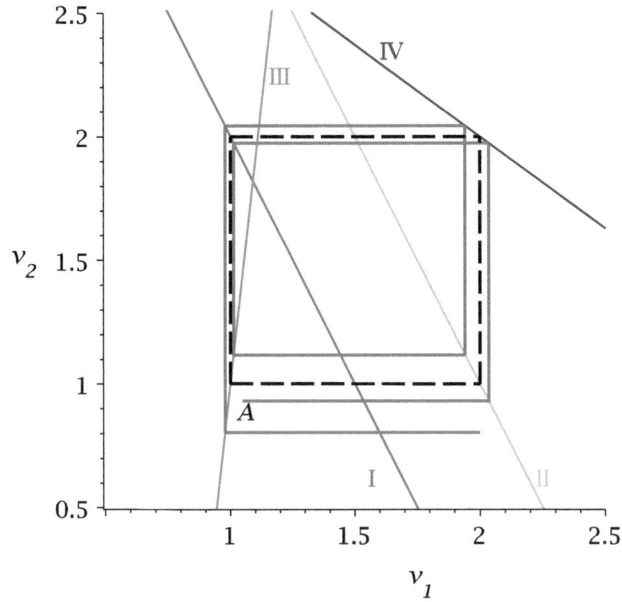

Fig. 2.21 The dashed rectangle defines a unique rectangular orbit which acts as a repeller. Starting from point A, the transient orbit alternates on the two sides of the steady state and increasingly deviates from it

Fig. 2.22 A limit cycle is established including a period in which both nodes slip. In the case illustrated, the limit cycle involves a double period of the loading cycle

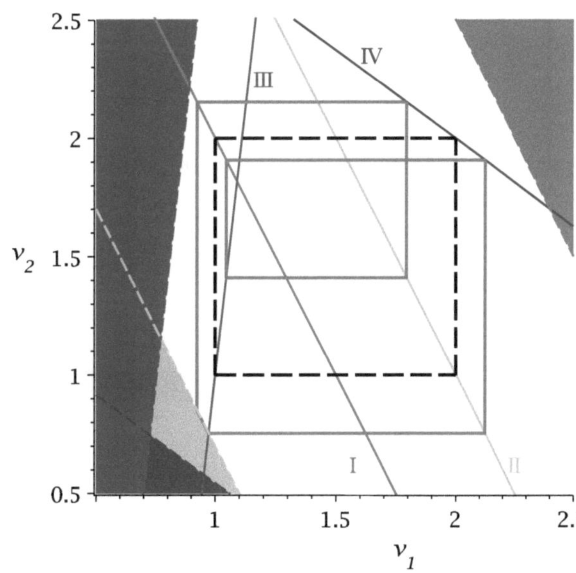

memory and the state is a unique function of the instantaneous loads. Subsequent evolution of the system follows the unique [and stable] ‘double’ orbit shown in Fig. 2.22.

Continuum Problems

Frictional problems for continuous elastic bodies can be reduced to discrete problems using [for example] finite element methods, but there remain important distinctions between the discrete and continuum formulations. To illustrate some of these, we consider the simple one-dimensional problem shown in Fig. 2.23.

A thin elastic strip of length L and cross-sectional area A and Young’s modulus E is pressed against a rigid plane by a force w per unit length, after which a time-varying force $F_0(t)$ is applied to the right end. We assume that Coulomb friction conditions apply at the interface, with friction coefficient f.

Figure 2.24 shows a small element of the strip of length δx. Axial equilibrium then requires that

$$F(x+\delta x) - F(x) + q\delta x = 0 \quad \text{and hence} \quad \frac{dF}{dx} + q = 0, \tag{2.39}$$

where q is the frictional force per unit length.

In a region that is slipping, the frictional force must oppose the relative motion, so

$$q = -fw\,\text{sgn}(\dot{u}) \quad \text{and} \quad \frac{dF}{dx} = fw\,\text{sgn}(\dot{u}), \tag{2.40}$$

where u is the local displacement in the positive x-direction. Since the strip is elastic, we also have

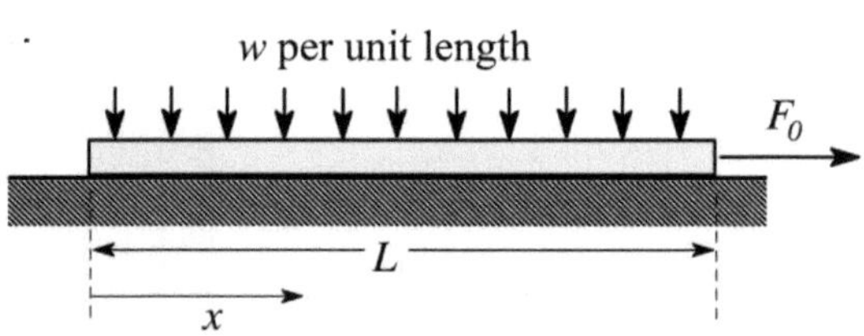

Fig. 2.23 An elastic strip pressed against a rigid plane and loaded by a time-varying force $F_0(t)$

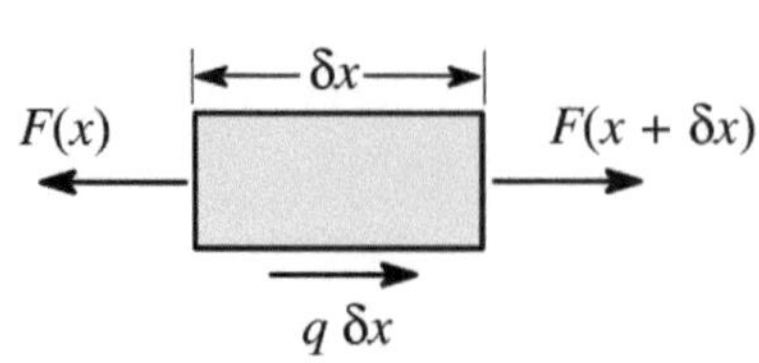

Fig. 2.24 Equilibrium of a small element of the strip

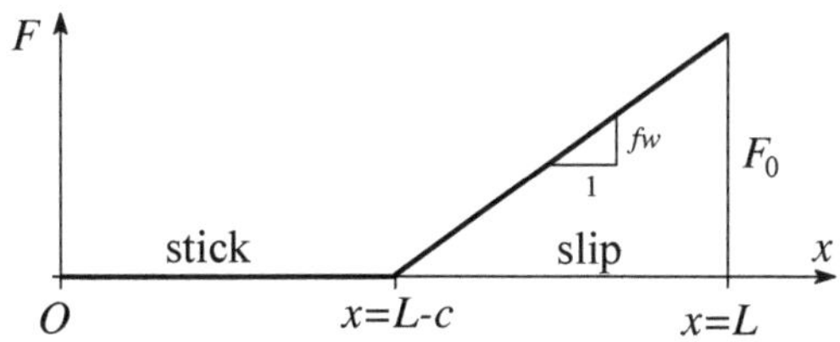

Fig. 2.25 Distribution of axial force in the strip under monotonically increasing loading

$$\sigma = \frac{F}{A} = E\frac{du}{dx} \quad \text{and hence} \quad F = EA\frac{du}{dx}. \tag{2.41}$$

Solving Eq. (2.40) for F, we obtain

$$F = fwx\,\mathrm{sgn}(\dot{u}) + C, \tag{2.42}$$

where C is a constant of integration.

If $F_0(t)$ increases monotonically with time t, slip will start on the right, whilst the left end of the strip remains stuck. In the stuck region, $u = 0$ and hence $F = 0$ from Eq. (2.41). This also implies that the frictional traction $q = 0$ in the stuck region.

These considerations show that the internal axial force $F(x)$ has the form shown in Fig. 2.25 and we conclude that the length c of the slip region is

$$c = \frac{F_0}{fw}. \tag{2.43}$$

Unloading. Suppose now that $F_0(t)$ increases to some value $F_{\max}$ and then decreases monotonically. If F_0 were held at $F_{\max}$ for some period before unloading, we would necessarily have $\dot{u}(x) = 0$, since we assume the loading rate to be slow enough for quasi-static assumptions to apply. Thus, the unloading process essentially starts from a state of stick, but one in which a distribution of residual stress exists, defined by Fig. 2.25.

During unloading, we anticipate that backward slip [$\dot{u} < 0$] will start on the right whilst the rest of the strip remains stuck, but the axial force in this stuck region will remain at the value reached during the loading phase, as shown in Fig. 2.26.

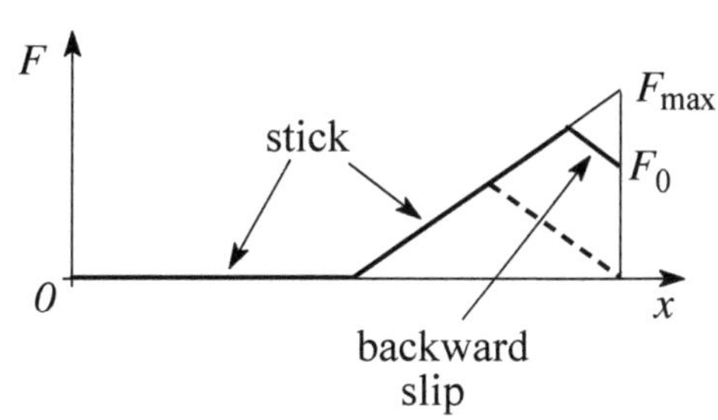

Fig. 2.26 Distribution of axial force in the strip during unloading

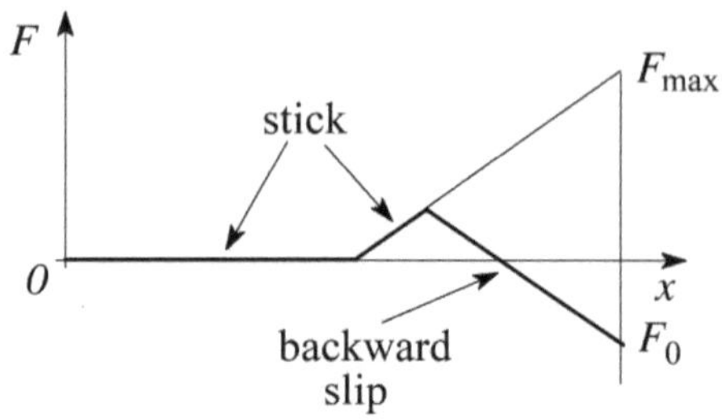

Fig. 2.27 Axial force distribution for $F_0(t) < 0$

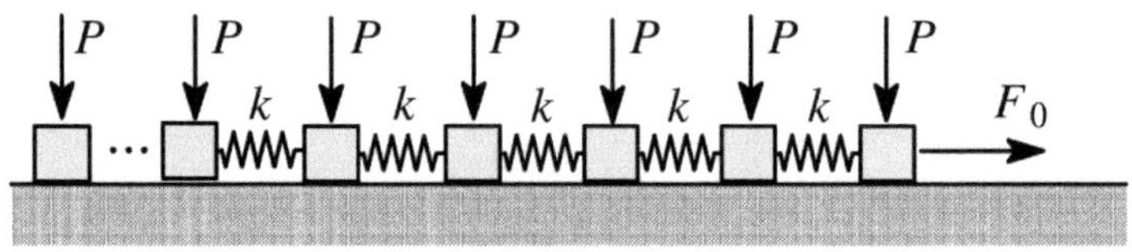

Fig. 2.28 Discrete model of the strip problem

Notice in particular that even if $F_0(t)$ is reduced to zero, the strip is left in a state of residual stress defined by the dashed line in Fig. 2.26. The unloading process can be extended to negative values of $F_0(t)$ as shown in Fig. 2.27. This shows that the distribution of axial force retains some memory of the initial loading phase [in particular of the maximum value F_{max}] as long as $|F_0(t)| < |F_{max}|$. In effect, the system remembers only the local extreme values of the time-varying load $F_0(t)$. Systems exhibiting this behaviour are also found in elastoplastic mechanics and are sometimes known as *Iwan models* (Iwan 1967).

The Inverse Problem

Suppose we wish to achieve a given state of residual stress $F(x)$ satisfying the 'unloaded' condition $F(0) = F(L) = 0$. Is it possible to devise a loading scenario $F_0(t)$ to achieve this, starting from the condition $F(x) = 0$ at $t = 0$? For the discrete problem, idealized as in Fig. 2.28, the answer is 'yes' provided the required frictional tractions on each element satisfy the frictional inequality $|Q| < fP$.

The residual stress distribution implies a prescribed set of nodal displacements u_i, assuming the first [leftmost] element is not moved [so $u_1 = 0$]. We first increase the magnitude of $F_0(t)$ monotonically [with an appropriate sign] until all the elements $i > 1$ slip and until u_2 reaches the desired value. We then reverse the direction of F_0 and change the magnitude until all the elements $i > 2$ slip in the opposite direction and until u_3 reaches the desired value. Alternating the direction of F_0 using appropriately chosen extrema will then yield the desired residual stress distribution.

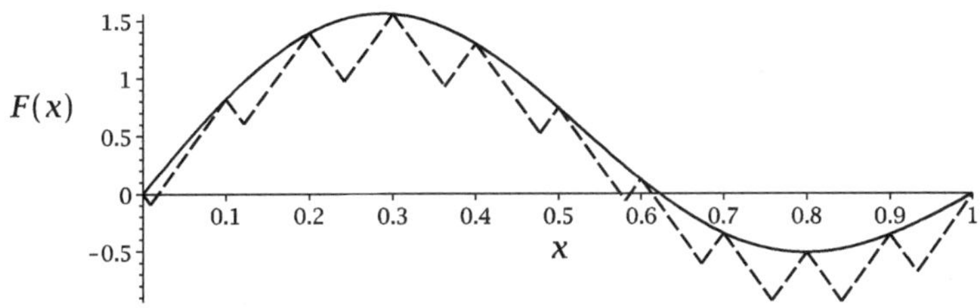

Fig. 2.29 The desired residual stress distribution is shown as a solid line, but we can only achieve the correct value at a denumerable set of points, as illustrated by the dashed line

The situation is very different for the continuous strip. A similar scenario will enable us to achieve the desired axial displacement $u(x)$ at any denumerable set of points, but between these points the derivative dF/dx is everywhere either $+fw$ or $-fw$. In Fig. 2.29, the dashed line shows one such 'approximation' to the desired residual stress distribution, shown as a solid line. Other scenarios can be devised that define residual stress fields that alternate above and below the desired curve. Also, we could choose the points to be non-uniformly spaced, so as to minimize an appropriate measure of the difference between the approximation and the desired curve.

Contact of Half Spaces

If the linear dimensions of the contact area are small compared with those of the contacting bodies, it is reasonable to approximate the bodies as half spaces. In this case, the normal and tangential problems are uncoupled [equivalent to $\boldsymbol{B} = \boldsymbol{0}$ in the discrete problem] if Dundurs' constant $\beta = 0$ [see Eq. (2.31)].

The Cattaneo–Mindlin Problem

Cattaneo (1938) solved the uncoupled problem where a Hertzian contact is loaded by a normal force P, which is then held constant whilst a tangential force Q is increased monotonically from zero. The same problem was later solved by Mindlin (1949), who was apparently unaware of Cattaneo's earlier publication.

The two-dimensional case is illustrated in Fig. 2.30. Since the problem is uncoupled, the contact area of width $2a$ remains constant during the tangential loading phase, but equal slip zones are developed on either side of a stick zone of width $2c$. The corresponding normal and tangential contact traction distributions are given by

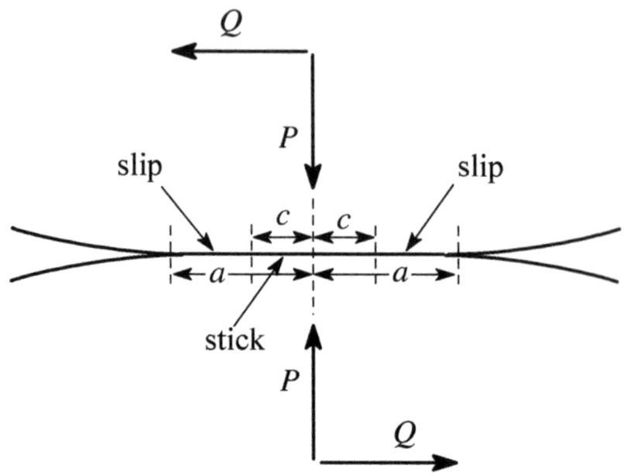

Fig. 2.30 The two-dimensional Cattaneo–Mindlin problem

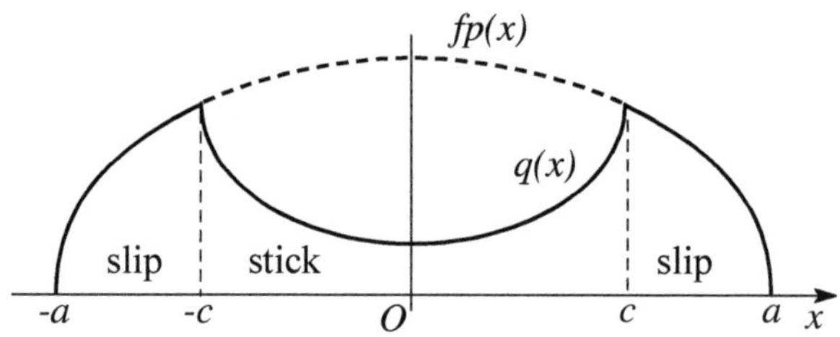

Fig. 2.31 Contact tractions for the Cattaneo–Mindlin problem

$$p(x) = \frac{E^*}{2R}\sqrt{a^2 - x^2} \tag{2.44}$$

$$q(x) = \frac{fE^*}{2R}\left(\sqrt{a^2 - x^2} - \sqrt{c^2 - x^2}\right) \tag{2.45}$$

and are illustrated in Fig. 2.31.

The tangential traction in Eq. (2.45) can be characterized as

$$q(x) = fp(x, P) - fp(x, P_S), \tag{2.46}$$

where $p(x, P)$ is the normal traction [contact pressure] when the normal load is P, and P_S is a fictitious load defined by

$$P_S = P - \frac{Q}{f}. \tag{2.47}$$

Cattaneo and Mindlin used this superposition for the general three-dimensional Hertzian problem, where the contact area is elliptical so the assumed stick region is a smaller ellipse of the same eccentricity. Equation (2.46) is then replaced by

$$q_x(x, y) = fp(x, y, P) - fp(x, y, P_S); \quad q_y(x, y) = 0. \tag{2.48}$$

They showed that the stick conditions are then satisfied in this central ellipse, and clearly $|\boldsymbol{q}| = fp$ in the surrounding slip region. However, Eq. (2.48) satisfies (2.1) and (2.2) only if the slip direction $\boldsymbol{e}_V$ is aligned with the negative x-direction throughout the slip region, and this condition is not satisfied, particularly near to the stick-slip boundary. However, Munisamy et al. (1994) used a numerical method to assess the effect of this error and found that it has very little effect on the actual traction distribution.

The Ciavarella–Jäger Theorem

Ciavarella (1998) and Jager (1998) have shown that the superposition (2.46) gives an exact solution to *all* two-dimensional uncoupled problems of the half space, loaded as shown in Fig. 2.32, which also shows a construction for the fictitious load P_S.

We also note that that the stick area

$$\mathcal{A}_{\text{stick}} = \mathcal{A}(P_S), \tag{2.49}$$

where $\mathcal{A}(P)$ comprises the extent of the contact area when the normal load has the value P. In other words, the first points to slip are the last ones that made contact during monotonic normal loading to P. These results apply for all initial contact geometries, including [for example] the case of a rough surface, where the highest asperities in the original surface will be the last to slip as Q is increased.

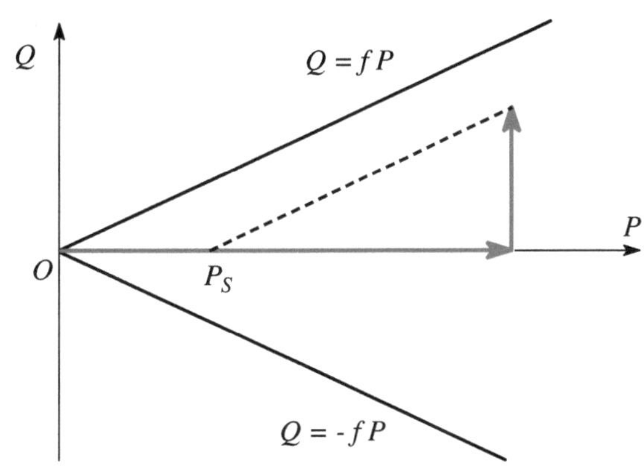

Fig. 2.32 Loading history for Cattaneo's problem. This figure also shows a geometrical construction for P_S of Eq. (2.47)

Periodic Loading

Consider the case where the steady-state loading is periodic, a typical case being defined by

$$P = P_0 + P_1 \cos(\omega t); \quad Q = Q_0 + Q_1 \cos(\omega t - \phi), \tag{2.50}$$

where P_0, P_1, Q_0, Q_1 are constants, ω is the frequency, t is time and ϕ is a phase lag.

This loading cycle is illustrated in Fig. 2.33. Notice that the point $\{P(t), Q(t)\}$ must satisfy the condition $-fP < Q < fP$ at all times, and we also need to define an initial loading path OA before the periodic loading starts.

During periods when

$$\frac{dP}{dt} > 0 \quad \text{and} \quad \frac{d|Q|}{dP} < f, \tag{2.51}$$

the contact area is increasing and all points are in a state of stick. The problem must be solved incrementally because slip displacements are 'locked in' as new areas come into contact. For all other parts of the loading cycle, the shear traction distribution can be obtained by superposition of appropriate Ciavarella–Jäger distributions (Barber et al. 2011).

Figure 2.34 identifies the conditions holding at various instants during the periodic cycle. The points B, C and E are defined such that the local tangent to the loading locus has slope of $\pm f$. Between C and E, a forward slip zone grows from the edges of the contact area, but the slip velocity at all points in this zone goes simultaneously to zero at E, which therefore corresponds to a state of instantaneous stick. A corresponding backward slip zone then grows between E and B. Between B and C, conditions (2.51) are satisfied, so the contact area grows in a state of complete stick.

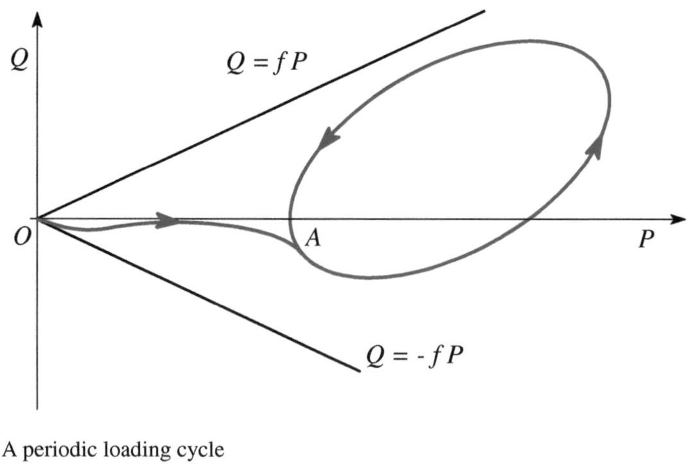

Fig. 2.33 A periodic loading cycle

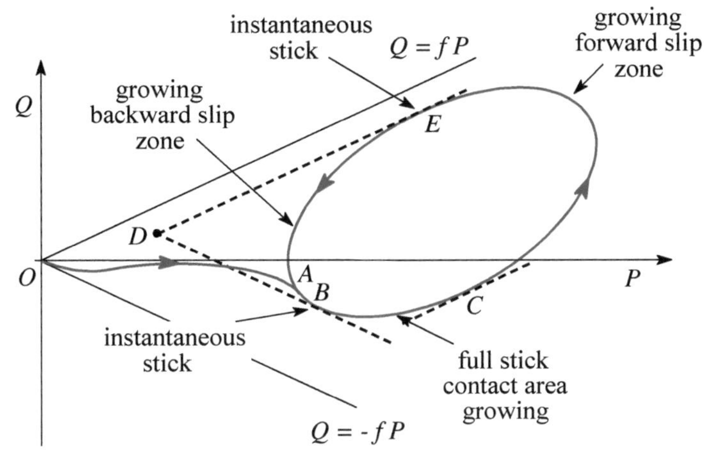

Fig. 2.34 Conditions obtaining during a steady-state periodic loading cycle

If the tangents at B and E are extended to intersect at the point D as shown, the permanent stick zone is then defined as $\mathcal{A}(P_D)$ (Barber et al. 2011). We notice that this point is independent of the initial loading segment OA, as required by "Ponter's theorem".

Example: A Flat and Rounded Punch

Consider the problem where a flat punch with rounded corners is pressed into a half plane, as shown in Fig. 2.35. Contact occurs throughout the flat region $\mathcal{A}_{\text{flat}} \equiv -b < x < b$ even when the normal load $P = 0$, so $\mathcal{A}_{\text{flat}} \in \mathcal{A}(P_S)$ for all P_S. It follows that $\mathcal{A}_{\text{flat}}$ never slips for any loading cycle of the form of Fig. 2.34.

Now suppose that microslip at the contact area causes wear governed by Archard's wear law

$$\dot{w}(x,t) = \alpha f p(x,t) V_s(x,t) \tag{2.52}$$

Fig. 2.35 Normal indentation by a flat and rounded punch

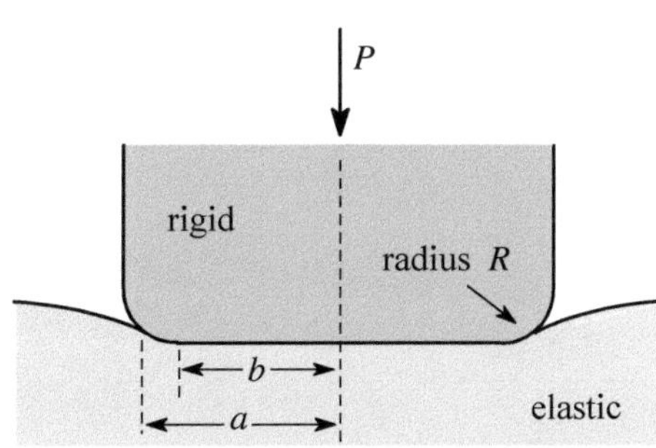

(Archard 1953), where $\dot{w}(x, t)$ is the time derivative of the wear depth, $V_s(x, t)$ is the local microslip velocity, and α is a constant. Since the region $\mathcal{A}_{\text{flat}}$ never slips, it will never wear and hence remains flat. Wear will occur in the rounded segments only.

More general cases. For indenters of more general shape, we can identify the permanent stick zone $\mathcal{A}(P_D)$ from the construction in Fig. 2.34. If wear then occurs according to Archard's law, the long-time solution will involve contact *only* in $\mathcal{A}(P_D)$ and there will be no slip.

Proof Consider the case where some wear has occurred outside $\mathcal{A}(P_D)$. Because of the change of profile, the solution of the contact problem $p(x, P)$, $\mathcal{A}(P)$ is changed for $P > P_D$, but it is not changed for $P \leq P_D$, since in this range only the profile inside $\mathcal{A}(P_D)$ affects the solution and no wear has occurred in this region. It follows that $\mathcal{A}(P_D)$ is not affected by wear and hence that all points $x \in \mathcal{A}(P_D)$ remain stuck at all times.

As wear occurs outside $\mathcal{A}(P_D)$, the contact pressure there decreases, so it must increase for $x \in \mathcal{A}(P_D)$ to preserve the same normal force $P(t)$ at any given point in the periodic loading cycle. As long as the solution remains elastic, the system tends asymptotically to one in which $\mathcal{A}(P_D)$ is the only contact area. But this implies the development of a square-root singularity at the boundary of $\mathcal{A}(P_D)$. In most cases, this will be limited by plastic deformation. A finite element solution (Hu et al. 2016) of the resulting elastic–plastic contact problem [including wear] shows that slip now starts to penetrate the original stick zone.

Coupled Problems

We recall that for the half space, contact problems are coupled if $\beta \neq 0$, where Dundurs constant β is defined in Eq. (2.31). The simplest problem of this kind comprises a rigid flat punch loaded by a normal force P and tangential force Q_x as shown in Fig. 2.36

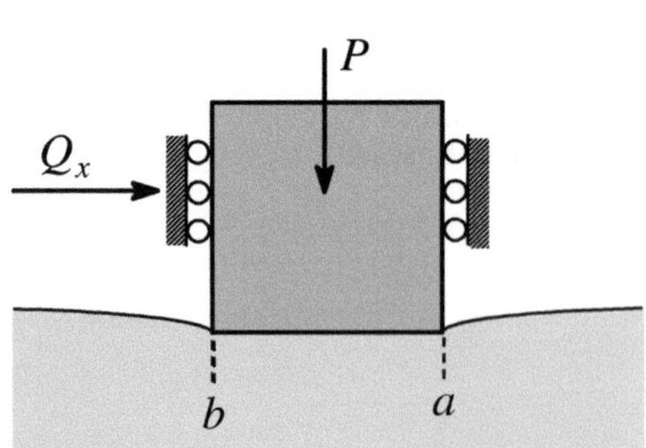

Fig. 2.36 Indentation and tangential loading of a flat rigid punch

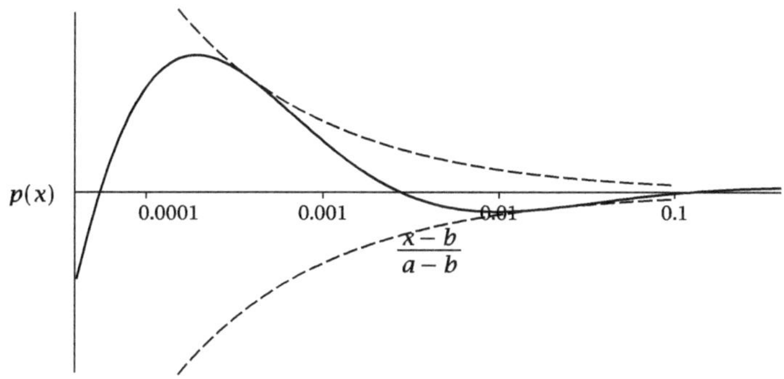

Fig. 2.37 Normal tractions near $x = b$ for purely normal loading $[Q_x = 0]$

If slip is everywhere prevented, the contact tractions can be determined as

$$p(x) + \imath q_x(x) = \frac{(P + \imath Q_x)}{\pi\sqrt{1-\beta^2}\sqrt{(a-x)(x-b)}}\left(\frac{x-b}{a-x}\right)^{\imath\epsilon}, \tag{2.53}$$

where

$$\epsilon = \frac{1}{2\pi}\ln\left(\frac{1+\beta}{1-\beta}\right). \tag{2.54}$$

Notice that if $\beta = 0$ then also $\epsilon = 0$.

At the edge of the contact area, the tractions exhibit an oscillatory singularity. The local normal traction $p(x)$ near $x = b$ is shown in Fig. 2.37 for the case where $Q_x = 0$. The dashed lines define positive and negative square-root singular distributions, but the contact pressure oscillates between these envelopes an infinite number of times as $x \to b$. This implies that the Signorini inequality [that the normal traction be everywhere compressive] is violated neat the edges of the contact region, except in the uncoupled case.

Finite coefficient of friction. Spence (1975) solved the normal loading problem for the problem of Fig. 2.36 and also the corresponding axisymmetric problem, assuming a finite coefficient of friction f. He found that a slip region is developed at the edge of the contact, the extent of which is independent of the magnitude of the normal force P. Remarkably, the size of the slip zone tends to a finite limit even as $f \to \infty$.

Unloading. If the punch in Fig. 2.36 is loaded by a normal force P which is first increased and then reduced, the traction during the unloading phase is considerably more complex (Turner 1979). Slip at the contact edge continues *in the same direction* at the beginning of the process, but eventually separate regions of radially inward and outward slip are obtained, separated by a stick region with moving boundaries.

Non-conformal coupled problems with Friction. If the contact area increases with P, points on the surface are free to move tangentially until they come after contact, after which they must move only vertically if friction is sufficient to prevent slip. This situation is illustrated in Fig. 2.38 for indentation by an axisymmetric rigid punch.

Fig. 2.38 Tangential displacement during non-conformal indentation

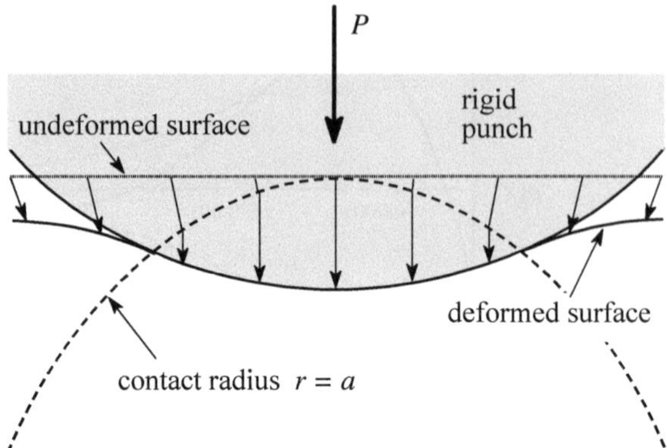

In the *incremental* problem [change from P to $P + \delta P$] with full stick, points in the instantaneous contact area $\mathcal{A}(P)$ all move downwards by the same distance. This is the same boundary condition as for the flat punch problem of planform $\mathcal{A}(P)$. Spence showed that for Hertzian contact with finite f, the slip zone in this incremental problem is also the same as that for a flat punch, and hence the proportion of the contact area that is slipping remains constant throughout the loading process.

Storakers and Elaguine (2005) extended this argument to show that the proportion of the contact area in a state of stick will remain constant during loading for any two-dimensional or axisymmetric indenter, provided only that the stick region increases monotonically with P—i.e. for which $\mathcal{A}_{\text{stick}}(P_1) \in \mathcal{A}_{\text{stick}}(P_2)$ if $P_2 > P_1$.

Elastodynamic Effects

We have seen that even for arbitrarily slow loading rates, frictional contact can result in rapid changes of configuration and these will occur on a time scale associated with one or more of the natural frequencies of the system. We shall also see that even the elementary Coulomb friction law can give rise to instabilities during arbitrarily slow sliding. To introduce this topic, we need to review some fundamental results in elastodynamics.

Bulk waves. In an infinite elastic body, two kinds of waves can propagate without dispersion [i.e. whilst retaining the same waveform].

If the particle displacements associated with the wave are aligned with the direction of propagation, it is known as a *dilatation wave* or *P-wave*. For example, if the wave propagates in the x-direction, we can write

$$u_x(x, y, z, t) = f(x - c_1 t); \quad u_y = u_z = 0 \tag{2.55}$$

and substitution into Hooke's law and the equations of motion shows that the *dilatational wave speed*

$$c_1 = \sqrt{\frac{2G(1-\nu)}{(1-2\nu)\rho}}, \tag{2.56}$$

where ρ is the density of the material.

By contrast, if the particle displacements are orthogonal to the direction of propagation, we have a *shear wave* or *S-wave*. For example,

$$u_y(x, y, z, t) = f(x - c_2 t); \quad u_x = u_z = 0 \tag{2.57}$$

where the *shear wave speed*

$$c_2 = \sqrt{\frac{G}{\rho}}, \tag{2.58}$$

For all materials, $c_2 < c_1$.

Rayleigh waves. For a half space with a traction-free plane surface, *Rayleigh waves* can propagate without dispersion along the surface at a speed $c_R < c_2$. These comprise displacement fields that decay with distance away from the surface.

Green's function for moving contact problems. In two dimensions, if a concentrated normal force P moves at constant speed V over the surface of the half plane $z > 0$, the normal surface displacement in the steady state is

$$u_z(x, 0) = \frac{P}{G}\left[F_1(V)\ln|x| + F_2(V)H(-x)\right], \tag{2.59}$$

where $H(\cdot)$ is the Heaviside step function and $F_1(V)$, $F_2(V)$ are functions of V that are illustrated in Fig. 2.39.

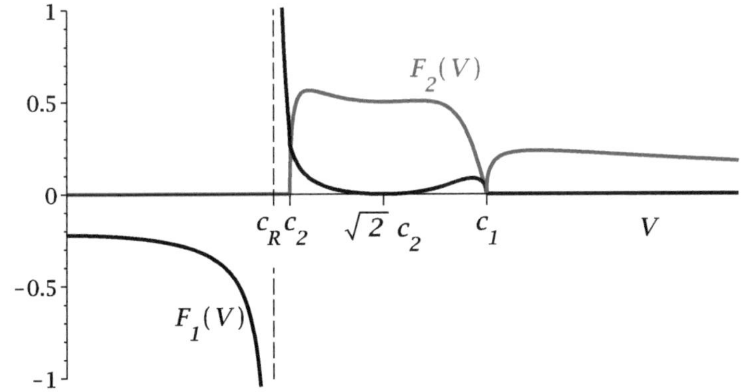

Fig. 2.39 The functions $F_1(V)$, $F_2(V)$ from Eq. (2.59) for the case where $\nu = 0.3$

For $V < c_2$, $F_2(V) = 0$ and the term including $F_1(V)$ is of the same form as the static Green's function. In this speed range, the motion has the same effect as a modification in the elastic modulus, but in other respects, solutions of two-dimensional contact problems proceed as in the quasi-static case. The special case $V = c_R$ can be seen as a resonance of the system in that a force moving at this speed produces an arbitrarily large displacement, or equivalently, a finite displacement can be produced by a negligibly small [or in the limiting case zero] force.

At the other extreme, if $V > c_1$, $F_1(V) = 0$ and the response is dictated by the step function term. Points ahead of the moving load experience no displacement. In a sense they cannot know that the load is approaching, since waves emanating from the load cannot travel faster than speed c_1. Behind the moving load, the displacement is constant.

Transmission of Waves Across an Interface

Contacting bodies rarely move at speeds close to or above the wave speeds, but contact problems of this kind arise if a bulk wave impinges on an inclined interface between two bodies. Consider the case of two identical half planes $z > 0$ and $z < 0$ that make contact over the common plane $z = 0$.

Figure 2.40 illustrates the case where a shear wave moving in the x'-direction strikes this interface. We suppose that the direction of particle motion lies in the y-direction and hence is orthogonal to the figure. Waves of this kind in an otherwise two-dimensional context are known as *SH waves*.

If the half planes were bonded together, they would comprise a single infinite body and could support an incident wave of the form

$$u_y(x', z', t) = u_0 \sin\left\{\omega(x' - c_2 t)\right\}. \tag{2.60}$$

This would imply the existence of time-dependent shear tractions on the interface defined as

$$q_y(x, t) = G u_0 \omega \cos\alpha \cos\{\widehat{\omega}(x - ct)\}, \tag{2.61}$$

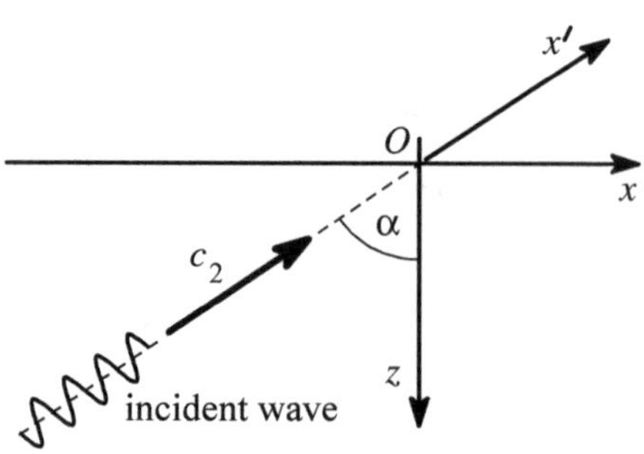

Fig. 2.40 Transmission of an SH wave across the interface $z = 0$ between two identical half planes

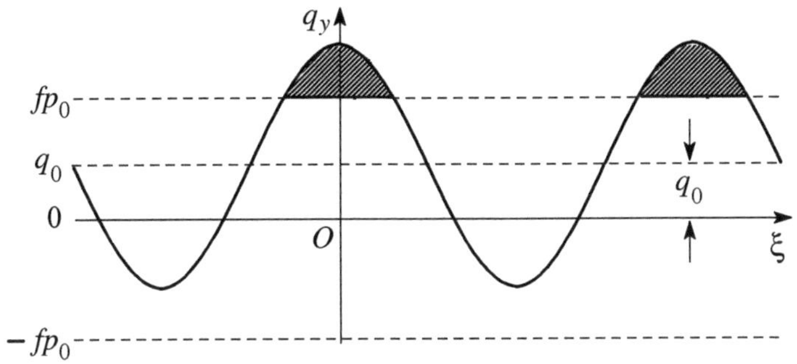

Fig. 2.41 Frictional tractions at the interface resulting from a transmitted SH wave, where $\xi = x - ct$

where

$$c = \frac{c_2}{\sin\alpha} > c_2 \quad \text{and} \quad \widehat{\omega} = \omega \sin\alpha, \tag{2.62}$$

so the disturbance appears to move along the interface at a speed greater than the wave speed.

Now suppose that the two half planes are actually not bonded, but are pressed together by a uniform traction p_0 and sheared by a uniform shear traction q_0. There will be no slip provided the coefficient of friction f satisfies the condition

$$|q_0 + Gu_0\omega\cos\alpha| < fp_0. \tag{2.63}$$

If this inequality is not satisfied, regions of slip and stick will propagate along the interface at speed $c > c_2$. This problem can be solved by superposing a 'corrective' distribution of out-of-plane forces moving along the interface at speed c (Chez et al. 1978). But since $c > c_2$, the corresponding Green's function is of step function form, so the correction is simply $fp_0 \pm q_y(x,t)$. In other words, it is purely local.

Figure 2.41 illustrates the resulting distribution for the case where the frictional inequality is violated only in the range $q_y > 0$. The shaded segments define the correction and the actual interfacial traction comprises the resulting truncated sinusoid. Notice also that the accumulated slip displacements cause the half planes to 'creep' relative to each other in the y-direction.

Dissimilar materials. If the two contacting half planes [1, 2] comprise different materials, they will generally have different shear wave speeds $c_2^{(1)}, c_2^{(2)}$. In this case, even if the bodies are bonded together we shall obtain a reflected wave and a refracted wave, as shown in Fig. 2.42. These must all correspond to the same velocity c along the interface, and hence the directions of propagation are determined from the equations

$$\beta_1 = \alpha; \quad \frac{c_2^{(1)}}{\sin\alpha} = \frac{c_2^{(2)}}{\sin\beta_2} = c. \tag{2.64}$$

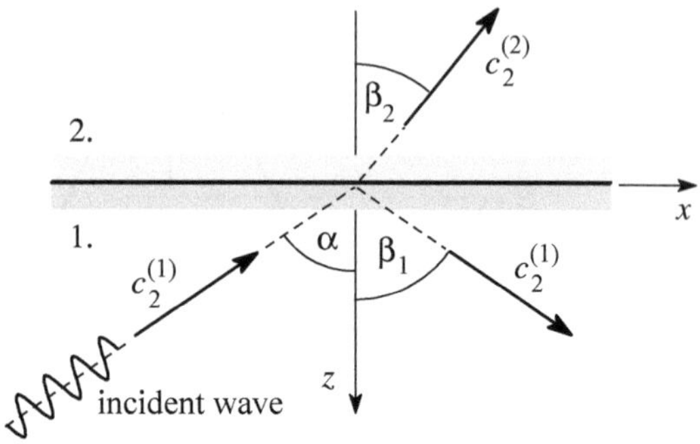

Fig. 2.42 Transmission of an incident SH wave across an interface between two dissimilar half planes

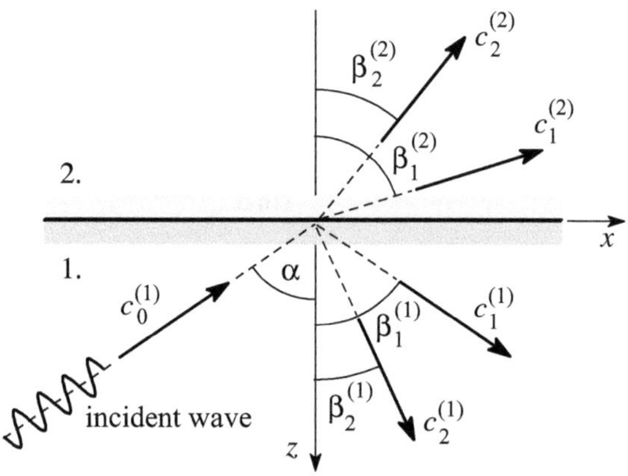

Fig. 2.43 Transmission of an incident in-plane wave across an interface between two dissimilar half planes. Notice that $c_0^{(1)}$ must be either $c_1^{(1)}$ or $c_2^{(1)}$, depending on whether the incident wave is a dilatational or a shear wave

Notice that if

$$c_2^{(2)} > \frac{c_2^{(1)}}{\sin\alpha} \quad \text{i.e.} \quad \sin\alpha > \frac{c_2^{(1)}}{c_2^{(2)}}, \tag{2.65}$$

there can be no refracted wave. Instead, an *evanescent wave* [whose amplitude decays with $|z|$, like that of a Rayleigh wave] propagates along the interface in half plane 2. In both cases, the problem for a finite friction coefficient can be solved by superposing an appropriate corrective solution involving moving point forces, but if the wave in body 2 is evanescent, the correction is subseismic [$c < c_2^{(2)}$] and the corresponding Green's function will be of logarithmic form.

In-plane waves. If the incident wave is in-plane [meaning that the periodic displacements lie in the plane of the figure], both shear and dilatational waves will be refracted, as shown in Fig. 2.43.

The angles defining the directions of propagation, and the velocity c at which the disturbance propagates along the interface are related by the equation

$$c = \frac{c_0^{(1)}}{\sin\alpha} = \frac{c_1^{(1)}}{\sin\beta_1^{(1)}} = \frac{c_2^{(1)}}{\sin\beta_2^{(1)}} = \frac{c_1^{(2)}}{\sin\beta_1^{(2)}} = \frac{c_2^{(2)}}{\sin\beta_2^{(2)}}. \tag{2.66}$$

The corresponding interface tractions now involve both normal and tangential components, so there can be propagating regions of separation as well as slip (Comninou and Dundurs 1979). As in the simpler SH-wave case, one or both of the refracted waves may become evanescent.

Interface Waves

If two dissimilar half spaces are bonded together at a common plane, *Stoneley waves* can propagate along the interface for some combinations of material properties. As with Rayleigh waves, the disturbance decays exponentially with distance from the interface. In fact, Rayleigh waves can be considered as the limiting case where the elastic modulus of one of the materials goes to zero. Notice that Stonely waves are *not* possible if the two half spaces are of similar materials, since in that case, there is essentially no interface.

Waves can also propagate along a frictionless interface between two half spaces. These are known as *slip waves*, since they involve relative tangential displacements at the interface. Slip waves can occur even when the materials of the two half spaces are similar. They would then take the form of identical Rayleigh waves moving along the surfaces of the two half spaces, oriented such that the normal displacements in each are always equal, ensuring complete contact, but without requiring tractions.

Slip waves with dissimilar materials. Slip waves can occur at a frictionless interface between dissimilar materials, but there will then generally be non-zero normal contact tractions. These must produce equal and opposite normal displacements on the two half planes, and this is satisfied for an arbitrary moving traction distribution if the propagation speed c is below the shear wave speed c_2 in both materials [so that $F_2(c)$ in Eq. (2.59) is zero] and if

$$f^{(1)}(c) = -f^{(2)}(c) \qquad \text{where} \qquad f(c) = \frac{F_1(c)}{G}. \tag{2.67}$$

We notice from Fig. 2.39 that $F_1(V)$ changes sign at $V = c_R$, so this condition requires that the propagation speed c be intermediate between the Rayleigh wave speeds of the two materials—e.g. $c_R^{(1)} < c < c_R^{(2)}$.

Stability of Steady Frictional Sliding

Consider now the case where two half planes are pressed together by a uniform pressure p_0 and caused to slide over each other by the application of uniform tangential tractions $q_0 = fp_0$. We suppose that the sliding speed V is very small compared with the wave speeds in the materials [e.g. $V \ll c_R$].

There is clearly a trivial solution in which the stresses are uniform throughout the two bodies. However, suppose we superpose the displacements and stresses associated with a frictionless slip wave. These will maintain contact everywhere and, if the materials are similar, they will not change the contact tractions, so the frictional sliding condition $q(x,t) = fp(x,t)$ is still satisfied everywhere. We conclude that this disturbance of the uniform solution is *neutrally stable*.

If the materials are *dissimilar*, the superposition of a frictionless slip wave would modify the normal tractions, but not the shear tractions, and the resulting solution would therefore violate the frictional condition $q(x) = fp(x)$. However, suppose that in place of the slip wave, we postulate an exponentially growing or decaying perturbation of the form

$$\boldsymbol{u}^{(i)}(x,z,t) = \Re\left[\boldsymbol{U}^{(i)} \exp\{\omega[\imath(x-ct) + at - \lambda_i z]\}\right] \tag{2.68}$$

in body i $[i = 1, 2]$. The governing equations define λ_i in terms of the other parameters and the material properties and, since the perturbation is not to change the condition $q(x) = fp(x)$, the boundary conditions lead to an eigenvalue problem for $(a - \imath c)$.

For similar materials, we obtain $a = 0$ indicating neutral instability, as argued above. However, for dissimilar materials, Adams (1995) has shown that if the material combination supports a frictionless slip wave, there always exists an eigenvalue with a positive exponential growth rate $a > 0$. In other words, uniform steady slip is unstable. Adams also showed that if the material combination does *not* support a frictionless slip wave, there exists some critical coefficient of friction above which uniform steady slip is unstable.

If the amplitude of an unstable wave grows exponentially, eventually there must be a transition in some region[s] to stick and/or separation. Adams (1998) identified steady-state [i.e. limit state] solutions for the sliding of dissimilar materials, which involve propagating regions of stick, slip and/or separation. He also showed that these propagating waves allow one body to creep over the other at a mean tangential traction lower than that required for gross slip.

Implications for finite element studies. The exponential growth rate a increases with wavenumber ω, so short wavelengths grow more rapidly. In finite element models, discretization errors act as initial perturbations and the 'effective wavelength' is linked to the smallest element size. This implies that mesh refinement will introduce more rapidly growing disturbances and hence require a shorter time step. Also, in FE studies, we usually assume that with sufficient mesh refinement we approach the continuum solution [in some sense]. This clearly fails if short wavelength perturbations are increasingly unstable. Even in a continuum formulation, *the transient elastodynamic frictional contact problem with Coulomb friction is mathematically ill-posed.*

Alternative Friction Laws

Numerous alternatives to the classical Coulomb friction law have been proposed. Usually, the motivation is to give a better approximation to observed physical behaviour, but as we have seen in the last section, an alternative motivation might be to avoid computational instabilities. Here we shall examine the following modifications to the Coulomb law:

- Differing static and dynamic friction coefficients.
- Velocity-dependent friction coefficient.
- 'Rate–state' friction laws.

Stick-Slip Friction

Figure 2.44 shows a mass M supported by a spring of stiffness k and a damper of coefficient c that is sliding against a surface that moves at constant speed V_0.

If the friction force is Q and the rightward displacement of the mass is u, the equation of motion is

$$M\ddot{u} + c\dot{u} + ku = Q. \tag{2.69}$$

During periods of stick, $\dot{u} = 0$, so $Q = cV_0 + ku$, and slip must start when $Q = f_s P$, where f_s is the coefficient of static friction.

At this instant

$$u = \frac{f_s P - cV_0}{k}; \quad \dot{u} = V_0, \tag{2.70}$$

and these define the initial conditions for a period of slip, during which we assume that $Q = f_d P$, where f_d [$< f_s$] is the coefficient of dynamic friction.

Figure 2.45 shows the subsequent evolution of the dimensionless velocity $\dot{u}/V_0$ for

$$\lambda \equiv \frac{\omega_n(f_s - f_d)}{kV_0} = 1 \quad \text{where} \quad \omega_n = \sqrt{\frac{k}{M}} \tag{2.71}$$

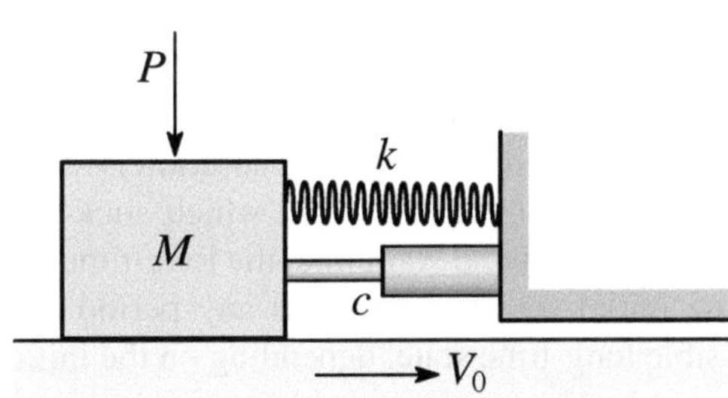

Fig. 2.44 An elastically supported mass sliding against a plane

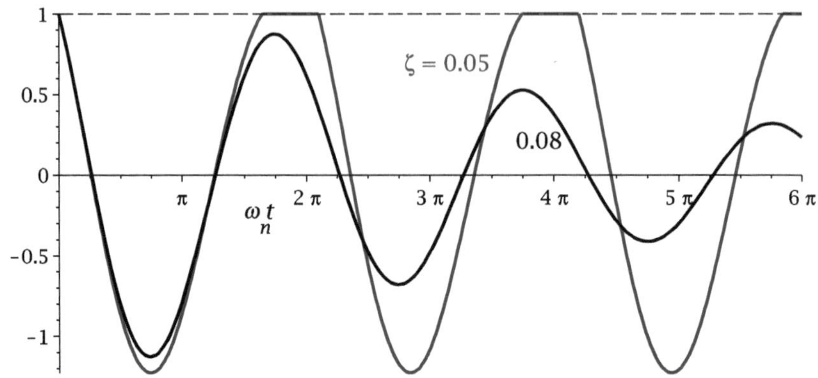

Fig. 2.45 Dimensionless slip velocity $\dot{u}/V_0$ following a period of stick for the system of Fig. 2.44 with $\lambda = 1$ and two values of ζ

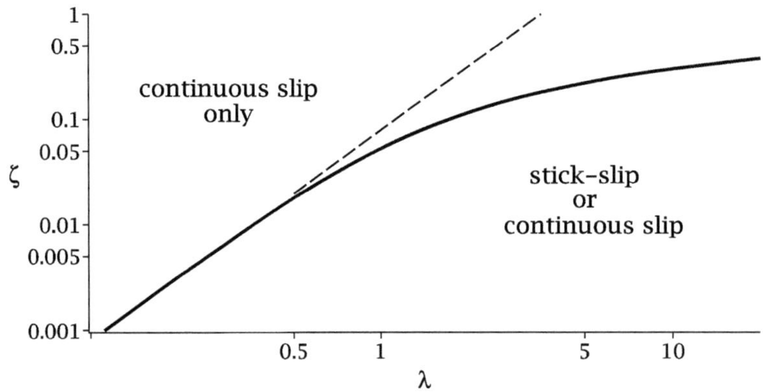

Fig. 2.46 Domain of the parameters λ, ζ in which stick-slip vibrations are possible

and two different values of the damping factor

$$\zeta = \frac{c\omega_n}{2k}. \tag{2.72}$$

For $\zeta = 0.05$, the speed returns to V_0 after one almost complete cycle of oscillation and the system reverts to a state of stick. The long-time solution is therefore a stick-slip vibration. However, for a higher level of damping $\zeta = 0.08$, the speed never returns to V_0 and the long-time solution is steady slip with $\dot{u} = 0$.

The range of parameters in which stick-slip vibrations can occur is defined in Fig. 2.46. Below the solid line, the long-time state will comprise stick-slip vibration if the initial transient involves any period of stick, but continuous slip is still a possible long-time state, depending on the initial conditions. For given values of the other parameters, stick-slip is more likely to occur when the sliding speed V_0 is low and hence λ is large.

Velocity-Dependent Friction

In lubricated systems, the coefficient of friction increases at high sliding speeds due to viscous effects, but at low speeds the lubricant film is thin compared with the surface roughness amplitude and solid–solid contact occurs, increasing friction. This typically leads to a dependence of f on V as shown in Fig. 2.47, known as the *Stribeck curve.*

At low speeds, the curve has a negative slope and this can give rise to instabilities. For example, suppose the system in Fig. 2.44 is sliding at constant speed V_0 so that $Q = Pf(V_0)$. We consider the stability of a small perturbation from this state, for which we can approximate the curve in Fig. 2.47 by the dashed straight line. The instantaneous sliding velocity is $V_0 - \dot{u}$, so the equation of motion (2.69) can be written

$$M\ddot{u} + c\dot{u} + ku = Pf(V) \approx Pf(V_0) - Pf'(V_0)\dot{u} \tag{2.73}$$

or

$$M\ddot{u} + \left[c + Pf'(V_0)\right]\dot{u} + ku = Pf(V_0) \tag{2.74}$$

The system will be unstable if the damping term is negative and hence if

$$f'(V_0) < -\frac{c}{P}. \tag{2.75}$$

The long-time state will then be either a periodic vibration [limit cycle] or stick-slip motion.

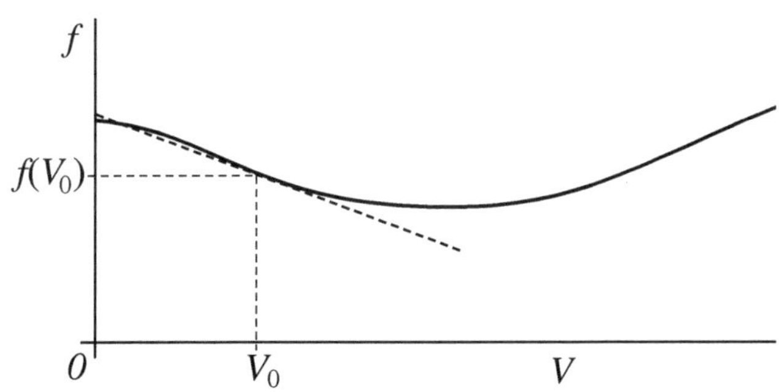

Fig. 2.47 Velocity-dependence of the coefficient of friction

Rate–State Friction Models

The friction force may depend on the history of loading or sliding as well as on instantaneous conditions. For example, if the shear traction is proportional to the actual contact area, an increase in this area due to an increased normal load may take time to develop because of viscoelasticity or creep. Also, if the static and dynamic coefficients of friction are different, some finite sliding distance may be required before the dynamic value is fully established, as shown in Fig. 2.48

This is an example of a *slip-weakening friction law*. In particular, we note that the work done per unit area required to 'break' the static friction bond is

$$W = p \int_0^\infty [f(u) - f_d]\, du = \frac{(f_s - f_d)\, p\delta}{2}, \tag{2.76}$$

where u is the sliding distance and p is the local contact pressure.

Rabinowicz (1951) devised a simple experiment to estimate W which is shown in Fig. 2.49. The inclination θ of the plane surface is chosen such that $f_s > \tan\theta > f_d$, so that the mass can rest in static equilibrium, but will slide down the plane once it is set in motion. The ball is then released from different heights h. If h is chosen as the minimum value needed to initiate permanent sliding down the plane, an energetic analysis, including the rebound distance of the ball, allows us to estimate the value of W.

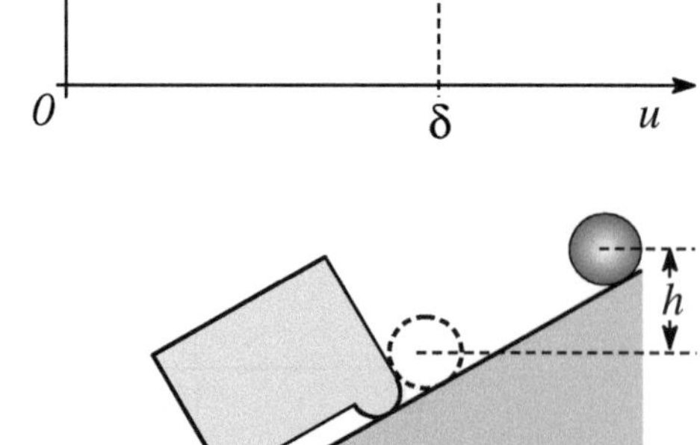

Fig. 2.48 A slip-weakening friction law

Fig. 2.49 Rabinowicz' experiment for estimating the parameter W in Eq. (2.76)

State variables. History-dependent friction laws of this kind can be characterized by defining a *state variable* S whose value represents some measure of the history. The tangential force is then defined by a function

$$Q = f(P, V, S), \tag{2.77}$$

where P is the normal force and V is the sliding speed. We also define an evolutionary equation for S, typically of the form

$$\frac{dS}{dt} = g(P, V, S). \tag{2.78}$$

This will typically predict an exponential approach to a new steady state when there is a step change in P and/or Q.

Example Ranjith and Rice (2001) proposed the simple rate–state model

$$Q = Q_0 \operatorname{sgn}(V); \quad \frac{dQ_0}{dt} = -\frac{|V|}{L}(Q_0 - fP), \tag{2.79}$$

where f is a constant coefficient of friction, Q_0 performs the function of a state variable and L is a length scale. They showed that this law is sufficient to regularize the elastodynamic problem at small length scales, hence avoiding the problem identified in section "Stability of Steady Frictional Sliding".

Estimating rate–state parameters. More generally, rate–state equations can be based on physical arguments, or we can simply use a 'black box' approach and determine the corresponding parameters with appropriate experiments. These might comprise:

- Steady sliding experiments [so $\dot{S} = 0$].
- Steady sliding with a superposed oscillation in speed or normal load (Cabboi et al. 2016).
- Response to a step change. Prakash (1998) generated step changes in frictional conditions by causing a bulk wave to impinge on an oblique interface.

Conclusions

These examples show that friction can give rise to quite complex phenomena in contact mechanics. Even the elementary Coulomb friction law can lead to non-existence and non-uniqueness of solution if the coefficient of friction is 'sufficiently high'. Also steady sliding of dissimilar materials can be elastodynamically unstable and numerical solutions of such problems may be ill-posed.

We have seen that coupling between normal and tangential loading has a critical effect on frictional behaviour. Uncoupled systems have limited dependence on loading history, they cannot become wedged and always lead to well-posed evolution problems. We also note that finite element codes involving frictional boundary conditions may fail or give erroneous results, particularly if the mechanical system is one that is prone to wedging.

References

Adams, G. G. (1995). Self-excited oscillations of two elastic half-spaces sliding with a constant coefficient of friction. *ASME Journal of Applied Mechanics*, *62*, 867–872.
Adams, G. G. (1998). Steady sliding of two elastic half-spaces with friction reduction due to interface stick-slip. *ASME Journal of Applied Mechanics*, *65*, 470–475.
Ahn, Y. J. (2010). Discontinuity of quasi-static solution in the two-node coulomb frictional system. *International Journal of Solids and Structures*, *47*, 2866–2871.
Ahn, Y. J., & Barber, J. R. (2008). Response of frictional receding contact problems to cyclic loading. *International Journal of Mechanical Sciences*, *50*, 1519–1525.
Ahn, Y. J., Bertocchi, E., & Barber, J. R. (2008). Shakedown of coupled two-dimensional discrete frictional systems. *Journal of the Mechanics and Physics of Solids*, *56*, 3433–3440.
Andersson, L.-E., Barber, J. R., & Ahn, Y.-J. (2013). Attractors in frictional systems subjected to periodic loads. *SIAM Journal of Applied Mathematics*, *73*, 1097–1116.
Andersson, L.-E., Barber, J. R., & Ponter, A. R. S. (2014). Existence and uniqueness of attractors in frictional systems with uncoupled tangential displacements and normal tractions. *International Journal of Solids and Structures*, *51*, 3710–3714.
Archard, J. F. (1953). Contact and rubbing of flat surfaces. *Journal of Applied Physics*, *24*, 981–988.
Archard, J. F. (1957). Elastic deformation and the laws of friction. *Proceedings of the Royal Society of London, A*, *243*, 190–205.
Barber, J. R. (2013). Multiscale surfaces and Amontons' law of friction. *Tribology Letters*, *49*, 539–543.
Barber, J. R., Davies, M., & Hills, D. A. (2011). Frictional elastic contact with periodic loading. *International Journal of Solids and Structures*, *48*, 2041–2047. https://doi.org/10.1016/j.ijsolstr.2011.03.008.
Bowden, F. P., & Tabor, D. (1950). *The friction and lubrication of solids*. Oxford: Clarendon Press.
Cabboi, A., Putelat, T., & Woodhouse, J. (2016). The frequency response of dynamic friction: Enhanced rate-and-state models. *Journal of the Mechanics and Physics of Solids*, *92*, 210–236.
Cattaneo, C. (1938). Sul contatto di due corpi elastici: Distribuzione locale degli sforzi. *Rendiconti dell'Accademia Nazionale dei Lincei*, *27*, 342–348, 434–436, 474–478.
Chez, E. L., Dundurs, J., & Comninou, M. (1978). Reflection and refraction of sh waves in presence of slip and friction. *Bulletin of the Seismological Society of America*, *68*, 999–1011.
Cho, H., & Barber, J. R. (1998). Dynamic behavior and stability of simple frictional systems. *Mathematical and Computer Modeling*, *28*, 37–53.
Ciavarella, M. (1998). The generalized Cattaneo partial slip plane contact problem. I—Theory, II—Examples. *International Journal of Solids and Structures*, *35*, 2363–2378.
Comninou, M., & Dundurs, J. (1979). Interaction of elastic waves with a unilateral interface. *Proceedings of the Royal Society of London, A*, *368*, 141–154. https://doi.org/10.1098/rspa.1979.0120.
Dundurs, J. (1969). Discussion on edge bonded dissimilar orthogonal elastic wedges under normal and shear loading. *ASME Journal of Applied Mechanics*, *36*, 650–652.
Greenwood, J. A., & Williamson, J. B. P. (1966). The contact of nominally flat surfaces. *Proceedings of the Royal Society (London), A 295*, 300–319.
Haslinger, J., & Nedlec, J. C. (1983). Approximation of the Signorini problem with friction, obeying the Coulomb law. *Mathematical Methods in the Applied Sciences*, *5*, 422–437.

Hassani, R., Hild, P., Ionescu, I. R., & Sakki, N. D. (2003). A mixed finite element method and solution multiplicity for Coulomb frictional contact. *Computer Methods in Applied Mechanics and Engineering*, *192*, 4517–4531.

Hild, P. (2004). Non-unique slipping in the Coulomb friction model in two-dimensional linear elasticity. *Quarterly Journal of Mechanics and Applied Mathematics*, *57*, 225–235.

Hu, Z., Lu, W., Thouless, M. D., & Barber, J. R. (2016). Effect of plastic deformation on the evolution of wear and local stress fields in fretting. *International Journal of Solids and Structures*, *82*, 1–8.

Iwan, W. D. (1967). On a class of models for the yielding behaviour of continuous and composite systems. *ASME Journal of Applied Mechanics*, *34*, 612–617.

Jager, J. (1998). A new principle in contact mechanics. *ASME Journal of Tribology*, *120*, 677–684.

Kikuchi, N., & Oden, J. T. (1988). *Contact problems in elasticity: A study of variational inequalities and finite element methods*. Philadelphia: SIAM.

Klarbring, A. (1990). Examples of non-uniqueness and non-existence of solutions to quasi-static contact problems with friction. *Ingenieur-Archiv*, *60*, 529–541.

Klarbring, A. (1999). Contact, friction, discrete mechanical structures and discrete frictional systems and mathematical programming. In P. Wriggers & P. Panagiotopoulos (Eds.), *New developments in contact problems* (pp. 55–100). Wien: Springer.

Klarbring, A., Ciavarella, M., & Barber, J. R. (2007). Shakedown in elastic contact problems with Coulomb friction. *International Journal of Solids and Structures*, *44*, 8355–8365.

Martins, J. A. C., Montiero Marques, M. D. P., Gastaldi, F., Simoes, F. M. F. (1992). A two degree-of-freedom "quasistatic" frictional contact problem with instantaneous jumps. In *Contact Mechanics International Symposium*

Melan, E. (1936). Theorie statisch unbestimmter Systeme aus ideal-plastischen Baustoff. *Sitzungsberichte der Akademie der Wissenschaften in Wien*, *145*, 195–218.

Mindlin, R. D. (1949). Compliance of elastic bodies in contact. *ASME Journal of Applied Mechanics*, *16*, 259–268.

Munisamy, R. L., Hills, D. A., & Nowell, D. (1994). Static axisymmetrical hertzian contacts subject to shearing forces. *ASME Journal of Applied Mechanics*, *61*, 278–283.

Persson, B. N. J. (2001). Theory of rubber friction and contact mechanics. *Journal of Chemical Physics*, *115*, 3840–3861.

Prakash, V. (1998). Frictional response of sliding interfaces subjected to time varying normal pressures. *ASME Journal of Tribology*, *120*, 97–102.

Rabinowicz, E. (1951). The nature of the static and kinetic coefficients of friction. *Journal of Applied Physics*, *22*, 1373–1379.

Ranjith, K., & Rice, J. R. (2001). Slip dynamics at an interface between dissimilar materials. *Journal of the Mechanics and Physics of Solids*, *49*, 341–361.

Spence, D. A. (1975). The Hertz problem with finite friction. *Journal of Elasticity*, *5*, 297–319. https://doi.org/10.1007/BF00126993.

Storakers, B., & Elaguine, D. (2005). Hertz contact at finite friction and arbitrary profiles. *Journal of the Mechanics and Physics of Solids*, *53*, 1422–1447.

Thaitirarot, A., Ahn, Y. J., Hills, D. A., Jang, Y. H., & Barber, J. R. (2014). The use of static reduction in the finite element solution of two-dimensional frictional contact problems. *Journal of Mechanical Engineering Science*, *228*, 1474–1487.

Turner, J. R. (1979). Frictional unloading problem on a linear elastic half-space. *Journal of the Institute of Mathematics and its Applications*, *24*, 439–469. https://doi.org/10.1093/imamat/24.4.439.

Chapter 3
Nonequilibrium Molecular Dynamics Simulations of Tribological Systems

James P. Ewen, Eduardo Ramos Fernández, Edward R. Smith and Daniele Dini

Abstract Nonequilibrium molecular dynamics (NEMD) simulations are increasingly being used to investigate the nanoscale behaviour of tribological systems. This chapter focuses on the application of classical NEMD simulations of liquid lubricants and additives confined between solid surfaces. Ab initio NEMD, which can be used to accurately model tribochemsitry, and coupled computational fluid dynamics (CFD)-NEMD are also introduced. Specific example systems and recommendations for future research are provided.

Introduction

In recent years, nonequilibrium molecular dynamics (NEMD) simulations have given unique insights into the nanoscale behaviour of fluids under shear. A detailed understanding of this behaviour is crucial in tribology since this often governs the macroscopic friction (Vanossi et al. 2013) and wear (Molinari et al. 2018) responses that are observed experimentally. MD was invented in the 1950s and can be used to study the dynamics of a system of interacting particles by numerically solving Newtons equations of motion using a finite difference scheme over a series of short time steps. In classical MD, the forces between the particles and their potential energies are calculated using interatomic potentials or molecular mechanics force fields (Ewen et al. 2018a). Initially, MD played a central role in corroborating theories of the liquid state (Barker and Henderson 1976). For example, Alder and Wainwright (1957) showed that 'hard sphere' liquids crystallized as the density were increased above a certain value while Rahman (1964) showed that the diffusion and structural evolu-

J. P. Ewen · E. R. Fernández · E. R. Smith · D. Dini (✉)
Department of Mechanical Engineering, Imperial College London, London SW7 2AZ, UK
e-mail: d.dini@imperial.ac.uk

E. R. Smith
Department of Mechanical and Aerospace Engineering, Brunel University London, Uxbridge, Middlesex UB8 3PH, UK

M. Paggi and D. Hills (eds.), *Modeling and Simulation of Tribological Problems in Technology*, CISM International Centre for Mechanical Sciences 593,
https://doi.org/10.1007/978-3-030-20377-1_3

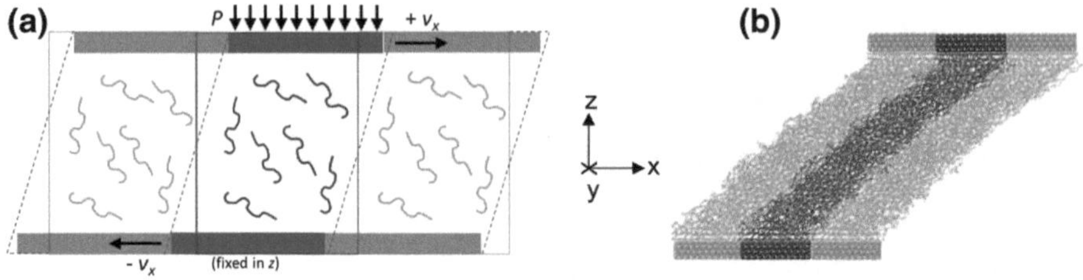

Fig. 3.1 Schematic of confined NEMD simulation (**a**) and an example snapshot from confined NEMD simulation (**b**). Adapted from Ewen et al. (2018a)

tion of liquid argon takes place by a series of small, highly coordinated motions of neighbouring atoms.

More recently, MD has been increasingly utilized to study liquid transport properties such as viscosity, initially without shear applied (Levesque et al. 1973) but using Green–Kubo formalism (Kubo 1957). This remains a popular equilibrium method to study such transport coefficient, but NEMD simulations, in which the fluid is sheared, is particularly useful for tribological applications. In early NEMD simulations, the system was sheared by applying an equal and opposite velocity to the regions of fluid atoms at the top and bottom of the simulation cell (Ashurst and Hoover 1975). Instead of moving the outer fluid atoms, shear is now more commonly applied by adding velocities to solid walls which confine the system (Bitsanis et al. 1987). Confinement can significantly influence the fluid behaviour (Granick 1991; Gubbins et al. 2011), which is discussed in more detail below.

An example system setup for confined NEMD simulations with moving solid surfaces is shown in Fig. 3.1. Periodic boundary conditions are applied in *x* and *y* directions.

Alternatively, bulk NEMD simulations can be performed with periodic boundary conditions in all three Cartesian directions. In bulk NEMD simulations, the periodic cell is deformed using the SLLOD equations of motion (Evans and Morriss 1984), Lees–Edwards boundary conditions (Lees and Edwards 1972) and temperature is controlled by a non-stochastic thermostat like Nosé–Hoover (Nosé 1984; Hoover 1985). It is important to note that excellent agreement between bulk and confined NEMD simulations has been observed when the surfaces are sufficiently separated such that there is a negligible confinement-induced viscosity increase in the latter (Liem et al. 1992; Todd and Daivis 2007). In this chapter, we will focus on recent applications of NEMD simulations in tribology. For the underlying theory, readers are directed to comprehensive books on the topic by Evans and Morriss (2008) and Todd and Daivis (2017).

Over the past four decades, advances in computational power, software parallelization and model sophistication have enabled NEMD simulations to progress significantly in terms of time and length scale accessibility, molecule complexity and accuracy compared to experiment (Ewen et al. 2018a). Thus, NEMD has now become capable of directly evaluating physical properties in industrially important systems. For example, NEMD can now accurately describe the high-pressure rheology of lubricant molecules (Jadhao and Robbins 2017), which is of significant industrial interest. Moreover, relatively new techniques, such as ab initio NEMD, are starting to be applied to study additive tribochemistry (Loehlé and Righi 2018; Kuwahara et al. 2019). As a result, NEMD can now be employed to test the applicability of macroscopic models and even facilitate the rational design of improved lubricants and additive molecules (Ewen et al. 2018a).

This chapter will first discuss key simulation methodology required to obtain accurate results from NEMD simulations. The subsequent sections will focus on three important areas where NEMD has been recently applied. The first section will discuss classical NEMD simulations of confined lubricants, the second and third sections are devoted, respectively, to NEMD and ab initio NEMD simulations of confined lubricant additives. The final section examines a multi-scale method by domain decomposition coupling CFD and NEMD and applied to tribologically relevant systems.

Simulation Details

Within the constraints of commonly available computational resources, only rather small time (ns) and length (nm) scales are generally accessible to MD simulations. As a consequence, an important consideration for any MD simulation is whether the behaviour of interest can be captured over these accessible scales (Elber 2016).

Examples of simulation methods used in tribology and their accessible time and length scales are shown in Fig. 3.2.

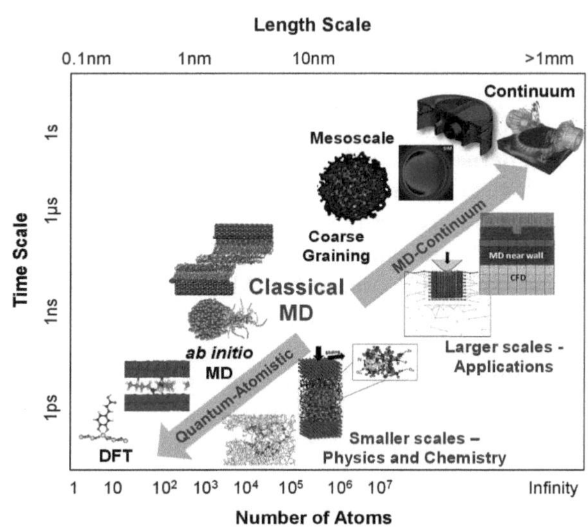

Fig. 3.2 Different methods and accessible scales for simulations of tribological systems. Adapted from Ewen et al. (2018a)

One important manifestation of the rather short accessible timescales in NEMD simulations are that high shear rates (generally $\geq 10^7$ s^{-1}) are required to reach a nonequilibrium steady state (Bair et al. 2002a; Jadhao and Robbins 2017). Simulating lower shear rates have been a long-term goal of NEMD simulations to facilitate direct overlap with experiments and real components. For comparison, the high-pressure viscosity of lubricants can generally only be measured up to shear rates of approximately $10^4\,s^{-1}$ (Bair et al. 2002a), tribology experiments can extend up to around $10^6\,s^{-1}$ (Spikes and Jie 2014), while high-performance engine components can reach up to $10^8\,s^{-1}$ (Taylor and de Kraker 2017). Direct NEMD simulations have been conducted for squalane at shear rates as low as $10^5\,s^{-1}$ through μs simulations on large supercomputers by Jadhao and Robbins (2017), although direct overlap between experiments and NEMD simulations has still not been obtained for realistic fluids. Techniques, such as the transient-time correlation function (TTCF) (Evans and Morriss 1988), facilitate accurate results for shear rates below $10^5\,s^{-1}$ using ensemble averaging correlation products from many independent NEMD trajectories (Pan and McCabe 2006). However, thousands of independent trajectories are required, making it a rather computationally expensive technique, and thus it has only been used to study rather small molecules (n-decane) up to now (Pan and McCabe 2006).

To obtain accurate results from confined NEMD simulations, which are representative of experiments, a number of methodological choices need to be carefully considered. For example, an appropriate thermostatting strategy and force field should be employed; these are briefly outlined below. In NEMD simulations, effective thermostatting is essential to reach a nonequilibrium steady state by removing heat produced during the shearing process. There are a number of possible thermostatting methods available for confined NEMD simulations, as reviewed by Bernardi et al. (2010) and Yong and Zhang (2013). For confined NEMD simulations, the thermostat is most commonly applied only to the wall atoms. This allows a thermal gradient to develop through the thickness of the lubricant film (Khare et al. 1997), as it occurs in tribology experiments (Lu et al. 2018). NEMD simulations with thermostats applied directly to the confined fluid molecules do not allow temperature gradients to develop and can artificially influence the structure, flow and friction behaviour (Martini et al. 2008; Bernardi et al. 2010; Yong and Zhang 2013). In confined NEMD simulations of tribological systems, wall thermostatting is performed using stochastic thermostatting algorithms, e.g. Langevin (Schneider and Stoll 1978), are generally utilized due to their efficient energy dissipation (Berro et al. 2011). For the relatively thin films usually studied in confined NEMD simulations (nm), the temperature rise, which is generally largest in the centre of the film, only becomes significant at high shear rates (Berro et al. 2011).

In MD simulations, force fields are used to describe the forces between the system of interacting atoms. The accuracy of any MD simulation is heavily dependent on the force field used, and thus significant effort should be spent choosing and testing before production simulations are conducted (Ewen et al. 2016b). The functional forms of most classical molecular force fields are quite similar, most include bonded (bond stretching, angle bending, and dihedral torsion) and nonbonded (van der Waals, electrostatic) terms. Bonded interactions are commonly represented with simple harmonic potentials while nonbonded terms are usually represented by the Lennard-Jones (LJ) and Coulomb potentials. LJ interactions are usually cut-off at a distance of around 10 Å whereas long-ranged Coulombic interactions are treated with a solver such as particle–particle, particle–mesh (PPPM) since truncation can lead to unphysical results (Feller et al. 1996). These potentials can be empirically parametrized to reproduce important experimental properties for a set of training compounds, derived from first principles calculations, or computed using a combination of the two. Force field parameterization is a complex and time-consuming process, and most of the force fields used to study liquid lubricants were originally developed for biological applications (Ewen et al. 2016b).

A key target of NEMD simulations of tribological systems is to yield realistic viscous behaviour of lubricant-sized molecules, which requires the use of highly accurate force fields. The computational expense of classical NEMD simulations is usually dominated by the nonbonded interactions. Consequently, most historic simulations of tribological systems which include alkane molecules have employed united-atom (UA) force fields where the nonpolar hydrogens are grouped with the carbon atoms to generate CH, CH_2 and CH_3 pseudo-atoms (Ewen et al. 2016b). This decreases the number of interaction sites by around 2/3 and computational expense by up to an order of magnitude compared to all-atom (AA) force fields (Martin and Siepmann 1999). However, UA force fields have been shown to lead to large viscosity under prediction for linear, long-chain alkanes (50% error for *n*-hexadecane) compared to experiment (Ewen et al. 2016b). For alkanes with multiple short branches, UA force fields can give reasonably accurate viscosity results (15% for squalane), but for those with fewer, longer branches they are less accurate (50% for 9-n-octyldocosane) (Moore et al. 2000). For such molecules, all-atom (AA) force fields are required to accurately reproduce viscous behaviour (Allen and Rowley 1997; Ewen et al. 2016b). For example, the L-OPLS-AA force field (Siu et al. 2012) within 10% for *n*-hexadecane at ambient and high-temperature–high-pressure conditions (Ewen et al. 2016b). Thus, for accurate viscosity prediction, AA force fields are required for longer, linear molecules, while for more branched, shorter molecules, UA force fields may be an acceptable trade-off.

Classical NEMD Simulations of Lubricants

Many lubricated engineering components include elements that roll and slide together, for example, rolling bearings, gears, constant velocity joints and cam/follower systems. In these components, a significant proportion of the friction loss is in the elastohydrodynamic lubrication (EHL) regime, where very thin (nm) fluid films are sheared at very high shear rate, $\dot{\gamma}$ (10^5–$10^8\,s^{-1}$) and pressure, P (GPa) (Spikes and Jie 2014). Such extreme conditions are clearly difficult to investigate with in situ experiments and thus prediction of EHL friction remains a considerable challenge.

Bulk NEMD Simulations

NEMD simulations have been widely utilized to study fluids under high-pressure and shear conditions. For example, Moore et al. (2000) studied the effect of high shear rate and temperature on the viscosity of C_{30} isomers. While the Newtonian viscosity was significantly underpredicted due to the UA force field used, the change in viscosity with temperature, also known as viscosity index (VI), was in excellent agreement with experiments. Similarly, McCabe et al. (2001) and Liu et al. (2015) performed NEMD simulations to study the change in viscosity with pressure of 9-octylheptadecane and 1-decene trimer. The Newtonian viscosity was again significantly underpredicted due to the UA force fields used; however, the change in viscosity with pressure, the α value was in good agreement with experiment in both cases. Accurate prediction of VI and α of new molecules is a potentially valuable application of NEMD since these properties are important to their performance under EHL conditions (Bair and Kottke 2003).

NEMD has been used to test the applicability of the Eyring (1936) and Carreau (1972) shear thinning models commonly applied to predict EHL friction. For example, Bair et al. (2002a) compared the high-pressure viscosity of a molecular lubricant (squalane) from a rheometer and bulk NEMD simulations. This was the first comparison of the nonlinear rheology predicted by NEMD with experiment, and was thus the first experimental test of NEMD simulations in the shear thinning regime. Although the experimental and simulation data were separated by several orders of magnitude in shear rate, they collapsed onto the same time–temperature superposition master curve (Bair et al. 2002a). This master curve was successfully fit using the power-law Carreau (1972) model. Similarly, Liu et al. (2017) used the Carreau (1972) model to describe the shear thinning behaviour of squalane as well as several types of poly-alpha-olefin (PAO) molecule. They correlated the shear thinning behaviour with changes to the conformation of the molecules, quantified through the radius of gyration. Recent bulk NEMD simulations by Jadhao and Robbins (2017) suggested that, although the viscosity-shear rate behaviour at lower pressure was better described by the Carreau (1972) model, the Eyring (1936) shear thinning model was more appropriate at higher pressure. Moreover, the 'incremental viscosity' of the Eyring model, measured for LJ fluids using a 'shear-kick' NEMD scheme, has been shown to be a special case of the Carreau model (Heyes et al. 2018).

Confined NEMD Simulation Results

A significant assumption of bulk NEMD simulations is that a linear velocity profile develops in the fluid between the sliding surfaces (Delhommelle et al. 2003; Cao and Likhtman 2012). However, deviations from this Couette case were suggested by Israelachvili (1986) from surface forces apparatus experiments and later confirmed by confined NEMD simulations (Thompson and Robbins 1990). Such behaviour in nanoconfinement has been attributed to phase transitions (vitrification or crystallization) and large viscosity increases that facilitate localization of the shear at the confining surfaces or within the fluid itself (Thompson and Robbins 1990). Robbins and Smith (1996) suggested that high pressures could induce similar phase transitions to those in the relatively thicker EHL films (≈100 nm) (Spikes and Jie 2014). Moreover, nonlinear flow has been used for more than half a century to explain observations from EHL experiments (Plint 1967; Ehret et al. 1998), but conclusive experimental proof has remained elusive for realistic lubricants. However, such behaviour has been observed in confined NEMD simulations, with increasing complexity, as discussed below.

Comprehensive confined NEMD simulations of atomic Lennard-Jones (LJ) fluids under EHL conditions (Heyes et al. 2012; Gattinoni et al. 2013; Maćkowiak et al. 2016) have identified the transition from Couette flow to various types of shear localization with increasing pressure. For example, central localization (CL), where the outer regions of the fluid move at the same velocity as the confining surfaces and only the central region is sheared. Another form of shear localization is plug slip (PS), where the outer regions of the fluid are sheared while the central region remains unsheared. In the LJ fluid systems, friction behaviour deviated from classical friction relations between macroscopic bodies, for example, the friction force was observed to decrease with increasing load and shear rate in some cases (Maćkowiak et al. 2016). More recently, similar NEMD simulations have been performed for molecular systems (Ewen et al. 2017b; Washizu et al. 2017; Porras-Vazquez et al. 2018).

The EHL friction coefficient from these NEMD simulations was in good agreement with extrapolations from experiments conducted at lower shear rates (Ewen et al. 2017b; Porras-Vazquez et al. 2018). For example, Fig. 3.3a shows the change in friction with logarithmic shear rate for two molecular fluids; 2,6,10,15,19,23-hexamethyl-tetracosane (squalane) and 2,3-dimethyl-2-[(3-methylbicyclo[2.2.1] hept-2-yl)methyl]bicyclo[2.2.1]heptane (DM2H). Squalane is a linear C_{24} alkane with six methyl branches that have commonly been employed as a model lubricant, designed to give low friction. DM2H is a rigid bicyclic molecule designed to give high friction for use in traction drives (Zhang et al. 2017).

In Fig. 3.3a-i, the friction coefficient for squalane increases logarithmically with shear rate and also generally increases with pressure (Ewen et al. 2017b). This type of behaviour is commonly observed in experiments and NEMD simulations of lubricants under EHL conditions (Spikes and Jie 2014). The slope of the friction coefficient with logarithmic shear rate decreases with increasing pressure, as was also observed in NEMD simulations of binary atomic LJ fluids (Gattinoni et al. 2013). At very high shear rate, the friction coefficient of squalane at 0.5 GPa exceeds that at 2.0 GPa, which has also been observed for LJ fluids (Maćkowiak et al. 2016).

Although Fig. 3.3a-ii shows that the friction coefficient of DM2H also increases logarithmically with shear rate, it does so with a much lower slope than for squalane. This behaviour is similar to that observed in NEMD simulations of single-component LJ fluids subjected to high pressures (Heyes et al. 2012; Gattinoni et al. 2013; Maćkowiak et al. 2016). Moreover, DM2H shows much higher friction than squalane,

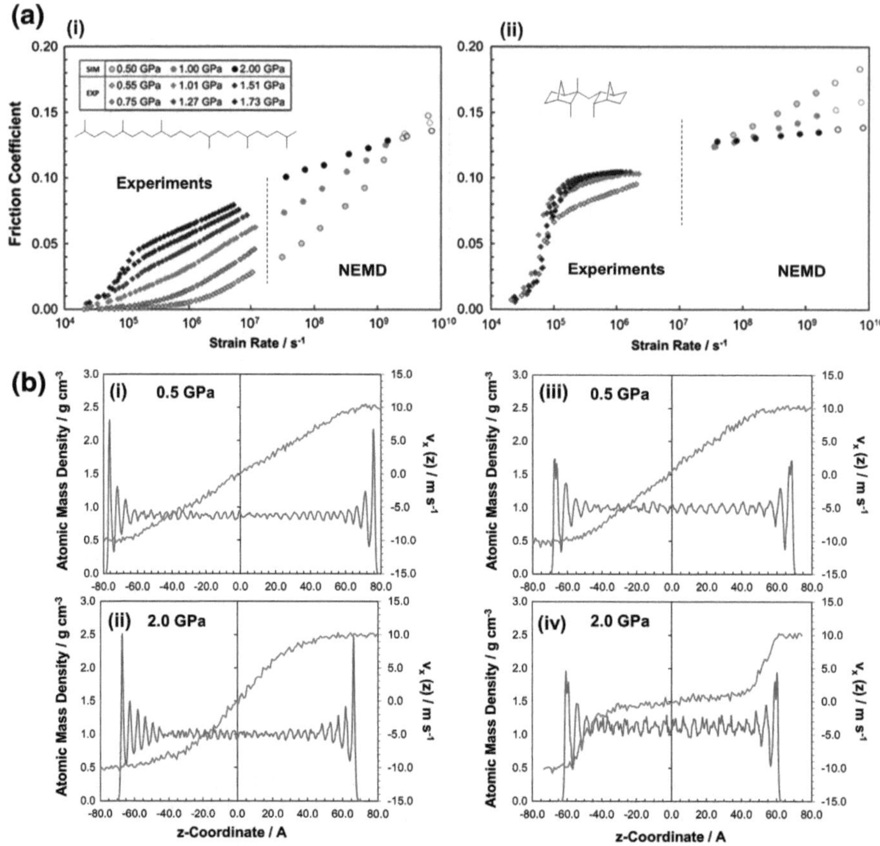

Fig. 3.3 **a** Friction coefficient versus log(shear rate) for the fluids at: 353 K and 0.5–2.0 GPa; squalane (i) a DM2H (ii). Thermally corrected experimental data shown as filled diamonds. Isothermal NEMD data shown as filled circles, NEMD data with a temperature rise shown as open circles. NEMD data time-averaged for the final 10 nm of sliding. **b** Flow profiles for squalane at 0.5 GPa (i), squalane at 2.0 GPa (ii), DM2H at 0.5 GPa (iii), DM2H at 2.0 GPa (iv). Adapted from Ewen et al. (2017b)

particularly at low shear rate. The high friction of DM2H can be attributed to interlocking of the bulky bicyclo[2.2.1]heptyl groups which coupled with its internal molecular stiffness, increases energy barriers for neighbouring molecules to slide over one another (Ewen et al. 2017b). The friction coefficient of DM2H at 0.5 GPa exceeds that at 2.0 GPa at much lower shear rate than for squalane.

Figure 3.3b shows the flow profiles for squalane at 0.5 GPa (i) and 2.0 GPa (ii) and DM2H at 0.5 GPa (iii) and 2.0 GPa (iv). In all cases, the outer molecular layer of fluid moves at the same velocity as the iron oxide surfaces, indicating that no boundary slip occurs (Ewen et al. 2017b). For squalane at 0.5 GPa, a Couette flow profile develops, with a linear velocity gradient in the fluid confined between the two surfaces. At 2.0 GPa, the mass density profile shows strong layering of the fluid close to the surfaces, suggesting the formation of an ordered, solid-like region close to the slabs, with a liquid-like region in the centre of the film. It is important to note that the strong layering extends much further into fluid than was directly influenced by the surface (LJ interactions cut-off at 12 Å). The flow profile for squalane at 2.0 GPa shows CL behaviour, in common with the simulations for LJ fluids (Heyes et al. 2012; Gattinoni et al. 2013; Maćkowiak et al. 2016). Experiments have also shown CL (Bair et al. 1993, 1994; Bair and McCabe 2004; Galmiche et al. 2016) under EHL conditions, although presently only for very viscous polymers.

The change in flow behaviour with pressure for DM2H is similar to that for squalane, but it shows greater divergence from Couette flow under equivalent conditions. Even at 0.5 GPa, DM2H shows weak CL, with layering of the molecules close to the surface resulting in a cubic velocity profile. At 2.0 GPa, DM2H shows PS, another phase identified in confined NEMD simulations of LJ fluids (Heyes et al. 2012; Gattinoni et al. 2013; Maćkowiak et al. 2016). Note that PS is distinct from the boundary slip behaviour, which is commonly observed in NEMD simulations of very thin films (Martini et al. 2008). PS has also been observed experimentally for very viscous polymers under EHL conditions (Ponjavic et al. 2014; Sperka et al. 2014; Jeffreys et al. 2019).

From these NEMD simulations, potential links have been drawn between the glass transition, nonlinear flow and the limiting shear stress. This is the point at which the friction coefficient becomes insensitive to shear rate and pressure (Martinie and Vergne 2016). Recent NEMD simulations have suggested that it is the glass transition of the confined fluid that drives both the unusual friction and flow behaviour (Porras-Vazquez et al. 2018).

Classical NEMD Simulations of Lubricant Additives

Organic friction modifiers (OFMs) are amphiphilic surfactant molecules that contain nonpolar hydrocarbon tail groups attached to polar head groups. Commercial OFMs generally contain unbranched aliphatic tail groups containing 12–20 carbon atoms as a result of their effective friction reduction, high base oil solubility and high availability from natural fats and oils. Many head groups have been employed, but the most commonly studied in the literature are carboxylic acids, amines, amides and glyceride esters. Extensive experimental evidence suggests that OFMs reduce friction through the adsorption of their polar head groups to metal-, ceramic- or carbon-based surfaces, with strong, cumulative van der Waals forces between proximal nonpolar tails leading to the formation of incompressible monolayers that prevent contact between solid surfaces (Spikes 2015).

Classical NEMD simulations can be used to simultaneously probe the nanoscale structure and friction of OFM films, making them a valuable complement to experiments (Ewen et al. 2018a; Apóstolo et al. 2019). Due to the relatively slow film formation process on MD timescales, NEMD simulations with preformed films, similar to those formed by Langmuir–Blodgett experiments (Briscoe and Evans 1982), are usually employed. Carboxylic acid Langmuir–Blodgett films on solid surfaces were first studied by MD in the 1990s (Moller et al. 1991). This was soon followed by the first NEMD simulations of such systems which showed that higher coverage films led to lower friction (Kong et al. 1997). Their simulations predicted that OFM molecules were approximately normal to the surface at high surface coverage ($4.8\,\mathrm{nm}^{-2}$) but became more tilted at lower surface coverage. The authors suggested that tilting maximizes the van der Waals attraction between the chains by increasing the packing efficiency. The tilting transition was attributed to the packing of the hydrogen atoms belonging to methylene groups on neighbouring molecules, meaning that the correct behaviour could only be accurately reproduced using AA and not UA force fields (Karaborni and Verbist 1994). NEMD simulations Ewen et al. (2016b) showed that the use of AA force fields is also critical to give accurate flow and particularly friction behaviour for OFM films under shear.

More recently, NEMD simulations of preformed OFM films have investigated the effect of head group type (Eder et al. 2013; Ewen et al. 2016a), tailgroup structure (Doig et al. 2014; Ewen et al. 2016a) and surface roughness (Eder et al. 2013; Ewen et al. 2017a) on their structure and friction under boundary lubrication conditions. For example, Fig. 3.4 shows the change in mass density and flow profiles (a) and friction coefficient (b) with surface coverage for a representative OFM, stearic acid (SA). The velocity profiles at all SA coverages in Fig. 3.4a show that there is no slip at the surface, as expected for the strongly absorbed OFM head groups. The OFM tail groups move at a similar velocity as the slab to which they are absorbed ($\pm 5\,\mathrm{m\,s}^{-1}$) until the region where they become interdigitated with the n-hexadecane lubricant. The low-coverage ($1.44\,\mathrm{nm}^{-2}$) flow profile (i) resembles Couette flow, with a near-linear fluid velocity profile between the slabs. The velocity profile contains step-like features, suggesting partial plug flow between the combined OFM-hexadecane layers. At medium coverage ($2.88\,\mathrm{nm}^{-2}$) (ii), the OFM tail groups are sheared in the region in which they are interdigitated with hexadecane. The velocity profile passes through zero at the centre of the hexadecane layer, with a steeper gradient than at low coverage. At high coverage ($4.32\,\mathrm{nm}^{-2}$) (iii), the fluid layers remain mostly unsheared and slip planes form between the well-defined OFM and hexadecane layers.

This flow behaviour correlates with the observed change in friction coefficient with surface coverage in Fig. 3.4b. Results are shown for SA, oleic acid (OA), stearamide (SAm), oleamide (OAm, glycerol monostearate (GMS), and glycerol monooleate (GMO). For all of the OFMs at $10\,\mathrm{m\,s}^{-1}$, the friction coefficient increases by around 5% for all OFMs between low and medium coverage (Fig. 3.10a) before decreasing by 30% between medium and high coverage. This trend is consistent with SFA experiments using monolayer films formed from other surfactants in which friction increased as the film moved from a liquid-like to an amorphous film and then decreased when a solid-like film was formed (Yoshizawa et al. 1993).

Comparisons have also been made between the friction-sliding velocity behaviour from confined NEMD simulations of OFMs and boundary friction experiments. Figure 3.5 shows the change in the friction coefficient with logarithmic sliding velocity from CETR UMT boundary friction experiments (Campen et al. 2012)

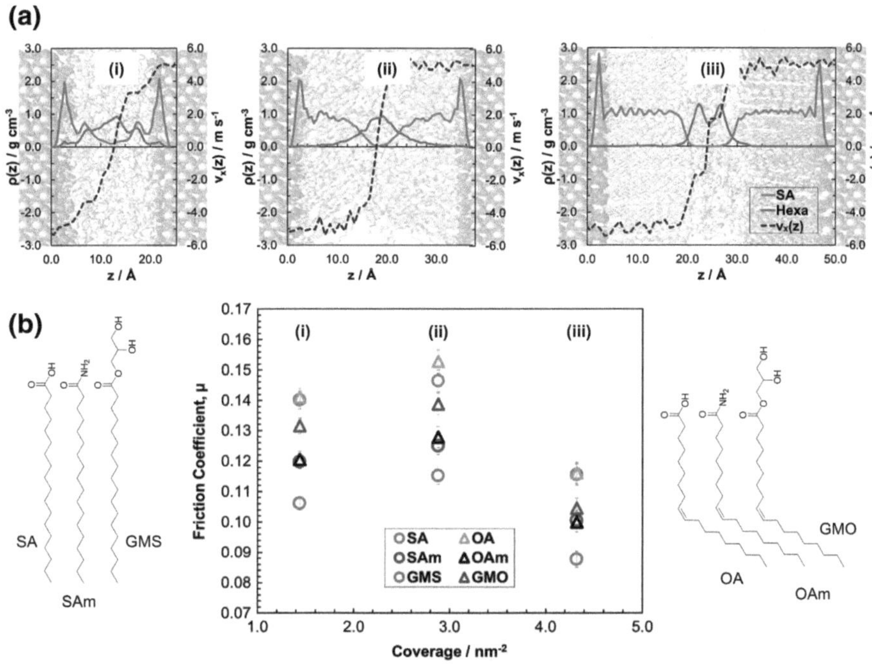

Fig. 3.4 **a** Flow and mass density profiles during sliding, for a representative OFM system (SA) at low (i), medium (ii) and high (iii) coverage. **b** Friction coefficient as a function of coverage for SA, OA, SAm, OAm, GMS and GMO. Both at 0.5 GPa and $10\,m\,s^{-1}$. Adapted from Ewen et al. (2016a)

and NEMD simulations (Ewen et al. 2016a, b). The NEMD simulations were performed at much higher sliding speeds ($1–10\,m\,s^{-1}$) than the tribology experiments ($10^{-7}–10^{-2}\,m\,s^{-1}$). The NEMD simulations were performed with iron oxide surfaces at 0.5 GPa and 300 K for low-coverage ($1.44\,nm^{-2}$) and high-coverage ($4.32\,nm^{-2}$) SA films. In the tribology experiments, OFM concentration, rather than coverage, is varied since this is far easier to measure and control. The tribology experiments were performed in a steel–steel contact at 0.7 GPa and 308–373 K for SA and OA at a concentration of $0.01\,mol\,dm^{-3}$. Desorption isotherm experiments have shown that SA forms films with higher maximum coverage ($\approx 4\,nm^{-2}$) compared to OA ($\approx 2\,nm^{-2}$) on iron oxide surfaces (Wood et al. 2016). This is due to the Z-alkene group in the OA tail group which leads to less efficient surface packing. Moreover, NEMD simulations (Fig. 3.4) have shown that, at equal surface coverage, there is negligible difference in friction coefficient between SA and OA (Ewen et al. 2016a). Therefore, at high concentration, the SA experimental results are comparable to the high-coverage ($4.32\,nm^{-2}$) NEMD simulations while the OA experimental results are comparable to the low-coverage ($1.44\,nm^{-2}$) NEMD simulations.

The tribology experiments in Fig. 3.5 show that OFMs with saturated tails (SA) give lower friction which increases logarithmically with sliding velocity. Conversely,

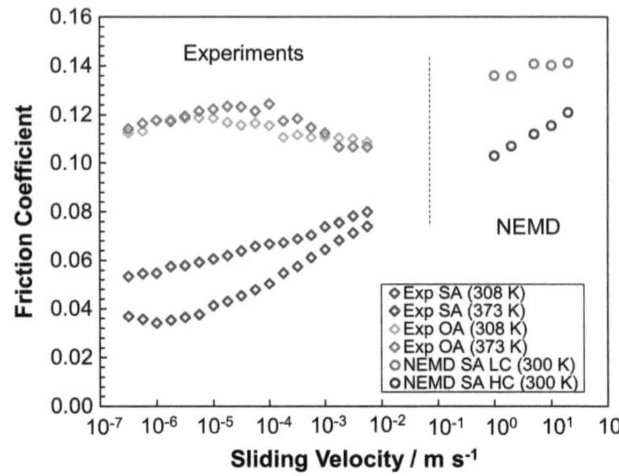

Fig. 3.5 Change in friction coefficient with logarithmic sliding velocity from experiments and NEMD simulations. NEMD results for SA at low coverage, 1.44 nm^{-2} (LC) and high coverage, 4.32 nm^{-2} (HC) from Ewen et al. (2016a, b) and experimental results for SA and OA from Campen et al. (2012)

OFMs with Z-unsaturated tail groups (OA) show higher friction, which is more weakly dependent on sliding velocity (Campen et al. 2012). Similarly, the NEMD simulations at low coverage give higher friction than at low coverage and increase with a shallower slope than at high coverage. This can be attributed to the more liquid-like film and widely spaced OFM molecules which are able to rearrange in order to reduce the energy barrier height during sliding (Ewen et al. 2016a). Moreover, the high-coverage NEMD results are in quantitative agreement with extrapolations from the SA experiments to higher sliding velocity. It is important to note that such agreement is only attainable with an accurate, AA force field (Ewen et al. 2016b). Although the low-coverage friction results are somewhat higher than extrapolations from the OA experiments to higher sliding velocity, and they show the same qualitative trend. Thus, combined these NEMD and experimental results provide strong evidence that OA forms lower coverage films than SA, which leads to higher friction which is less dependent on sliding velocity.

Some NEMD studies have studied film formation from concentrated OFM solutions rather than using preformed films. OFM molecules are unlikely to exist in isolation in nonpolar solvents and most are expected to form dimers (Jaishankar et al. 2019). Moreover, some OFMs, such as glyceride esters, are known to form reverse micelles in nonpolar solvents (Bradley-Shaw et al. 2015). NEMD simulations have also been used to study the resilience of these reverse micelles to pressure and shear (Bradley-Shaw et al. 2016, 2018). They have shown that, while they are stable without sliding, the reverse micelles usually disintegrate under shear to form surface films. Thus, it is expected that such OFMs eventually form films similar to those shown in Fig. 3.4 inside tribological contacts.

In addition to reducing boundary friction, OFMs have also been postulated to reduce friction in the hydrodynamic lubrication regime by inducing liquid slip (Choo et al. 2007). This has also been investigated in recent NEMD simulations (Ewen et al. 2018b). The simulations showed that a measurable slip length was only observable for OFM films with a high surface coverage, which provide smooth interfaces between well-defined OFM and n-hexadecane layers. Slip commenced above a critical shear rate, beyond which the slip length first increased with increasing shear rate and then asymptoted towards a constant value. This was consistent with previous NEMD simulations of n-alkane slip on flexible solid surfaces (Martini et al. 2008). The maximum slip length increased significantly with increasing pressure. Systems and conditions which showed a larger slip length typically gave a lower friction coefficient. Generally, the friction coefficient increased linearly with logarithmic shear rate; however, it showed a much stronger shear rate dependency at low pressure than at high pressure. Relating slip and friction, slip only occurred above a critical shear stress (Spikes and Granick 2003), after which the slip length first increased linearly with increasing shear stress and then asymptoted. This behaviour was well-described using the slip model due to Wang and Zhao (2011) which is based on Eyring's molecular kinetic theory. The NEMD simulations supported that high-coverage OFM films can significantly reduce friction by promoting slip, even when the surfaces are well separated by a lubricant (Ewen et al. 2018b).

Ab Initio NEMD Simulations of Lubricant Additives

A detailed understanding of tribochemical reactions is of paramount importance for designing new, more effective and environmental-friendly lubricant additives. Density functional theory (DFT) calculations can provide insights into the adsorption and dissociation of additives on solid surfaces from first principles (Gattinoni and Michaelides 2015; Loehlé and Righi 2017; Gattinoni et al. 2018). However, in most cases, these calculations are static, i.e. no shear is applied.

Sliding surfaces are ubiquitous in tribology, and thus 'mechanochemistry' is of critical importance (Spikes and Tysoe 2015; Spikes 2018). Classical NEMD simulations can give insights regarding tensile forces on the bonds of molecules under compression and shear (Adams et al. 2015). However, the harmonic form of the bonding term in most classical force fields prevents the study of tribochemical processes. Force fields which approximate chemical reactivity through bond order potentials, for example, ReaxFF (Senftle et al. 2016) can be used to study tribochemistry in NEMD simulations. ReaxFF NEMD simulations of phosphoric acid confined and sheared between solid surfaces were used to explain its liquid superlubricity behaviour (Yue et al. 2013). Previous experiments by Li et al. (2011) showed that phosphoric acid and water mixtures can exhibit liquid superlubricity when confined between sapphire surfaces. Confined NEMD simulations of phosphoric acid molecules using ReaxFF (Yue et al. 2013) showed that the variation of hydrogen bond interaction strength and change of velocity accommodation affect the frictional response. At high temperature, tribochemical reactions, specifically, polymerization of phosphoric acid molecules, generation of water molecules and formation of slip planes occurred. The generation of water molecules, and their accumulation at the sliding interface, leads to weaker interfacial hydrogen bond interactions and velocity accommodation between the interfacial water layers.

ReaxFF is more computationally expensive than most classical force fields, and thus simulations are limited to shorter timescales ($\approx$1 ns) and higher sliding velocities ($\approx$100 m s^{-1}) (Yue et al. 2013). As well as ReaxFF, many other 'reactive' force fields are available for MD simulations, as discussed in a recent review by Harrison et al. (2018). A considerable drawback of all reactive force fields is the limited availability of reliable parameter sets to study the materials, conditions and properties of interest in tribology. The generation of high-quality parameter sets is a very difficult and time-consuming task due to the huge number of parameters which need to be fitted in comparison to conventional molecular force fields (Ewen et al. 2018a). However, when properly fitted, reactive force fields facilitate simulation studies of relatively large systems for appreciable timescales, which can significantly improve our understanding of additive tribochemistry.

Ab initio MD can be used to model additive tribochemistry from first principles and thus does not require force field parameterization. However, they increase the computational expense by more than an order of magnitude compared to MD simulations with reactive force fields, meaning that the accessible length (usually single molecules) and timescales (ps) are more limited. This can be mitigated somewhat by employing less computationally intensive ab initio techniques such as Car–Parrinello and tight binding.

Despite the limited length and timescales accessible, recent ab initio MD simulations have given useful insights in tribochemistry of lubricant additives. For example, Mosey et al. (2005) performed compression/decompression cycles up to very high pressures (2.5–32.5 GPa) using ab initio MD simulations to study the tribochemistry of zinc dialkyldithiophosphate (ZDDP) antiwear additives. The Car–Parrinello method (Car and Parrinello 1985) was used for the ab initio calculations. The simulations revealed the molecular origins of ZDDP film formation, wear resistance and energy dissipation. These effects were shown to originate from pressure-induced changes in the coordination number of atoms acting as cross-linking agents to form chemically connected networks (Mosey et al. 2005). More recently, shear has also been applied, for example, by Loehlé and Righi (2018) who performed ab initio NEMD simulations of trimethylphosphite (TMPi) molecules confined between Fe surfaces. The ab initio calculations were performed using full DFT. Gas phase lubrication (GPL) experiments suggested that TMPi dissociates on Fe surfaces leading to the formation of Fe–P tribofilms that significantly reduces friction and wear (Gao et al. 2004). The ab initio molecular dynamics simulations (Loehlé and Righi 2018) uncovered the atomistic mechanisms that lead to P release under boundary lubrication conditions. These P atoms are critical to passivate the Fe surface and reduce friction. Simulation snapshots of a TMPi molecule confined between Fe surfaces, thermostatted at 300 K, pressurized at 2.0 GPa, and moved at 200 m s^{-1} are shown in Fig. 3.6. The activation time for molecular dissociation observed in the simulations was orders of magnitudes smaller than that expected for open surfaces under static conditions on the basis of the calculated activation barriers. This observation and the observed dependence of reaction rates on the applied load constitute clear evidence that mechanical stress is able to activate tribochemical reactions, even at room temperature (Loehlé and Righi 2018). These findings are consistent with exper-

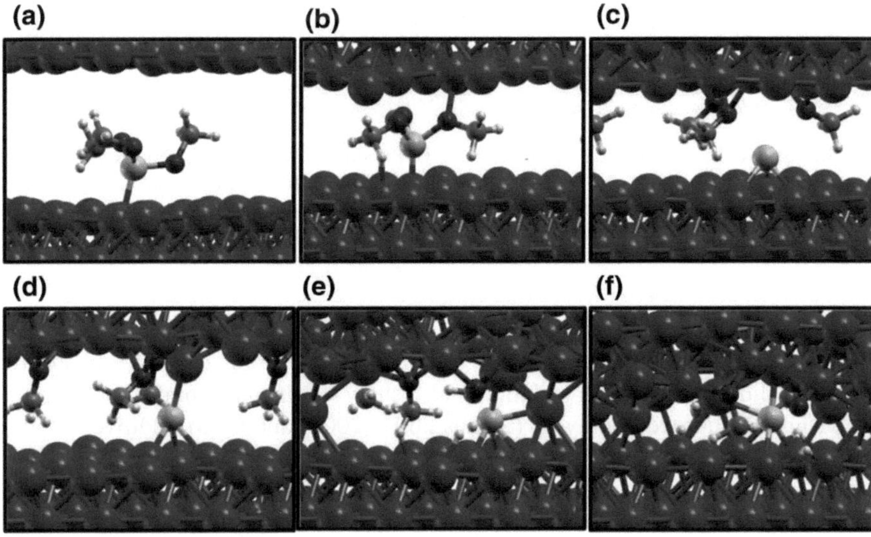

Fig. 3.6 Snapshots acquired during ab initio NEMD simulation of a TMPi molecule confined between sliding iron surfaces at 200 m s^{-1}, 2.0 GPa, and 300 K. The simulation time increases from 0 ps (**a**) to 7 ps (**f**). Fe is coloured in blue, P in yellow, O in red, C in grey, and H in white. Adapted from Loehlé and Righi (2018)

imental evidence for ZDDP film formation on solid surfaces (Gosvami et al. 2015; Zhang and Spikes 2016). Combined, experimental and simulation evidence strongly supports a stress-augmented thermal activation (SATA) model for antiwear additive tribochemistry (Spikes 2018).

Ab initio NEMD simulations have also been used by Kuwahara et al. (2019) to study the mechanochemical decomposition of OFMs with multiple reactive centres. This is important since such additives have shown superlow friction when confined and sheared between tetrahedral amorphous carbon (ta-C) surfaces (Kano et al. 2005). The ab initio calculations were performed using the tight binding approximation (Sutton et al. 1988). The simulations (Kuwahara et al. 2019) revealed that, due to the simultaneous presence of two reactive centres (carboxylic acid and alkene groups), unsaturated fatty acids can concurrently chemisorb on both of the ta-C surfaces. When sliding was initiated, mechanical strain triggers a cascade of molecular fragmentation reactions releasing passivating hydroxyl, keto, epoxy and olefinic groups. Simulation snapshots of an OA molecule confined between ta-C surfaces, thermostatted at 300 K, pressurized at 5.0 GPa, and moved at 100 m s^{-1} are shown in Fig. 3.7. Similarly, glycerol, which has three hydroxyl groups, reacts simultaneously with both ta-C surfaces, causing complete mechanochemical fragmentation and formation of aromatic passivation layers with superlow friction. Conversely, OFMs with only one reactive centre, such as SA, can only adsorb to one of the ta-C surfaces, and are thus less reactive, are less efficient in passivating the surfaces, and thus show higher friction (Kuwahara et al. 2019).

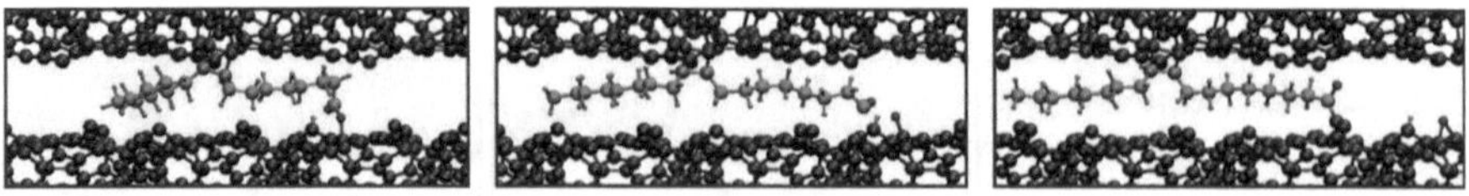

Fig. 3.7 Snapshots acquired during ab initio NEMD simulation of an OA molecule confined between sliding ta-C surfaces at $100\,\mathrm{m\,s^{-1}}$, 5.0 GPa, and 300 K. The simulation time increases from 0 ps (left) to 10 ps (right). Surface C is coloured in black, OA saturated C in green, OA unsaturated C in orange, O in red and H in white. Adapted from Kuwahara et al. (2019)

These examples showcase how ab inito NEMD simulations can provide mechanistic insights into the tribochemistry of lubricant additives confined and sheared between solid surfaces. Such techniques also show significant promise for the rational design of improved antiwear (Spikes 2004, 2008) and OFM additives (Spikes 2015) in the near future.

Multi-scale Modelling of Hydrodynamic Lubrication. Coupling MD-CFD Using Domain Decomposition

Introduction

It is evident, from the previous sections, that classical and ab initio NEMD methods are of paramount importance to the understanding of tribological mechanisms. Despite this, computational fluid dynamics (CFD) remains the de facto method for modelling hydrodynamic lubrication. By solving the Navier–Stokes equations or, most commonly, their thin film approximation known as Reynolds equation, a wide range of tribological problems can be studied. Accurate continuum simulation of tribological systems requires accurate constitutive relations, transport coefficients and boundary conditions for the system of interest. In some cases, this information may not be readily available nor even measurable with the current experimental technology. In addition, many physical processes that can be modelled explicitly at the atomic level (e.g. adsorption, phase changes, slip, etc.) cannot be easily incorporated into continuum methods. These aspects arise naturally in atomistic modelling, from the interatomic interactions defined by the force field in use. Why then not use NEMD pervasively? The big drawback of atomistic simulations is their enormous computational cost. Despite the huge increase in available computer power since the advent of high-performance computing (HPC), length scales over a few nanometres and timescales beyond nanoseconds are currently unfeasible to simulate. This limitation can be found in the systems reviewed in previous sections, where lubricant films modelled do not go beyond a few tens of nanometres at most. In the case where the scales of interests are beyond our computational power a *multi-scale method* can, for certain cases, be employed as a trade-off between computational efficiency

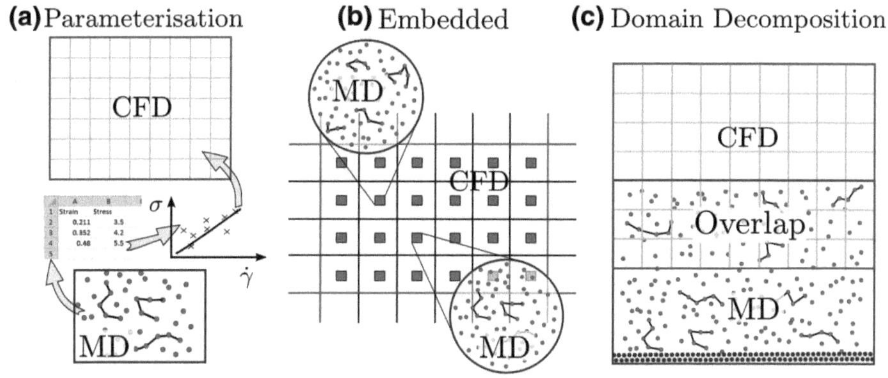

Fig. 3.8 Three types of coupling, **a** parameterizing data by running MD simulations to build up lookup tables or design closure models, **b** embedded coupling using MD simulation on-the-fly driven by the local CFD field, with resulting behaviour passed back to the CFD, **c** domain decomposition where both CFD and MD share a single simulation domain, exchanging information at the overlapping boundaries

and accuracy. An extensive overview of tribological modelling at different scales (including multi-scale modelling) can be found in Vakis et al. (2018). A multi-scale simulation method is a combination of two or more methods which operate at distinct scales. It is useful to categorize multi-scale methods into three types:

1. Parameterization of macroscopic models through microscopic data, see Fig. 3.8a. Constitutive relations like stress-velocity in fluids are computed and tabulated from a microscopic model and incorporated into a macroscopic one. This strategy is also feasible to obtain non-trivial boundary conditions (e.g. slip condition) for PDE-based macroscopic models (Holland et al. 2015).
2. *Heterogeneous Multi-scale Method* (Ren and Weinan 2005) (also termed embedded method) represents a general framework for tackling multi-scale problems, see Fig. 3.8b. In this method, the macroscopic solver is used throughout the whole simulation domain and the microscopic solvers are used to perform 'local refinement' on demand (e.g. computing viscosity using NEMD-SLLOD algorithm (Hoover et al. 2008) to feedback a CFD solver). This strategy is useful when the parameter space is too big to pre-compute and tabulate the data.
3. Methods which deal with isolated *defects*, see Fig. 3.8c. In this context, a defect means a certain region of the simulation domain where the macroscopic model is invalid. We can find several methods in this category (Weinan et al. 2004) but, among them, it is worth highlighting *coupled continuum-NEMD* strategies through domain decomposition (Mohamed and Mohamad 2009). The defect in fluid dynamics is often at an interface between two flow regimes where the continuum becomes invalid, for example, a solid–liquid or liquid–vapour interface or a moving shock wave between fast and slow moving fluid.
   Domain decomposition along an interface will be the method of interest in this section.

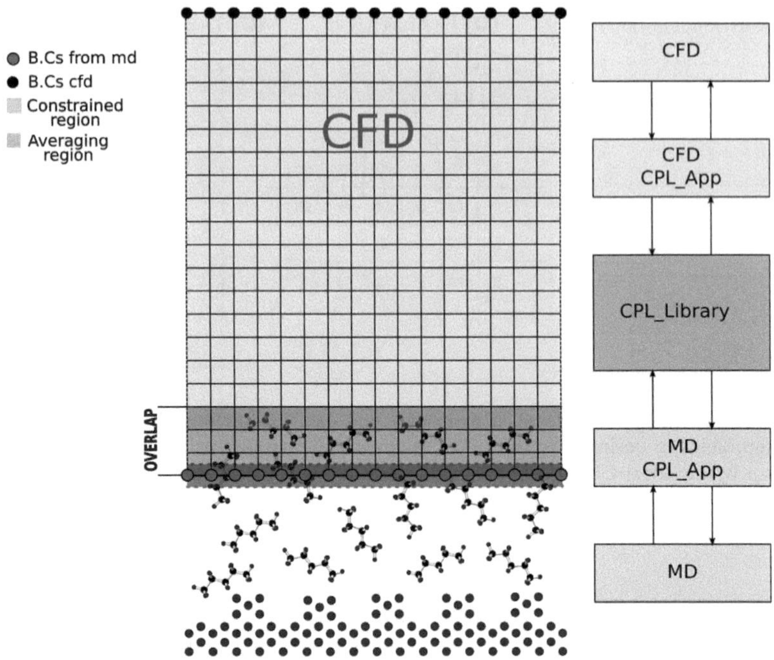

Fig. 3.9 A detailed typical setup for MD-CFD domain decomposition is depicted. The bottom domain where the liquid interacts with the surface is modelled using an atomistic description. Far from the surface, where bulk behaviour of the fluid is found, CFD is employed. The mesh (grid) discretizing the domain has been plotted with black lines. Two kinds of boundary conditions can be found in the CFD. The ones provided by the MD domain to the CFD (pink dots) and the user-provided (black dots). The constrained region is the region within the MD domain where the CFD constrains/fluxes are applied. In the averaging region, the coupled fields are averaged and passed from the MD to the CFD solver. The box diagram on the right is a schematic of how the software framework is set up. CPL_Library (Smith et al. 2016b) acts as a communication interface through the CFD and MD solver

Another family of multi-scale methods which deserve a mention are the 'equation-free' type. They are used to compute macroscopic properties without having to fit the data to a model when this is too complicated or unknown in advance (Weinan et al. 2004). The NEMD simulations presented in the previous sections of fluid lubricant films under shear are limited to the nanoscale. Simulations are limited to a few hundred nanometres in length scale with the longest timescales generally of order nanoseconds. The timescale limitation cases are hard to tackle, but length scale acceleration is naturally suited to a domain decomposition multi-scale method (Fig. 3.9). The idea is to use NEMD near the surface to capture interfacial phenomena and CFD far from it, where the liquid behaviour can be well described by bulk approximations. This coupled approach allows the exploration of larger fluid (lubricant) films by enlarging the CFD domain at virtually no extra cost, since the bottleneck, in terms of computational resources, is always in the MD domain which remains fixed in size.

The complexity lays in how to couple both solvers consistently which is explained in this section, after a brief introduction to computational fluid dynamics. The rest of the chapter is devoted to the theory and implementation details of the domain decomposition method for fluids with special consideration of tribological systems.

The Domain Decomposition Method

The continuum hypothesis is an idealization of a real material, which replaces molecular detail with a mean-field approximation. Intuitively, a mean field considers so many molecules they can be treated as a continuous flowing substance, with the individual molecular motion and structure no longer important. The evolution of this mean field is then governed by identical physical laws to molecular dynamics, namely, mass and energy conservation along with a continuum analogue of Newton's Law. This mean description allows the evolution of countless molecules to be simulated with affordable computational effort.

The key equation for fluid simulation is the Navier–Stokes equations, a set of nonlinear partial differential equations describing how the fluid velocity field evolves in time,

$$\frac{\partial \boldsymbol{u}}{\partial t} + (\boldsymbol{u} \cdot \nabla) \boldsymbol{u} = -\frac{1}{\rho} \nabla P + \nu \nabla^2 \boldsymbol{u}, \tag{3.1}$$

where $\boldsymbol{u}$ is velocity field, P the pressure field, ρ density and ν the coefficient of kinematic viscosity. The use of the Navier–Stokes equation in this form is valid for many fluids, where the complex evolution of molecular structures can be reduced to a single viscosity coefficient (Gad-el Hak 2006). Often in tribological applications, this assumption is not valid and more complex descriptions would be required. The analytical solution of Eq. (3.1) turns out to be impossible in all but the simplest of cases. Numerical simulations, called computational fluid dynamics (CFD), are therefore the main engineering approach to predicting fluid motion. Various methods are employed to solve the equations numerically. Perhaps the simplest is the finite differences method, where derivatives in Eq. (3.1) are approximated using a truncated Taylor expansion and the higher the truncation, the higher the accuracy. More common for fluid simulation is the finite volume approach, where the continuum equations of motion are solved in a weakened form as fluxes over the surface of a volume. The finite volume form has the advantage of being conservative as flow from one volume goes directly into the adjacent one, as well as allowing arbitrary-shaped grid cells. The time evolution at each point on the grid is therefore obtained from a combination of the values at the previous time and flow from adjacent points on the grid, described by the terms in Eq. (3.1). The values at the edges, known as boundary conditions, have to be explicitly provided and determine the evolution of the simulation. It is these values that coupled domain decomposition simulation aims to provide from the average behaviour of the molecular simulation.

The continuum representation of a fluid often fails to capture solid-liquid interfacial behaviour. Nucleation events for micelle and bubble formation, cavitation, liquid–vapour interfaces and the non-Newtonian liquids prevalent in tribological applications, all require the addition of extra models with a set of increasingly tenuous empirical assumptions. In addition, the Navier–Stokes equations do not ensure energy conservation without additional models, a factor which can become decisive in high-pressure and shear systems. MD captures all this and more with no additional models, extending to chemical reactions and interfacial interactions with the addition of quantum detail. Domain decomposition aims to model these key events by using a molecule model only where molecular detail is important.

The first example of domain decomposition for fluid dynamics is given in the work of O'Connell and Thompson (1995), which established the key features of this form of two-way concurrent coupling. Many of these features have a historical precedent from techniques used in solid mechanics for coupling of particles and continuum, dating back to the 1970s (Curtin and Miller 2003). The simulation domain is decomposed into two subdomains which overlap in a certain region, as shown in Fig. 3.9. This region acts as a 'handshake' zone, also called the hybrid simulation interface (HSI), where the two subdomains interchange data in real time. The bottom subdomain where the liquid interacts with a flat/rough surface is modelled atomistically, shown with molecules in Fig. 3.9. In the top subdomain, CFD is employed and the mesh (grid) discretizing the domain into a grid of finite volumes is denoted by the black lines. Two boundary conditions are shown in Fig. 3.9 for the CFD, the bottom boundary condition provided by the MD subdomain (pink dots) and the top boundary condition specified by the user (black dots). The left and right boundaries are set to periodic, which is consistent with the periodic boundaries in the MD part of the domain. In the MD region, the constrained region applies a force to guide the molecular to a value which agrees with the CFD subdomain, while in the average region, the data from the MD system is accumulated to provide the CFD boundary.

The challenges for domain decomposition coupling include the following:

1. A termination of the MD subdomain. A restraint mechanism is required at the boundary of the molecular region to prevent molecules escaping. This can be a generic force (O'Connell and Thompson 1995; Nie et al. 2004), one based on a previous simulation or calculated from the radial distribution function (Werder et al. 2005). Another method uses a reflective boundary with a correction for density fluctuations based on exponential smoothing of the unbalanced forces due to the missing fluid at the termination boundary (Issa and Poesio 2014). Alternatively, a buffer zone of molecules can be used (Hadjiconstantinou 1999; Delgado-Buscalioni 2012).
2. A method of inserting and removing molecules is required to match the mass flux from the continuum. For simple molecules, the most common method is a steepest descent energy search approach called USHER (Delgado-Buscalioni and Coveney 2003). For complex molecules, energy insertion location can be impossible to find, so techniques exist which try to insert slowly, gradually increasing the complexity of the inserted molecules (Praprotnik et al. 2005).

3. A procedure for averaging the MD region to obtain the continuum boundary conditions (blue region in Fig. 3.9). The MD simulation can be thought of as simply providing a boundary condition to the CFD solver, with as many boundary conditions as fields required. In the simplest case, this is just a velocity boundary value, but for more complex CFD problems, required fields can include density, stress, temperature, concentration, phase or many others. This requires summing over time and space to establish averaged values in discrete locations (Allen and Tildesley 1987; Irving and Kirkwood 1950). The removal of statistical noise is a key issue (Kotsalis et al. 2007), as is the ratio of timesteps used in the two solvers and size of averaged region in terms of statistics (Hadjiconstantinou et al. 2003). The domain decomposition coupling literature is divided into state coupling (mass, momentum, energy) and flux coupling (mass flux, stress, energy flux) (Mohamed and Mohamad 2009), with different statistical properties for the different averaging methods (Hadjiconstantinou et al. 2003).
4. A constraint must be applied to the molecular region to match properties to the continuum (green region in Fig. 3.9). This can be performed by an applied force derived as a constraint using a variational principle formulation of mechanics (Goldstein et al. 2002; O'Connell and Thompson 1995; Nie et al. 2004), an applied force based on stresses (Flekkøy et al. 2000) or Maxwell's demon-type approach (Hadjiconstantinou 1998).
5. Software to exchange the information between the two descriptions, ideally designed for distributed simulations on high-performance computers (HPC).

The last three points are the focus of this section.

We start with point three, the averaging of the MD to give a CFD boundary condition. A rigorous link between the continuum and molecular descriptions is obtained from Irving and Kirkwood (1950) in a derivation of the equations of fluid motion in terms of statistical mechanics. The equations of Irving and Kirkwood (1950) are equivalent to the pointwise continuum, with the limit of the continuum infinitesimal resulting in a Dirac delta function in the molecular description. As a result, an equivalent description for the whole domain results in a number of inherent problems for coupled simulations. As a mathematical idealization of an infinitely thin point, the Dirac delta functional cannot be used in practice in a numerical simulation, so some relaxed function must be employed, such as a Gaussian Kernel or other weighting function (Hardy 1982; Lucy 1977). For NEMD, in general, this is perfectly valid as we obtain an average field for the molecular system which can be used to interpret the system and measure quantities. For domain decomposition coupling, we need both systems to be expressed in the same form, as they exist at the same time and length scales. Any choice of relaxed function departs from the rigorous Irving and Kirkwood (1950) limit and the descriptions are no longer equivalent, with some arbitrary-averaged region assumed to be equivalent to a point in the continuum. To avoid these problems, both molecular and continuum equations should be expressed in terms of control (or finite) volumes (Smith et al. 2012). This replaces the Dirac delta function with an integral over a volume, called the control volume function, defined as $\vartheta_i \equiv \int_V \delta(\boldsymbol{r}_i - \boldsymbol{r}) dV$. Mathematically, this gives an expression for ϑ_i in terms of

Heaviside functions, useful as it is amenable to both theoretical manipulation and numerical implementation (for details, please see Smith et al. 2012). More important is the conceptual shift, we relax the assumption that it is possible to define quantities at every point in space, accepting that only an average in a volume is possible to obtain. This is exactly the assumption made when solving CFD using the finite volume method, and as many CFD solvers express the continuum equations in finite volume form, the exact flux conservation can be matched between the two descriptions. The relationship between molecular and continuum can then be formally expressed as follows, for any face of a control volume,

$$\int_{S_f} [-\rho \boldsymbol{u}\boldsymbol{u} + \boldsymbol{\Pi}] \cdot d\mathbf{S}_f = -\sum_{i=1}^{N} m_i \dot{\boldsymbol{r}}_i \dot{r}_{fi} dS_{fi} + \frac{1}{2}\sum_{i,j}^{N} \boldsymbol{f}_{ij} dS_{fij}, \tag{3.2}$$

where the left-hand side has the continuum convective term, $\rho \boldsymbol{u}\boldsymbol{u}$ and stress $\boldsymbol{\Pi}$ over the surface of a control volume, equal to the fluxes over a surface f of an equivalent (or overlapping) molecular control volume. Molecular fluxes are obtained by taking the sum of all molecular motions and selecting only the surface-motions fluxes using the function dS_{fi}, as well as the sum over all cumulative intermolecular interactions and selecting only the ones crossing the surface using the function dS_{fij}. The form of these surface-crossing terms are obtained by applying the same process used in the continuum control volume derivation to the function ϑ_i through standard manipulations of the Delta function. The resulting form of stress is also well known in the literature as the method of planes (Todd et al. 1995). The sum of this stress on every surface of an enclosed volume can be shown to entirely define the momentum evolution inside. In this way, we have obtained a consistent description of both continuum and molecular systems.

Next, we consider point 4, applying a constraint to ensure the MD region agrees with the CFD. In this context, the mathematical control volume operator is useful as it can be incorporated directly in a minimization constraint. As described in section "Introduction", much of the theory of NEMD focuses on techniques for periodic systems (i.e. SLLOD) or aims to constrain to ensembles, such as the NVT with the Nosé–Hoover thermostat. The use of local thermostatting is valid when applied to wall molecules as no flow is induced. When applied to a part of the fluid, a heat gradient will be created and careful control over molecules entering and leaving the thermostatted region may be required. For coupled simulation, the constraint must be local, as shown in Fig. 3.9, and so we have to apply these local constraints with care. This unique challenge for coupling can still use NEMD theory: applying minimization principles, which include Gauss' principle of least constraint (Hoover 1991). When minimization is applied to a region in space as selected by the control volume function (Smith et al. 2015), a constrained equation is obtained, Eq. (3.3), localized to a volume in space. This is simply a statement that the fluxes must be the same in both continuum and molecular systems,

$$m_i \ddot{\boldsymbol{r}} = \sum_{i,j}^{N} \boldsymbol{f}_{ij} - \sum_{f}^{N_{faces}} F_f, \tag{3.3}$$

where the force F_f ensures Eq. (3.2) is satisfied for each face f, which over all faces enforces the constraint that the sum of surface crossing is subtracted for a volume and replaced by the continuum surface fluxes. This provides a differential constraint which ensures the momentum evolution $(d/dt \int \rho \boldsymbol{u} dV)$ in a molecular control volume is identical to the continuum. In order to achieve this, the constraint is iterated as molecules enter and leave the volume in order to ensure the correct change in momentum to machine precision. Careful implementation can be used to provide exact control of momentum in any arbitrary volume in the molecular system.

Finally, we address point 5, the information exchange between the two descriptions in parallel. The development of CPL_library (Smith et al. 2016b) started with the coupling of two in-house codes, aiming to provide optimal scaling by ensuring minimal point-to-point communication between overlapping processes. This evolved to become an open-source shared library with a minimal interface to facilitate the coupling of any CFD to any MD program. This is done by providing a minimal set of functions in Fortran, C++ and Python to set up the mapping between two overlapping subdomains, ensuring all information is sent and received on the right processes. By providing a minimal shared library, it becomes easy to couple new codes, as well as allowing the user to divide up the problem so both CFD and MD codes can be tested in isolation. This divide and test philosophy is a major part of the design of CPL_library, with a wide range of automated unit and integration tests provided for both the library and for coupled case with OpenFOAM and LAMMPS. Deployment is provided using Docker and Anaconda, with a range of minimal examples and quickstart guides detailed on the project website http://www.cpl-library.org.

We now demonstrate the combination of control volume averaging, constrained dynamics Eq. (3.3) and CPL_library applied to the problem of domain decomposition. One of the simplest useful simulations for tribological application is pictured in

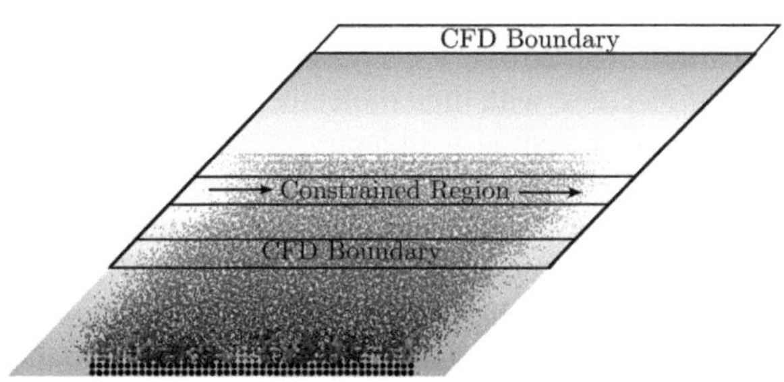

Fig. 3.10 A schematic of a coupled Couette flow with polymer brushes. The top wall of the continuum is driven at a velocity of one, and the colours in the background show the velocity field in the coupled system (with molecules coloured by the same field). The constraint regions and CFD boundaries are denoted on the figure to show the way the coupling is applied

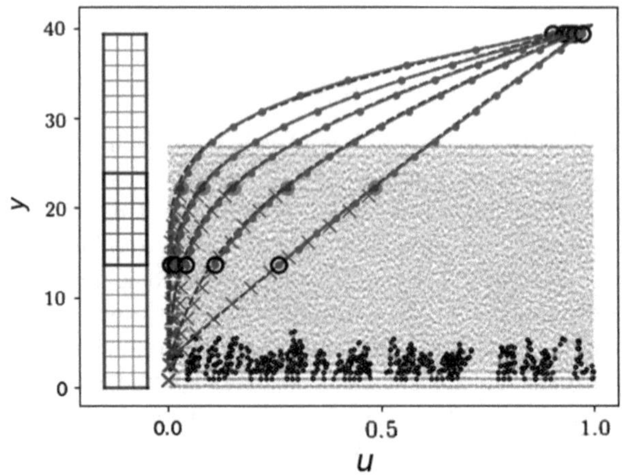

Fig. 3.11 Couette flow analytical solution compared to the coupled CFD-MD results overlapping a snapshot of the liquid molecules (black) and polymer brushes (pink). The MD velocity is shown by green crosses with a green spot showing the constrained cell, red lines and points denote the CFD values with black circles the boundary values and all values are matched to the analytical solution shown by the dotted blue line at times $t = \{15, 25, 37.5, 65, 200\}$

Fig. 3.10, with the effect of shear shown using the same schematic as in Fig. 3.1. This comparison emphasizes an important application of coupled modelling, namely, the reduction of MD system size by use of a continuum model. By modelling polymer brushes attached to the wall, the complex molecular detail is isolated to a region in space and the remainder of the domain is well described by a Newtonian fluid. For this reason, the use of a continuum solver is ideal as it replaces the complexity and cost of modelling a large molecular system, while retaining the effect of the polymer brushes on the flow itself, which persists into the continuum region. This can be seen in Fig. 3.11 where the Couette flow solution is effectively shifted up from the wall by the presence of the polymer brushes. The MD solver is Flowmol, which is fully verified (Smith 2014) and has been demonstrated in a range of fluid dynamics studies (Smith et al. 2016a; Trevelyan and Zaki 2016). The MD simulation consists of 356,864 molecules, with fluid molecules at a density of $\rho = 0.8$ interacting via the Lennard-Jones potential with a cut-off of $2^{1/6}$. The walls are set to the same density of $\rho = 0.8$, tethered using the anharmonic potential of (Petravic and Harrowell 2006) with the polymer brushes grafted onto them at a density of 0.1 as chains of 10 FENE molecules with maximum separation of $R_0 = 30.0$ and spring constant $k_{FENE} = 1.5$. The outer half of the walls is thermostatted to $T = 1.0$ using the Nosé–Hoover thermostat with a heat bath size obtained from the product of 0.1, the number of thermostatted molecules and the timestep $\Delta t = 0.005$. The MD subdomain is $140.2 \times 27.4 \times 116.3$ in reduced LJ units split into four cells in x each of size 35.1, 16 in y of size 1.71, and in z of size 29.1. The CFD has the same number of cells and the same size, with both subdomains overlapping by eight cells and the timestep ratio between the CFD and MD codes is set to one (i.e. $\Delta t_{CFD} = \Delta t_{MD}$), so both systems evolve together. The constraint (Eq. 3.3) is applied to cell 14 of the MD subdomain, and the top two cells are left as a buffer. The control volume averaging and constraint methodologies discussed above are implemented in Flowmol with runtime tests to ensure exact momentum control at each step. In Fig. 3.11, the constrained region can be seen to agree exactly with the CFD value at that location, with control to machine precision.

The continuum solver is based on OpenFOAM's icoFOAM (version 3.0.1; Weller et al. 1998), a flow solver for Eq. (3.1) with a single viscosity coefficient $\nu = 2.1$, estimated from the current MD density $\rho = 0.8$ and temperature $T = 1.0$ through a parameter study obtained in previous work (Smith 2015). The icoFOAM solver is adapted to receive information from the MD through CPL_library and apply this to the bottom cells of the domain as a boundary condition, all other features of the solver are kept identical. The bottom CFD boundary can be seen to result from an average of the MD region, while the top boundary is set to a velocity of one, with left and right boundaries periodic. Both MD and CFD codes are run on four processes each, with optimal parallel exchange between the overlapping processes managed by CPL_Library.

Beyond Lennard-Jones fluids

Lennard-Jones fluids are ubiquitous in MD-CFD-coupled simulations due to their simplicity. They consist of point particles which do not exhibit non-Newtonian behaviour nor significant shear heating at moderate shear rates. Furthermore, measurable quantities (velocity, stress, temperature, etc.) are less noisy than in molecular fluids, since they do not contain any internal degree of freedom leading to high-frequency motion modes. Hence, they are perfect candidates for test cases and proof of concept simulations. It is difficult to find in the literature studies which actually exploit the benefit of domain decomposition to solve a real problem. To achieve this, particularly in tribology, it would be at least necessary to couple molecular fluids using a realistic potential.

A fully coupled simulation using OPLS benzene is shown in Fig. 3.12. The simulation is performed at $\gamma = 10^{10}\,\mathrm{s}^{-1}$, $T = 300\,\mathrm{K}$ and $\rho = 0.89\,\mathrm{kg/m}^3$, where the molecule exhibit Newtonian behaviour (Lee 2004). Excellent agreement is shown during the transient and (pseudo-) steady state and shows promising evidence that moving from simple to molecular fluids is achievable without much modification of the method. It is yet to be proven if this is also true for more complicated molecules like linear and branched hydrocarbons.

Applications of Domain Decomposition in Tribology

NEMD shear simulations applied to tribology are limited by several factors which make the comparison against experimental data quite difficult: (a) high shear rates of O(10^8–$10^{11}\,\mathrm{s}^{-1}$), due to small system sizes and high shear velocities (Bair et al. 2002a, b); (b) number of atoms, O(10^6) which for dense fluids means characteristic

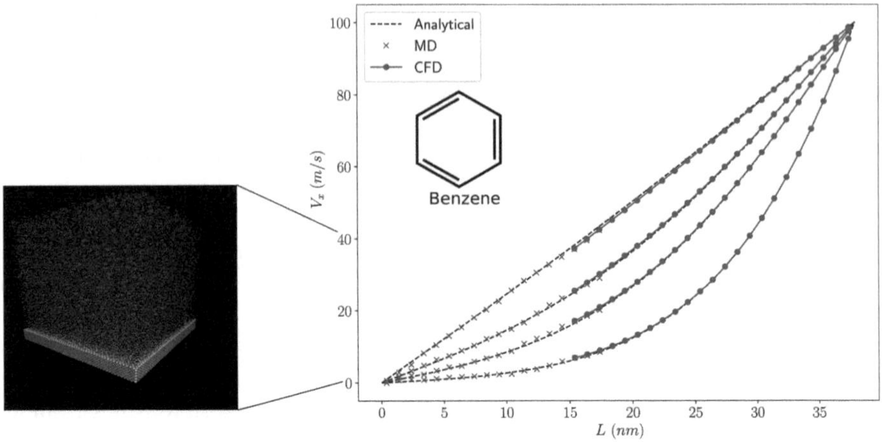

Fig. 3.12 Time evolution of velocity field for a CFD-MD coupled simulation of 40,691 molecules of OPLS (Jorgensen et al. 1996) benzene. Initial configuration equilibrated at $T = 300\,\mathrm{K}$ and $P = 1000\,\mathrm{atm}$ with a density $\rho = 0.89\,\mathrm{kg/m^3}$. The NEMD viscosity (SLLOD method) of benzene measured at $\gamma = 10^{10}\,\mathrm{s^{-1}}$ at the given conditions is $\eta = 0.58\,\mathrm{mPa\,s}$. Shearing velocity at the top wall $v = 100\,\mathrm{m/s}$ with a total domain length $L = 40\,\mathrm{nm}$. Fluid–surface interactions are strong enough to avoid slip. A shear rate of $\gamma = 2.5 \times 10^{10}\,\mathrm{s^{-1}}$ is registered at the steady state and a linear profile obtained is a sign of the Newtonian behaviour of benzene under simulated conditions

system sizes of no more than O(10 nm); (c) simulation time, O(10 ns) due to timestep sizes of O(1 fs).

Multi-scale domain decomposition allows the removal of limitation (b). By modelling most of the domain using a CFD solver and keeping the NEMD solver close to the surface (Fig. 3.9), we can arbitrarily enlarge the direction perpendicular to it. This method provides the opportunity to tackle (at least) two type of problems:

1. Comparing simulations with experimental data. The possibility of simulating liquid lubricant films of hundreds of nanometres to micrometres would permit the comparison of time-evolving velocity fields obtained from numerical simulations, with the ones obtained from thin-film experiments such as fluorescence lifetime-based techniques (Galmiche et al. 2016; Ponjavic et al. 2015). Furthermore, lubricant layers of this size are technologically relevant since they can be found in lubricated engine parts (Tung and McMillan 2004) and in biological systems (Myant et al. 2012).
2. Prediction of *no-slip* condition. Experimental data is fitted to models which usually assume no-slip condition (Ponjavic et al. 2015; Ponjavic and Wong 2014) but it has been experimentally known for almost two decades (Pit et al. 2000) that this is not always true. The question to answer is: for a particular interface, at a certain thermodynamic point and fixed shear velocity, what is the minimum film thickness that makes the no-slip assumption valid (Asproulis et al. 2012)? Therefore, being able to compute scaling laws for slip can guide experimentalists to obtain more accurate measurements by choosing appropriate models.

The applications presented here are interesting enough to justify the implementation of a domain decomposition framework despite the difficulties this might entail. A caveat of multi-scale methods is the complexity to understand, develop and test them properly. Typically, a decent amount of software development is required since it is unlikely to find a readily available software package for this specific purpose. In the author experience, the effort on this regard should not be taken lightly. This is discussed in the next sections along with some recommendations for someone who wants to develop a domain decomposition framework.

Implementation Caveats of Domain Decomposition

The complexity of using/developing a multi-scale method where two (or more) methods are coupled, is greater than the sum of the complexities of using/developing each individual method alone. This is true because the coupling method itself introduces a new layer of complexity both at the conceptual and implementation level. A list of some caveats of domain decomposition for MD-CFD coupling is presented (most of are also, in general, applicable to any multi-scale method). Along with them, some recommendations are made on how to tackle each one.

General

1. *Training on each of the methods to be coupled.* This can be a daunting task itself due to the disparity in the theory and technical considerations between them. This is an effort that has to be assumed.
2. *Development considerations I.* The first decision to make is to either develop *in-house* solvers or to use existing software packages that allow source code modification or development through a scripting interface. In this regard, it is advisable to follow the DRY (Do not Repeat Yourself) principle from software development. It is tempting to write your own code for the sake of learning and self-prise, but widespread third-party software is in general well tested and stable. It is a good idea to use as much as it is already available and add features to it as needed.
3. *Development considerations II.* Coupling two solvers at the software level means passing back and forth information between them in some way. This can be done in (A) a *monolithic* fashion, by compiling and linking them together at the binary level (calling one from inside the other) or (B) decoupling them by using a *coupling library* designed for the purpose, in which case the solvers are developed and compiled independently and data is exchanged in a smart way (commonly) through MPI communication. In this regards, there are several options. From more general frameworks like MUI library (Tang et al. 2015) and PRECICE (Bungartz et al. 2016), to more user-friendly options targeted for coupled simulation like CPL_Library (Smith et al. 2016b). In general, while strategy (A) is more efficient, strategy (B) is preferable. Some notable advantages of strategy (B) are:

(a) Independent development and testing of each coupled solver is easier. Version upgrading of a certain solver is less involved.
(b) Avoid solver incompatibility due to dependencies. A complicated issue arises if two solvers depend directly (or indirectly) on a certain library, but different versions of it. If those versions are not compatible between them we are in trouble.
(c) Arbitrary parallel domain decomposition of each solver. Each solver can independently decompose its domain, whereas in a monolithic approach, one decomposition has to fit both solvers.

4. *Difficult to identify sources of errors and quantify them*. Errors in the outcome of a multi-scale simulation can be difficult to track back to their origin. Is it an error in the inputs/outputs of the individual solvers or in the coupling methodology itself? Furthermore, codes used in multi-scale simulations are designed usually to run in parallel which makes debugging much harder.

Tribology related

1. *Pre-computing transport coefficients and Equation of State from molecular dynamics simulations*. This is necessary to achieve consistent coupling between MD and CFD (time evolution of momentum, density and energy are consistent in both solvers). In the case of tribologically relevant fluids (e.g. hydrocarbon mixtures) under high pressure, these fluids exhibit strong non-Newtonian behaviour (Jadhao and Robbins 2017). This requires a careful selection of a non-Newtonian model which represent the viscous behaviour of the molecular fluid in the MD. As far as we know, the non-Newtonian regime is still fertile ground to explore in the field of coupled simulations.
2. *Coupling density field for complex molecules*. There is currently no easy procedure to tackle this problem. If the density field varies significantly over time, removal or insertion of molecules is needed at the overlap region. This can be successfully performed for point-particle fluids like Lennard-Jones using the USHER method (Delgado-Buscalioni and Coveney 2003). On the other hand, molecule insertion can be trickier than point particles and a triple-scale scheme (AdResS) (Praprotnik et al. 2005) has been developed for this purpose to coarse-grain all-atom molecules into blobs near the insertion/removal zone. Other less physically meaningful methods like FADE (Borg et al. 2014) could, in principle, do the job but there are concerns about the effect non-instantaneous insertion on time-evolving flows.

This section has been written with the idea of facilitating the newcomer to grasp the potential complexities of implementing a multi-scale method, in particular, CFD-MD domain decomposition. It is our hope that by attracting research momentum into these techniques, eventually they will find their way to the mainstream engineering use.

Conclusions

In this chapter, we have reviewed recent advances in methodologies made available as modelling tools to tribologists who want to investigate molecular-scale effects and their link to structural and electronic properties at the lowest scales and to the macroscopic response at the larger scales. Some recent applications of NEMD simulations in tribology are first discussed and then critically reviewed. Such simulations have given unique insights into the nanoscale structure, flow and friction behaviour of lubricants and additives. Classical NEMD simulations of lubricants have shown how these fluids behave under high-pressure and shear conditions, which is particularly useful to study EHL. Classical NEMD simulations of lubricant additives have demonstrated the importance of surface coverage on the friction behaviour. Ab initio NEMD simulations of lubricant additives have started to reveal tribolochemical degradation mechanisms of lubricant additives under compression and shear. Coupled CFD-NEMD simulations of lubricants provide access to larger time and length scales, which are inaccessible to standard NEMD simulations. An overview of how to tackle issues related to the use of coupling techniques has been provided, together with a critical outlook.

References

Adams, H. L., Garvey, M. T., Ramasamy, U. S., Ye, Z., Martini, A., & Tysoe, W. T. (2015). Shear induced mechanochemistry: Pushing molecules around. *Journal of Physical Chemistry C, 119*(13), 7115–7123.

Alder, B. J., & Wainwright, T. E. (1957). Phase transition for a hard sphere system. *Journal of Chemical Physics, 27*, 1208–1209.

Allen, M. P., & Tildesley, D. J. (1987). *Computer simulation of liquids* (1st ed.). Oxford: Clarendon Press.

Allen, W., & Rowley, R. L. (1997). Predicting the viscosity of alkanes using nonequilibrium molecular dynamics: Evaluation of intermolecular potential models. *Journal of Chemical Physics, 106*(24), 10273–10281.

Apóstolo, R. F. G., Tsagkaropoulou, G., & Camp, P. J. (2019). Molecular adsorption, self-assembly, and friction in lubricants. *Journal of Molecular Liquids, 277*, 606–612.

Ashurst, W. T., & Hoover, W. G. (1975). Dense-fluid shear viscosity via nonequilibrium molecular dynamics. *Physical Review A, 11*(2), 658–678.

Asproulis, N., Kalweit, M., & Drikakis, D. (2012). A hybrid molecular continuum method using point wise coupling. *Advances in Engineering Software, 46*(1), 85–92.

Bair, S., & Kottke, P. (2003). Pressure–viscosity relationships for elastohydrodynamics. *Tribology Transactions, 46*(3), 289–295.

Bair, S., & McCabe, C. (2004). A study of mechanical shear bands in liquids at high pressure. *Tribology International, 37*(10), 783–789.

Bair, S., Qureshi, F., & Winer, W. O. (1993). Observations of shear localization in liquid lubricants under pressure. *Journal of Tribology, 115*(3), 507–514.

Bair, S., Qureshi, F., & Khonsari, M. (1994). Adiabatic shear localization in a liquid lubricant under pressure. *Journal of Tribology, 116*(4), 705.

Bair, S., McCabe, C., & Cummings, P. T. (2002a). Comparison of nonequilibrium molecular dynamics with experimental measurements in the nonlinear shear-thinning regime. *Physical Review Letters*, *88*(5), 058302.
Bair, S., McCabe, C., & Cummings, P. T. (2002b). Calculation of viscous EHL traction for squalane using molecular simulation and rheometry. *Tribology Letters, 13*(4), 251–254.
Barker, J. A., & Henderson, D. (1976). What is "liquid"? Understanding the states of matter. *Reviews of Modern Physics*, *48*(4), 587–671.
Bernardi, S., Todd, B. D., & Searles, D. J. (2010). Thermostating highly confined fluids. *Journal of Chemical Physics*, *132*(24), 244706.
Berro, H., Fillot, N., Vergne, P., Tokumasu, T., Ohara, T., & Kikugawa, G. (2011). Energy dissipation in non-isothermal molecular dynamics simulations of confined liquids under shear. *Journal of Chemical Physics*, *135*, 134708.
Bitsanis, I., Magda, J. J., Tirrell, M., & Davis, H. T. (1987). Molecular dynamics of flow in micropores. *Journal of Chemical Physics*, *87*(3), 1733–1750.
Borg, M. K., Lockerby, D. A., & Reese, J. M. (2014). The FADE mass-stat: A technique for inserting or deleting particles in molecular dynamics simulations. *Journal of Chemical Physics*, *140*(7).
Bradley-Shaw, J. L., Camp, P. J., Dowding, P. J., & Lewtas, K. (2015). Glycerol monooleate reverse micelles in nonpolar solvents: Computer simulations and small-angle neutron scattering. *Journal of Physical Chemistry B*, *119*(11), 4321–4331.
Bradley-Shaw, J. L., Camp, P. J., Dowding, P. J., & Lewtas, K. (2016). Molecular dynamics simulations of glycerol monooleate confined between mica surfaces. *Langmuir*, *32*(31), 7707–7718.
Bradley-Shaw, J. L., Camp, P. J., Dowding, P. J., & Lewtas, K. (2018). Self-assembly and friction of glycerol monooleate and its hydrolysis products in bulk and confined non-aqueous solvents. *Physical Chemistry Chemical Physics*, *20*, 17648–17657.
Briscoe, B. J., & Evans, D. C. B. (1982). The shear properties of Langmuir–Blodgett layers. *Proceedings of the Royal Society of London A*, *380*, 389.
Bungartz, H. J., Lindner, F., Gatzhammer, B., Mehl, M., Scheufele, K., Shukaev, A., et al. (2016). preCICE—A fully parallel library for multi-physics surface coupling. *Computers & Fluids*, *141*, 250–258.
Campen, S., Green, J., Lamb, G., Atkinson, D., & Spikes, H. (2012). On the increase in boundary friction with sliding speed. *Tribology Letters*, *48*, 237–248.
Cao, J., & Likhtman, A. E. (2012). Shear banding in molecular dynamics of polymer melts. *Physical Review Letters*, *108*(2), 028302.
Car, R., & Parrinello, M. (1985). Unified approach for molecular dynamics and density-functional theory. *Physical Review Letters*, *55*(22), 2471–2474.
Carreau, P. J. (1972). Rheological equations from molecular network theories. *Journal of Rheology*, *16*(1972), 99.
Choo, J. H., Forrest, A. K., & Spikes, H. A. (2007). Influence of organic friction modifier on liquid slip: A new mechanism of organic friction modifier action. *Tribology Letters*, *27*(2), 239–244.
Curtin, W. A., & Miller, R. E. (2003). Atomistic/continuum coupling in computational material science. *Modelling and Simulation in Materials Science and Engineering*, *11*, 33.
Delgado-Buscalioni, R. (2012). Tools for multiscale simulation of liquids using open molecular dynamics. *Lecture Notes in Computational Science and Engineering*, *82*, 145.
Delgado-Buscalioni, R., & Coveney, P. V. (2003). Usher: An algorithm for particle insertion in dense fluids. *Journal of Chemical Physics*, *119*, 978.
Delhommelle, J., Petravic, J., & Evans, D. J. (2003). On the effects of assuming flow profiles in nonequilibrium simulations. *Journal of Chemical Physics*, *119*(21), 11005–11010.
Doig, M., Warrens, C. P., & Camp, P. J. (2014). Structure and friction of stearic acid and oleic acid films adsorbed on iron oxide surfaces in squalane. *Langmuir*, *30*, 186–195.
Eder, S. J., Vernes, A., & Betz, G. (2013). On the Derjaguin offset in boundary-lubricated nanotribological systems. *Langmuir*, *29*(45), 13760–13772.
Ehret, P., Dowson, D., & Taylor, C. M. (1998). On lubricant transport conditions in elastohydrodynamic conjunctions. *Proceedings of the Royal Society of London A*, *454*, 763–787.

Elber, R. (2016). Perspective: Computer simulations of long time dynamics. *Journal of Chemical Physics, 144*, 060901.
Evans, D. J., & Morriss, G. P. (1984). Nonlinear-response theory for steady planar Couette flow. *Physical Review A, 30*(3), 1528–1530.
Evans, D. J., & Morriss, G. P. (1988). Transient-time-correlation functions and the rheology of fluids. *Physical Review A, 38*(8), 4142–4148.
Evans, D. J., & Morriss, G. P. (2008). *Statistical mechanics of nonequilibrium liquids* (2nd ed.). Cambridge: Cambridge University Press.
Ewen, J. P., Gattinoni, C., Morgan, N., Spikes, H. A., & Dini, D. (2016a). Nonequilibrium molecular dynamics simulations of organic friction modifiers adsorbed on iron oxide surfaces. *Langmuir, 32*, 4450.
Ewen, J. P., Gattinoni, C., Thakkar, F. M., Morgan, N., Spikes, H., & Dini, D. (2016b). A comparison of classical force-fields for molecular dynamics simulations of lubricants. *Materials, 9*(8), 651.
Ewen, J. P., Echeverri Restrepo, S., Morgan, N., & Dini, D. (2017a). Nonequilibrium molecular dynamics simulations of stearic acid adsorbed on iron surfaces with nanoscale roughness. *Tribology International, 107*(18), 264–273.
Ewen, J. P., Gattinoni, C., Zhang, J., Heyes, D. M., Spikes, H. A., & Dini, D. (2017b). On the effect of confined fluid molecular structure on nonequilibrium phase behaviour and friction. *Physical Chemistry Chemical Physics, 19*(27), 17883–17894.
Ewen, J. P., Heyes, D. M., & Dini, D. (2018a). Advances in nonequilibrium molecular dynamics simulations of lubricants and additives. *Friction, 6*, 349–386.
Ewen, J. P., Kannam, S. K., Todd, B. D., & Dini, D. (2018b). Slip of alkanes confined between surfactant monolayers adsorbed on solid surfaces. *Langmuir, 34*, 3864–3873.
Eyring, H. (1936). Viscosity, plasticity, and diffusion as examples of absolute reaction rates. *Journal of Chemical Physics, 4*, 283–291.
Feller, S. E., Pastor, R. W., Rojnuckarin, A., Bogusz, S., & Brooks, B. R. (1996). Effect of electrostatic force truncation on interfacial and transport properties of water. *Journal of Physical Chemistry, 100*(42), 17011–17020.
Flekkøy, E. G., Wagner, G., & Feder, J. (2000). Hybrid model for combined particle and continuum dynamics. *Europhysics Letters, 52*, 271.
Gad-el Hak, M. (2006). Gas and liquid transport at the microscale. *Heat Transfer Engineering, 27*(4), 13.
Galmiche, B., Ponjavic, A., & Wong, J. S. S. (2016). Flow measurements of a polyphenyl ether oil in an elastohydrodynamic contact. *Journal of Physics: Condensed Matter, 28*(13), 134005.
Gao, F., Furlong, O., Kotvis, P. V., & Tysoe, W. T. (2004). Reaction of tributyl phosphite with oxidized iron: Surface and tribological chemistry. *Langmuir, 20*, 7557–7568.
Gattinoni, C., & Michaelides, A. (2015). Understanding corrosion inhibition with van der Waals DFT methods: The case of benzotriazole. *Faraday Discussions, 180*, 439–458.
Gattinoni, C., Heyes, D. M., Lorenz, C. D., & Dini, D. (2013). Traction and nonequilibrium phase behavior of confined sheared liquids at high pressure. *Physical Review E, 88*(5), 052406.
Gattinoni, C., Ewen, J. P., & Dini, D. (2018). Adsorption of surfactants on α-Fe_2O_3(0001): A density functional theory study. *Journal of Physical Chemistry C, 122*, 20817–20826.
Goldstein, H., Poole, C., & Safko, J. (2002). *Classical mechanics* (3rd ed.). Boston: Addison Wesley.
Gosvami, N. N., Bares, J. A., Mangolini, F., Konicek, A. R., Yablon, D. G., & Carpick, R. W. (2015). Mechanisms of antiwear tribofilm growth revealed in situ by single-asperity sliding contacts. *Science, 348*(6230), 102–106.
Granick, S. (1991). Motions and relaxations of confined liquids. *Science, 253*(5026), 1374–1379.
Gubbins, K. E., Liu, Y.-C., Moore, J. D., & Palmer, J. C. (2011). The role of molecular modeling in confined systems: Impact and prospects. *Physical Chemistry Chemical Physics, 13*, 58–85.
Hadjiconstantinou, N. G. (1998). *Hybrid atomistic–continuum formulations and the moving contact-line problem*. Ph.D. thesis, MIT, USA.
Hadjiconstantinou, N. G. (1999). Hybrid atomistic–continuum formulations and the moving contact-line problem. *Journal of Computational Physics, 154*, 245.

Hadjiconstantinou, N. G., Garcia, A. L., Bazant, M. Z., & He, G. (2003). Statistical error in particle simulations of hydrodynamic phenomena. *Journal of Computational Physics, 187*, 274.

Hardy, R. J. (1982). Formulas for determining local properties in molecular dynamics simulations: Shock waves. *Journal of Chemical Physics, 76*, 622.

Harrison, J. A., Schall, J. D., Maskey, S., Mikulski, P. T., Knippenberg, M. T., & Morrow, B. H. (2018). Review of force fields and intermolecular potentials used in atomistic computational materials research. *Applied Physics Reviews, 5*, 031104.

Heyes, D. M., Smith, E. R., Dini, D., Spikes, H. A., & Zaki, T. A. (2012). Pressure dependence of confined liquid behavior subjected to boundary-driven shear. *Journal of Chemical Physics, 136*(13), 134705.

Heyes, D. M., Dini, D., & Smith, E. R. (2018). Incremental viscosity by non-equilibrium molecular dynamics and the Eyring model. *Journal of Chemical Physics, 148*, 194506.

Holland, D. M., Lockerby, D. A., Borg, M. K., Nicholls, W. D., & Reese, J. M. (2015). Molecular dynamics pre-simulations for nanoscale computational fluid dynamics. *Microfluidics and Nanofluidics, 18*(3), 461–474.

Hoover, W. G. (1985). Canonical dynamics: Equilibrium phase-space distributions. *Physical Review A, 31*(3), 1695–1697.

Hoover, W. G. (1991). *Computational statistical mechanics* (1st ed.). Oxford: Elsevier Science.

Hoover, W. G., Hoover, C. G., & Petravic, J. (2008). Simulation of two- and three-dimensional dense-fluid shear flows via nonequilibrium molecular dynamics: Comparison of time-and-space-averaged stresses from homogeneous doll's and sllod shear algorithms with those from boundary-driven shear. *Physical Review E, 78*, 046701.

Irving, J. H., & Kirkwood, J. G. (1950). The statistical mechanics theory of transport processes. IV. The equations of hydrodynamics. *Journal of Chemical Physics, 18*, 817.

Israelachvili, J. N. (1986). Measurement of the viscosity of liquids in very thin films. *Journal of Colloid and Interface Science, 110*, 263–271.

Issa, K. M., & Poesio, P. (2014). Algorithm to enforce uniform density in liquid atomistic subdomains with specular boundaries. *Physical Review E, 89*, 043307.

Jadhao, V., & Robbins, M. O. (2017). Probing large viscosities in glass-formers with nonequilibrium simulations. *Proceedings of the National Academy of Sciences of the United States of America, 114*(30), 7952–7957.

Jaishankar, A., Jusufi, A., Vreeland, J. L., Deighton, P., Pellettiere, J. R., & Schilowitz, A. M. (2019). Adsorption of stearic acid at the iron oxide/oil interface—Theory, experiments and modeling. *Langmuir*.

Jeffreys, S., di Mare, L., Liu, X., Morgan, N., & Wong, J. S. S. (2019). Elastohydrodynamic lubricant flow with nanoparticle tracking. *RSC Advances, 9*, 1441–1450.

Jorgensen, W. L., Maxwell, D. S., & Tirado-Rives, J. (1996). Development and testing of the OPLS all-atom force field on conformational energetics and properties of organic liquids. *Journal of the American Chemical Society, 118*(45), 11225–11236.

Kano, M., Yasuda, Y., Okamoto, Y., Mabuchi, Y., Hamada, T., Ueno, T., et al. (2005). Ultralow friction of DLC in presence of glycerol mono-oleate (GMO). *Tribology Letters, 18*(2), 245–251.

Karaborni, S., & Verbist, G. (1994). Effect of chain conformation on the tilt behaviour in Langmuir monolayers. *European Letters, 27*, 467.

Khare, R., de Pablo, J., & Yethiraj, A. (1997). Molecular simulation and continuum mechanics study of simple fluids in non-isothermal planar couette flows. *Journal of Chemical Physics, 107*(7), 2589.

Kong, Y. C., Tildesley, D. J., & Alejandre, J. (1997). The molecular dynamics simulation of boundary-layer lubrication. *Molecular Physics, 92*(1), 7–18.

Kotsalis, E. M., Walther, J. H., & Koumoutsakos, P. (2007). Control of density fluctuations in atomistic–continuum simulations of dense liquids. *Physical Review E, 76*, 016709.

Kubo, R. (1957). Statistical–mechanical theory of irreversible processes. I. General theory and simple applications to magnetic and conduction problems. *Journal of the Physical Society of Japan, 12*(6), 570–586.

Kuwahara, T., Romero, P. A., Makowski, S., Weihnacht, V., Moras, G., & Moseler, M. (2019). Mechano-chemical decomposition of organic friction modifiers with multiple reactive centres induces superlubricity of ta-C. *Nature Communications*, *10*, 151.

Lucy, L. B. (1977). A numerical approach to the testing of the fission hypothesis. *Astronomical Journal*, *82*(12).

Lee, S. H. (2004). Shear viscosity of benzene, toluene, and p-xylene by non-equilibrium molecular dynamics simulations. *Bulletin of the Korean Chemical Society*, *25*(2), 321–324.

Lees, A. W., & Edwards, S. F. (1972). The computer study of transport processes under extreme conditions. *Journal of Physics C: Solid State Physics*, *5*(15), 1921–1929.

Levesque, D., Verlet, L., & Kurkijar, J. (1973). Computer experiments on classical fluids. IV. Transport properties and time-correlation functions of the Lennard-Jones liquid near its triple point. *Physical Review A*, *7*(5), 1690–1700.

Li, J., Zhang, C., & Luo, J. (2011). Superlubricity behavior with phosphoric acid–water network induced by rubbing. *Langmuir*, *27*(15), 9413–9417.

Liem, S. Y., Brown, D., & Clarke, J. H. R. (1992). Investigation of the homogeneous-shear nonequilibrium-molecular-dynamics method. *Physical Review A*, *45*(6), 3706–3713.

Liu, P., Lu, J., Yu, H., Ren, N., Lockwood, F. E., & Wang, Q. J. (2017). Lubricant shear thinning behavior correlated with variation of radius of gyration via molecular dynamics simulations. *Journal of Chemical Physics*, *147*(8), 084904.

Liu, P. Z., Yu, H. L., Ren, N., Lockwood, F. E., & Wang, Q. J. (2015). Pressure–viscosity coefficient of hydrocarbon base oil through molecular dynamics simulations. *Tribology Letters*, *60*(3), 9.

Loehlé, S., & Righi, M. C. (2017). First principles study of organophosphorus additives in boundary lubrication conditions: Effects of hydrocarbon chain length. *Lubrication Science*, *29*, 485–491.

Loehlé, S., & Righi, M. C. (2018). Ab initio molecular dynamics simulation of tribochemical reactions involving phosphorus additives at sliding iron interfaces. *Lubricants*, *6*(2), 31.

Lu, J., Reddyhoff, T., & Dini, D. (2018). 3D measurements of lubricant and surface temperatures within an elastohydrodynamic contact. *Tribology Letters*, *66*, 7.

Maćkowiak, Sz., Heyes, D. M., Dini, D., & Brańka, A. C. (2016). Non-equilibrium phase behavior and friction of confined molecular films under shear: A non-equilibrium molecular dynamics study. *Journal of Chemical Physics*, *145*(16), 164704.

Martin, M. G., & Siepmann, J. I. (1999). Novel configurational-bias Monte Carlo method for branched molecules. Transferable potentials for phase equilibria. 2. United-atom description of branched alkanes. *Journal of Physical Chemistry B*, *103*(21), 4508–4517.

Martini, A., Hsu, H. Y., Patankar, N. A., & Lichter, S. (2008). Slip at high shear rates. *Physical Review Letters*, *100*(20), 206001.

Martinie, L., & Vergne, P. (2016). Lubrication at extreme conditions: A discussion about the limiting shear stress concept. *Tribology Letters*, *63*(2), 21.

McCabe, C., Cui, S. T., Cummings, P. T., Gordon, P. A., & Saeger, R. B. (2001). Examining the rheology of 9-octylheptadecane to giga-pascal pressures. *Journal of Chemical Physics*, *114*(4), 1887–1891.

Mohamed, K. M., & Mohamad, A. A. (2009). A review of the development of hybrid atomistic–continuum methods for dense fluids. *Microfluidics and Nanofluidics*, *8*, 283.

Molinari, J.-F., Aghababaei, R., Brink, T., Frérot, L., & Milanese, E. (2018). Adhesive wear mechanisms uncovered by atomistic simulations. *Friction*, *6*(3), 245–259.

Moller, M. A., Tildesley, D. J., Kim, K. S., & Quirke, N. (1991). Molecular dynamics simulation of a Langmuir–Blodgett film. *Journal of Chemical Physics*, *94*(12), 8390–8401.

Moore, J. D., Cui, S. T., Cochran, H. D., & Cummings, P. T. (2000). Rheology of lubricant basestocks: A molecular dynamics study of C-30 isomers. *Journal of Chemical Physics*, *113*(19), 8833–8840.

Mosey, N. J., Müser, M. H., & Woo, T. K. (2005). Molecular mechanisms for the functionality of lubricant additives. *Science*, *307*(5715), 1612–1615.

Myant, C., Underwood, R., Fan, J., & Cann, P. M. (2012). Lubrication of metal-on-metal hip joints: The effect of protein content and load on film formation and wear. *Journal of the Mechanical Behavior of Biomedical Materials*, *6*, 30–40.
Nie, X. B., Chen, S. Y., E, W. N., & Robbins, M. O. (2004). A continuum and molecular dynamics hybrid method for micro- and nano-fluid flow. *Journal of Fluid Mechanics*, *500*, 55.
Nosé, S. (1984). A molecular-dynamics method for simulations in the canonical ensemble. *Molecular Physics*, *52*(2), 255–268.
O'Connell, S. T., & Thompson, P. A. (1995). Molecular dynamics-continuum hybrid computations: A tool for studying complex fluid flow. *Physical Review E*, *52*, R5792.
Pan, G., & McCabe, C. (2006). Prediction of viscosity for molecular fluids at experimentally accessible shear rates using the transient time correlation function formalism. *Journal of Chemical Physics*, *125*(19), 194527.
Petravic, J., & Harrowell, P. (2006). The boundary fluctuation theory of transport coefficients in the linear-response limit. *Journal of Chemical Physics*, *124*, 014103.
Pit, R., Hervet, H., & Léger, L. (2000). Direct experimental evidence of slip in hexaecane: Solid interfaces. *Physical Review Letters*, *85*(5), 980–983.
Plint, M. A. (1967). Traction in elastohydrodynamic contacts. *Proceedings of the Institution of Mechanical Engineers*, *182*(14), 300–306.
Ponjavic, A., & Wong, J. S. S. (2014). The effect of boundary slip on elastohydrodynamic lubrication. *RSC Advances*, *4*(40), 20821–20829.
Ponjavic, A., di Mare, L., & Wong, J. S. S. (2014). Effect of pressure on the flow behavior of polybutene. *Journal of Polymer Science Part B: Polymer Physics*, *52*(10), 708–715.
Ponjavic, A., Dench, J., Morgan, N., & Wong, J. S. S. (2015). In situ viscosity measurement of confined liquids. *RSC Advances*, *5*, 99585.
Porras-Vazquez, A., Martinie, L., Vergne, P., & Fillot, N. (2018). Independence between friction and velocity distribution in fluids. *Physical Chemistry Chemical Physics*, *20*, 27280–27293.
Praprotnik, M., Delle Site, L., & Kremer, K. (2005). Adaptive resolution molecular-dynamics simulation: Changing the degrees of freedom on the fly. *Journal of Chemical Physics*, *123*, 224106.
Rahman, A. (1964). Correlations in the motion of atoms in liquid argon. *Physical Review*, *136*, 405–411.
Ren, W., & Weinan, E. (2005). Heterogeneous multiscale method for the modeling of complex fluids and micro fluidics. *Journal of Computational Physics*, *204*, 1–26.
Robbins, M. O., & Smith, E. D. (1996). Connecting molecular-scale and macroscopic tribology. *Langmuir*, *12*(19), 4543–4547.
Schneider, T., & Stoll, E. (1978). Molecular-dynamics study of a three-dimensional one-component model for distortive phase-transitions. *Physical Review B*, *17*(3), 1302–1322.
Senftle, T. P., Hong, S., Islam, M. M., Kylasa, S. B., Zheng, Y., Shin, Y. K., et al. (2016). The ReaxFF reactive force-field: Development, applications and future directions. *npj Computational Materials*, *2*, 15011.
Siu, S. W. I., Pluhackova, K., & Bockmann, R. A. (2012). Optimization of the OPLS-AA force field for long hydrocarbons. *Journal of Chemical Theory and Computation*, *8*(4), 1459–1470.
Smith, E. R. (2014). *On the coupling of molecular dynamics to continuum computational fluid dynamics*. Ph.D. thesis, Imperial College London. http://hdl.handle.net/10044/1/15610.
Smith, E. R. (2015). A molecular dynamics simulation of the turbulent Couette minimal flow unit. *Physics of Fluids*, *27*, 115105.
Smith, E. R., Heyes, D. M., Dini, D., & Zaki, T. A. (2012). Control-volume representation of molecular dynamics. *Physical Review E*, *85*, 056705.
Smith, E. R., Heyes, D. M., Dini, D., & Zaki, T. A. (2015). A localized momentum constraint for non-equilibrium molecular dynamics simulations. *Journal of Chemical Physics*, *142*(7), 074110.
Smith, E. R., Mller, E. A., Craster, R. V., & Matar, O. K. (2016a). A langevin model for fluctuating contact angle behaviour parametrised using molecular dynamics. *Soft Matter*, *12*, 9604–9615.
Smith, E. R., Trevelyan, D., & Ramos Fernandez, E. (2016b). *cpl-library*. https://doi.org/10.5281/zenodo.46573.

Sperka, P., Krupka, I., & Hartl, M. (2014). Evidence of plug flow in rolling-sliding elastohydrodynamic contact. *Tribology Letters*, *54*(2), 151–160.
Spikes, H. (2004). The history and mechanisms of ZDDP. *Tribology Letters*, *17*(3), 469–489.
Spikes, H. (2008). Low- and zero-sulphated ash, phosphorus and sulphur anti-wear additives for engine oils. *Lubrication Science*, *20*, 103–136.
Spikes, H. (2015). Friction modifier additives. *Tribology Letters*, *60*, 5.
Spikes, H., & Granick, S. (2003). Equation for slip of simple liquids at smooth solid surfaces. *Langmuir*, *19*(12), 5065–5071.
Spikes, H., & Jie, Z. (2014). History, origins and prediction of elastohydrodynamic friction. *Tribology Letters*, *56*(1), 1–25.
Spikes, H., & Tysoe, W. (2015). On the commonality between theoretical models for fluid and solid friction, wear and tribochemistry. *Tribology Letters*, *59*(1), 14.
Spikes, H. A. (2018). Stress-augmented thermal activation: Tribology feels the force. *Friction*, *6*(1), 1–31.
Sutton, A. P., Finnis, M. W., Pettifor, D. G., & Ohta, Y. (1988). The tight-binding bond model. *Journal of Physics C: Solid State Physics*, *21*, 35–66.
Tang, Y.-H., Kudo, S., Bian, X., Li, Z., & Karniadakis, G. E. (2015). Multiscale universal interface: A concurrent framework for coupling heterogeneous solvers. *Journal of Computational Physics*, *297*, 13–31.
Taylor, R. I., & de Kraker, B. R. (2017). Shear rates in engines and implications for lubricant design. *Proceedings of the Institution of Mechanical Engineers, Part J: Journal of Engineering Tribology*, *231*(9), 1106–1116.
Thompson, P. A., & Robbins, M. O. (1990). Shear flow near solids: Epitaxial order and flow boundary conditions. *Physical Review A*, *41*(12), 6830–6837.
Todd, B. D., & Daivis, P. J. (2007). Homogeneous non-equilibrium molecular dynamics simulations of viscous flow: Techniques and applications. *Molecular Simulation*, *33*(3), 189–229.
Todd, B. D., & Daivis, P. J. (2017). *Nonequilibrium molecular dynamics: Theory, algorithms and applications*. Cambridge: Cambridge University Press.
Todd, B. D., Evans, D. J., & Daivis, P. J. (1995). Pressure tensor for inhomogeneous fluids. *Physical Review E*, *52*, 1627.
Trevelyan, D. J., & Zaki, T. A. (2016). Wavy taylor vortices in molecular dynamics simulation of cylindrical couette flow. *Physical Review E*, *93*, 043107.
Tung, S. C., & McMillan, M. L. (2004). Automotive tribology overview of current advances and challenges for the future. *Tribology International*, *37*(7), 517–536.
Vakis, A. I., Yastrebov, V. A., Scheibert, J., Nicola, L., Dini, D., Minfray, C., et al. (2018). Modeling and simulation in tribology across scales: An overview. *Tribology International*, *125*, 169–199.
Vanossi, A., Manini, N., Urbakh, M., Zapperi, S., & Tosatti, E. (2013). Colloquium: Modeling friction: From nanoscale to mesoscale. *Reviews of Modern Physics*, *85*(2), 529–552.
Wang, F.-C., & Zhao, Y.-P. (2011). Slip boundary conditions based on molecular kinetic theory: The critical shear stress and the energy dissipation at the liquidsolid interface. *Soft Matter*, *7*(18), 8628.
Washizu, H., Ohmori, T., & Suzuki, A. (2017). Molecular origin of limiting shear stress of elastohydrodynamic lubrication oil film studied by molecular dynamics. *Chemical Physics Letters*, *678*, 1–4.
Weinan, E., Li, X., & Vanden-Eijnden, E. (2004). Some recent progress in multiscale modeling. In S. Attinger, & P. Koumoutsakos (Eds.), *Multiscale modelling and simulation* (pp. 3–21). Berlin, Heidelberg: Springer Berlin Heidelberg. ISBN 978-3-642-18756-8.
Weller, H. G., Tabor, G., Jasak, H., & Fureby, C. (1998). A tensorial approach to computational continuum mechanics using object-oriented techniques. *Computers in Physics*, *12*(6), 620–631.
Werder, T., Walther, J. H., & Koumoutsakos, P. (2005). Hybrid atomistic continuum method for the simulation of dense fluid flows. *Journal of Computational Physics*, *205*, 373.

Wood, M. H., Casford, M. T., Steitz, R., Zarbakhsh, A., Welbourn, R. J. L., & Clarke, S. M. (2016). Comparative adsorption of saturated and unsaturated fatty acids at the iron oxide/oil interface. *Langmuir*, *32*, 534.

Yong, X., & Zhang, L. T. (2013). Thermostats and thermostat strategies for molecular dynamics simulations of nanofluidics. *Journal of Chemical Physics*, *138*(8), 084503.

Yoshizawa, H., Chen, Y. L., & Israelachvili, J. (1993). Fundamental mechanisms of interfacial friction. 1. Relation between adhesion and friction. *Journal of Physical Chemistry*, *97*(16), 4128–4140.

Yue, D. C., Ma, T. B., Hu, Y. Z., Yeon, J., van Duin, A. C. T., Wang, H., et al. (2013). Tribochemistry of phosphoric acid sheared between quartz surfaces: A reactive molecular dynamics study. *Journal of Physical Chemistry C*, *117*(48), 25604–25614.

Zhang, J., & Spikes, H. (2016). On the mechanism of ZDDP antiwear film formation. *Tribology Letters*, *63*(2), 24.

Zhang, J., Tan, A., & Spikes, H. (2017). Effect of base oil structure on elastohydrodynamic friction. *Tribology Letters*, *65*(1), 13.

Chapter 4
Computational Methods for Contact Problems with Roughness

Marco Paggi, Alberto Bemporad and José Reinoso

Abstract This chapter provides a self-consistent introduction to computational methods for the solution of contact problems between bodies separated by rough interfaces. Both frictional and frictionless contact problems are examined. The mathematical formulation of the boundary element method is presented first, with details on the possible algorithmic implementation strategies and their computational efficiency. In the second part of the chapter, the fundamentals of the finite element method for the solution of contact problems are presented, along with an overview on the different strategies available in the literature to accurately discretize the multiscale features of roughness. A synopsis of the major advantages and disadvantages provided by the computational methods based on the boundary element method or the finite element method concludes the chapter, illustrating also perspective research directions.

Introduction

Contact mechanics between rough surfaces is a very active area of theoretical and applied research in physics and engineering (Vakis et al. 2018). Due to roughness, when two bodies separated by rough boundaries are brought into contact, they exchange forces through the so-called contact spots, which correspond to the tips of the asperities, i.e., the local maxima of the surfaces. As a consequence, the real area of contact is usually a small percentage of the nominal one, which would be attained only if the surfaces were perfectly flat. The evolution of the contact domain, which includes all the contact spots, the size of the real area of contact, and the normal contact stiffness, do depend on the applied normal load level, see Borri-Brunetto et al.

M. Paggi (✉) · A. Bemporad
IMT School for Advanced Studies Lucca, Piazza San Francesco 19, 55100 Lucca, Italy
e-mail: marco.paggi@imtlucca.it

J. Reinoso
School of Engineering, University of Seville, Camino de los Descubrimientos s/n, 41092 Seville, Spain

M. Paggi and D. Hills (eds.), *Modeling and Simulation of Tribological Problems in Technology*, CISM International Centre for Mechanical Sciences 593,
https://doi.org/10.1007/978-3-030-20377-1_4

(1999), Ciavarella et al. (2000, 2004, 2008a), Campaña et al. (2001), Barber (2003), Nosonovsky and Bhushan (2005), Persson (2006), Hyun and Robbins (2007), Carbone and Bottiglione (2008), Paggi and Ciavarella (2010), Paggi and Barber (2011), Paggi et al. (2014), and Yastrebov et al. (2015) for a selection of studies. Similarly, when a shearing load is applied, there is a progressive transition of the contact spots from a full stick condition, with perfect adhesion and no relative displacement between the bodies, to full slip, when sliding takes place. Such a transition is ruled by the Coulomb friction law at the asperity level, while the emerging quantities, such as for instance the total shearing load versus the size of the contact area in stick or slip conditions, are the result of a collective response emerging from the complex local interactions (Carpinteri and Paggi 2005, 2009; Paggi et al. 2014).

In this context, semi-analytical micromechanical contact theories relying on the statistical distribution of the elevation of the asperities and their radii of curvature have been proposed and widely explored in the engineering community (see McCool 1986; Zavarise et al. 2004a for comprehensive review articles), following the pioneering approach by Greenwood and Williamson (1966) and extending it to more complex statistical distributions of elevations and curvatures (Ciavarella et al. 2006; Greenwood 2006; Paggi and Ciavarella 2010), considering also elastic interactions between asperities (Ciavarella et al. 2008b) that were not included in the original pioneering formulations. Since the 1990s, research focused on the multiscale features of roughness, exploiting the use of fractal geometry for the understanding of its role on the contact behavior (Majumdar and Bhushan 1990; Borri-Brunetto et al. 1999; Carpinteri and Paggi 2005; Persson et al. 2005).

More recently, it has been found that neither the random process theory, which is the theoretical framework for the derivation of micromechanical contact theories, nor the fractal description of roughness is able to reproduce the complex morphology of surfaces (Greenwood and Wu 2001), as recently proved for natural or engineering surfaces with functionalized textures (Borri and Paggi 2015, 2016). Therefore, the predictions of semi-analytical contact models based on random process theory or fractal assumptions should be checked with care and led to a wide range of comparisons and validation studies (Mueser et al. 2017). On the other hand, experimental investigations are challenging to be performed and involve approximations too (Woo and Thomas 1980). For example, very often the contact quantities can only be estimated by indirect measurements of thermal or electric resistances of compressed rough joints (Sridhar and Yovanovich 1994), or they are mostly concerned with the measurement of the real area of contact under special conditions allowing for its inspection (O'Callaghan and Probert 1970; Hendriks and Visscher 1995).

Therefore, due to the general considerations above, numerical methods able to deal with realistic surface topologies without making approximations and assumptions on their shape, and with any constitutive response of the continuum and of the interface, are very important to predict the contact response and infer general conclusions on the observed trends.

In the linear elastic regime, if the multiscale character of roughness covering a wide range of wavelengths is one of the most prominent research topics, then the use of the boundary element method (BEM) has been historically preferred over the finite element method (FEM) (Andersson 1981; Man 1994). This is essentially due to the fact that only the surface must be discretized in the boundary element method, and not the surrounding continuum, as required by the finite element method. Moreover, it is not necessary to adopt surface interpolation techniques, like Bezier curves, to discretize the interface (see, e.g., the approach in Wriggers and Reinelt (2009)) and make it amenable for the application of contact search algorithms. This avoids an undesired smoothing of the fine-scale geometrical features of roughness.

In the application of the boundary element method, the core of the procedure is based on the knowledge of the so-called Green functions that relate the displacement of a generic point of the half-plane to the action of a concentrated force on the surface caused by contact interactions. An integral convolution of the effects of all the contact tractions provides the deformed contact configuration. After introducing a discretization of the half-plane consisting of a grid of boundary elements, the problem of point-force singularity is solved numerically by using the closed-form solution for a patch load acting on a finite-size boundary element (Johnson 1985, Chaps. 3, 4). The contact problem is then set in terms of equalities and inequalities stemming from the unilateral contact constraints and it can be solved by constrained optimization algorithms, see Polonsky and Keer (1999), Bemporad and Paggi (2015). The basic version of the boundary element method can be also extended to solve contact problems with friction (Li and Berger 2003; Pohrt and Li 2014) and between viscoelastic materials, see Carbone and Putignano (2013) and the references therein given.

However, standard boundary element formulations are based on the fundamental assumptions of linear elasticity and homogeneity of the materials, and their extension to inhomogeneities (Leroux et al. 2010), finite-size geometries (Putignano et al. 2015), or interface constitutive nonlinearities such as adhesion (Rey et al. 2017; Popov et al. 2017; Li et al. 2018) are sometimes possible but are not so straightforward. For these problems, the finite element method would be conceived as a more versatile computational approach to pursuit in order to overcome all the major limitations of the boundary element method. The finite element method can in fact take into account any material or interface constitutive nonlinearity, and it can easily treat finite-size geometries of practical interest in industrial applications. Moreover, it is prone to be extended for the solution of nonlinear multi-field problems involved in heat transfer or in reaction–diffusion systems (Zavarise et al. 2019; Sapora and Paggi 2014; Lenarda et al. 2018), for which the boundary element method has not been applied so far.

In spite of the different appealing aspects of the finite element method over the boundary element method, this approach has been limited to few remarkable studies concerning contact problems with roughness, especially in relation to elastoplasticity (Pei et al. 2005; Hyun et al. 2004). The motivation is primarily due to the need for discretizing the bulk and also the rough interface, which is not an easy task from the mathematical standpoint and it also gives a rise in computation costs. As shown in Wriggers and Reinelt (2009), Bezier interpolation techniques can be employed to regularize rough interfaces to become amenable for contact search algorithms. Nevertheless, smoothing should be applied with care in order to avoid artificial filtering of finer surface features relevant to the physics of contact. To overcome such limitations, a recent approach which does not explicitly discretize roughness, but it embeds its analytical form in a nominally flat interface finite element has proved to be very efficient in solving contact problems with roughness in the finite element method (Paggi and Reinoso 2018), significantly simplifying the issue of roughness discretization.

In this chapter, an overview of computational methods for solving contact problems with roughness is proposed. Section "The Boundary Element Method" focuses on the fundamentals of the boundary element method, first in relation to the frictionless normal contact problem. Based on the results in Bemporad and Paggi (2015), special attention is given to the review of the computational challenges of the method, which regard two main aspects: (i) efficiently solve the system of linear equations; (ii) impose the satisfaction of the unilateral contact constraints (contact inequalities). Regarding the first issue, iterative methods like the conjugate gradient algorithm or the Gauss–Seidel method (Francis 1983; Borri-Brunetto et al. 1999, 2001), or the capabilities of multigrid or multilevel methods (Raous 1999; Polonsky and Keer 1999), or even the solution of the linear system of equations in the Fourier space (Nogi and Kato 1997; Polonsky and Keer 2000a, b; Batrouni et al. 2002; Scaraggi et al. 2013; Prodanov et al. 2014; Vollebregt 2014) are possible strategies.

Regarding the imposition of the contact inequalities, on the other hand, a greedy approach where the boundary elements bearing tensile loads are iteratively excluded can be exploited (Kubo et al. 1981; Borri-Brunetto et al. 1999, 2001; Karpenko and Akay 2001; Batrouni et al. 2002), although it has been demonstrated by Bemporad and Paggi (2015) that it often fails for very compact contact domains. The constrained conjugate gradient method proposed by Polonsky and Keer (1999) and based on the theory in Hestenes (1980, Chaps. 2, 3) to solve the linear system of equations and rigorously impose the satisfaction of the contact constraints is also discussed, along with its developments (Polonsky and Keer 2000a, b). Finally, other optimization algorithms based on the solution of the corresponding quadratic program, such as nonnegative least squares (NNLS) and the alternative direction method of multipliers (ADMM), proposed in Bemporad and Paggi (2015), are detailed. A careful comparison of the available methods in terms of computation cost is also provided, along with other more advanced acceleration strategies.

The presentation moves then to the contact problem with friction, which is significantly complicated by the coupling between the normal and the tangential deformations, and by the Coulomb friction law to be locally satisfied in the tangential direction.

Section "The Finite Element Method" presents the basis of the finite element method, starting from the variational formulation of the problem, including the strong form, the governing equations for the continuum and the interface, and also the weak form. Next, finite element procedures based on the explicit discretization of roughness are briefly summarized in relation to seminal work published in the literature. Finally, the zero-thickness interface finite element with embedded profile for joint roughness (MPJR interface finite element, recently proposed in Paggi and Reinoso (2018)) is presented, along with its implementation details in the research finite element analysis program FEAP (Zienkiewicz and Taylor 2000). A comparison between the different finite element discretization strategies for the solution of a benchmark Hertzian contact problem concludes the section and it shows the great potential of the MPJR interface finite element for future research.

The last section provides an overall summary of the chapter with a synopsis reporting the major advantages and disadvantages of the computational methods herein reviewed, along with an overview of perspective research directions in the field of contact mechanics between rough surfaces.

The Boundary Element Method

The boundary element method (BEM) is an efficient technique to solve the contact problem between two linear elastic bodies, say Ω_1 and Ω_2, with rough boundaries Γ_1 and Γ_2, respectively. The first step for the application of the method is the knowledge of the topographies Γ_i $(i = 1, 2)$, which are nowadays acquired using confocal profilometers or atomic force microscopes. They are stored in a matrix containing the x_i, y_i, and z_i coordinates with respect to an arbitrary datum. Such surfaces are nonconforming, i.e., they do not match when in contact. Hence, the shape of the contact area changes with the applied load and the contact area is an unknown of the problem, which is a source of complexity for the solution procedure.

Due to linear elasticity of the continuum, the actual contact problem can be simplified by reducing it to the solution of a fictitious contact problem between a rigid microscopically rough surface with a composite topography and a flat linear elastic half-space with composite elastic moduli E and G dependent on the Young's moduli E_i and the Poisson's ratios ν_i of the two materials $(i = 1, 2)$. The composite topography is simply obtained by summing up the elevations of the two rough surfaces Γ_i $(i = 1, 2)$. The composite elastic moduli, on the other hand, are given by the following formulae:

$$E = \left(\frac{1-\nu_1^2}{E_1} + \frac{1-\nu_2^2}{E_2}\right)^{-1}, \tag{4.1a}$$

$$G = \left(\frac{2-\nu_1}{4G_1} + \frac{2-\nu_2}{4G_2}\right)^{-1}, \tag{4.1b}$$

where $G_i = E_i/[2(1+\nu_i)]$. The composite Poisson ratio ν is related to E and G via $\nu = E/(2G) - 1$.

In this simplified setting, the next step for the application of the method is the knowledge of the so-called Green functions, which relate the displacements of any point belonging to the deformable half-space to the applied tractions on its surface. This allows formulating the problem by involving only contact tractions and surface displacements, getting rid of the surrounding continua. While the simplest expressions for the Green functions are those for a cylindrical punch on a homogenous and isotropic half-space, other forms for more complex material configurations can be of interest in engineering applications. For instance, multilayered half-spaces, deformable elastic bodies with finite size (Putignano et al. 2015), or half-spaces with spatial inclusions or voids (Leroux et al. 2010) have been investigated in the literature. Quite recently, adhesion effects in tension have been included in BEM for frictionless normal contact problems, see Rey et al. (2017), Popov et al. (2017), Li et al. (2018).

For more complex heterogeneous material compositions due to the random presence of inclusions or voids of arbitrary shape, or functionally graded compositions, the Green functions can be solely determined in numerical form (Paggi and Zavarise 2011). For that, the finite element method (FEM) can be used to preliminary extract the Green functions. This is done by discretizing the bulk of the deformable body and its internal microstructure. Then, unit tractions are applied at any point over the half-plane boundary, and the induced surface displacements are computed. Therefore, point-by-point, the Green functions can be numerically reconstructed. In the sequel, we shall restrict the attention to linear elastic homogeneous and isotropic half-spaces, for the sake of simplicity. More complex material compositions can be dealt with by modifying the expression for the Green functions.

It is important to remark here that the knowledge of the relation between tractions and surface displacements is a key point for the simplification of the complexity of the contact problem, since it allows avoiding the discretization of the bulk. This is indeed a significant gain over the finite element method in terms of computation costs, especially for problems with rough boundaries where the discretization of the rough interface has to be very fine to capture its essential multiscale features. On the other hand, BEM is rigorously exact only for linear elastic contact problems, since the principle of superposition is applied to convolute the effect of any distribution of surface tractions applied over the half-plane. For elastoplastic contact problems with isotropic or kinematic hardening laws, Chen et al. (2008) generalized the BEM formulation by considering modified discrete expressions for the Green functions and the residual displacements caused by plastic deformation. For other relevant publications on this topic, the reader is referred to Kogut and Etsion (2002), Chang et al. (1987), Nelias et al. (2006). For other problems involving material or geometrical (finite elasticity) nonlinearities of the continuum, or for multi-field problems, the finite element method is indeed preferable over the boundary element method and it would be the ideal framework for their investigation.

The Frictionless Normal Contact Problem

Back to the classical framework of BEM for frictionless normal contact problems in linear elasticity, the normal displacements $u(\mathbf{x})$ at any point of the half-plane identified by the position vector $\mathbf{x}$ are related to the contact tractions $p(\mathbf{y})$ at other points as follows (Johnson 1985; Barber 2018):

$$u(\mathbf{x}) = \int_S H(\mathbf{x}, \mathbf{y}) p(\mathbf{y}) \mathrm{d}\mathbf{y}, \tag{4.2}$$

where $H(\mathbf{x}, \mathbf{y})$ represents the displacement at a point $\mathbf{x}$ due to a surface contact pressure p acting at $\mathbf{y}$, and S is the elastic half-plane. For homogeneous, isotropic, linear elastic materials, the influence coefficients (Green function) are given by Boussinesq:

$$H(\mathbf{x}, \mathbf{y}) = \frac{1 - \nu^2}{\pi E} \frac{1}{\| \mathbf{x} - \mathbf{y} \|}, \tag{4.3}$$

where $\| \cdot \|$ denotes the standard Euclidean norm. The total contact force P is the integral of the contact pressure field:

$$P = \int_S p(\mathbf{x}) \mathrm{d}A. \tag{4.4}$$

By referring to Fig. 4.1, in the following we define for each surface point $\mathbf{x} \in S$ its elevation $\xi(\mathbf{x})$, measured with respect to a reference frame, and set $\xi_{\max} \triangleq \max_{\mathbf{x} \in S} \xi(\mathbf{x})$ the maximum elevation. The indentation of the half-plane at the points in contact is denoted by $\bar{u}$, whereas a generic displacement along the surface is u.

For a given far-field displacement $\Delta \geq 0$ in the direction perpendicular to the undeformed half-plane, the problem is to find the solution of the normal contact $u(\mathbf{x})$, $p(\mathbf{x})$ satisfying (4.2) and the unilateral contact (linear complementarity) conditions

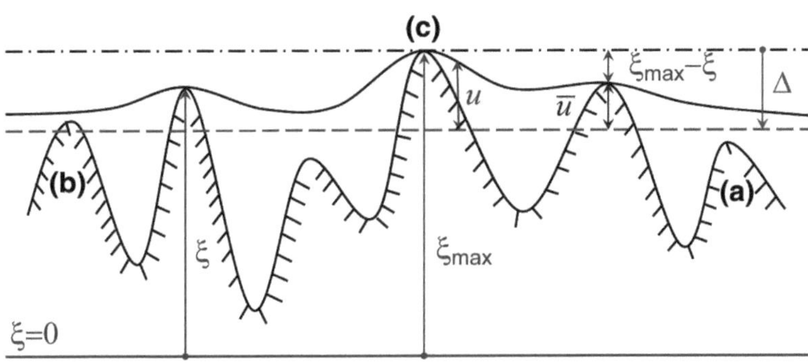

Fig. 4.1 Sketch of the contact problem between a rigid rough surface and an elastic half-plane. Its deformed configuration corresponding to the imposed far-field displacement Δ is depicted with a solid line. The dashed line corresponding to a rigid body motion of the half-plane identifies the heights to be included in the initial trial contact domain. Once the contact problem is solved, one may have (i) heights certainly not in contact from the beginning, type (**a**); (ii) heights loosing contact due to elastic interactions, type (**b**); (iii) heights in contact, type (**c**). Adapted from Bemporad and Paggi (2015)

$$u(\mathbf{x}) - \bar{u}(\mathbf{x}, \Delta) \geq 0, \tag{4.5a}$$

$$p(\mathbf{x}) \geq 0, \tag{4.5b}$$

$$(u(\mathbf{x}) - \bar{u}(\mathbf{x}, \Delta))p(\mathbf{x}) = 0, \tag{4.5c}$$

for all points $\mathbf{x} \in S$, where contact tractions are considered positive when compressive.

Introducing the quantity $w(\mathbf{x}, \Delta) = u(\mathbf{x}) - \bar{u}(\mathbf{x}, \Delta)$, Eq. (4.5) can be rewritten as

$$w(\mathbf{x}, \Delta) \geq 0, \tag{4.6a}$$

$$p(\mathbf{x}) \geq 0, \tag{4.6b}$$

$$w(\mathbf{x}, \Delta)p(\mathbf{x}) = 0. \tag{4.6c}$$

The above contact problem is an infinite-dimensional linear complementarity problem. A finite-dimensional approximate solution can be sought by discretizing the surface as a square grid of spacing δ consisting of $N \times N$ average heights. Let S_{ij} be the cell of area δ^2 indexed by $i, j \in I_N$, with $I_N \triangleq \{1, \ldots, N\} \times \{1, \ldots, N\}$. Let $\mathbf{x}_{i,j} \triangleq \frac{1}{S_{ij}} \int_{\mathbf{x} \in S_{ij}} \mathbf{x} \mathrm{d}A$, $\xi_{i,j} \triangleq \frac{1}{S_{ij}} \int_{\mathbf{x} \in S_{ij}} \xi(\mathbf{x}) \mathrm{d}A$, $p_{i,j} \triangleq \int_{\mathbf{x} \in S_{ij}} p(\mathbf{x}) \mathrm{d}A$, and $u_{i,j} \triangleq \frac{1}{S_{ij}} \int_{\mathbf{x} \in S_{ij}} u(\mathbf{x}) \mathrm{d}A$ be, respectively, the barycentric coordinate, average height, resultant of the contact tractions, and the corresponding average displacement on the surface element S_{ij}. Consider the following discretized version of (4.2)

$$u_{i,j} = \sum_{k=1}^{N} \sum_{l=1}^{N} H_{i-k,j-l}\, p_{k,l} \tag{4.7}$$

for all $(i, j), (k, l) \in I_N$, $p_{k,l} \geq 0$, where the term $H_{i-k,j-l}$ is the Green function in (4.3) averaged over the elementary area δ^2. For instance, Borri-Brunetto et al. (1999) used the following approximation related to a uniform pressure acting on a rounded punch of radius $\delta/2$:

$$H_{i-k,j-l} = \begin{cases} \dfrac{2}{E \pi \delta}, & \text{if } i = k \text{ and } j = l \\ \dfrac{2}{E \pi \delta} \arcsin \dfrac{\delta}{2\|\mathbf{x}_{i,j} - \mathbf{x}_{k,l}\|}, & \text{if } i \neq k, j \neq l \end{cases} \tag{4.8}$$

but other formulae for a square punch can also be taken as in Pohrt and Li (2014).

Let $\bar{I}_C \triangleq \{(i, j) \in I_N : \xi_{i,j} < \xi_{\max} - \Delta\}$ be the set of indices corresponding to elements S_{ij} that are certainly not in contact (cf. Fig. 4.1), and hence

$$p_{k,l} = 0, \forall (k, l) \in \bar{I}_C. \tag{4.9}$$

Let $m = \#\bar{I}_C$ be the number of elements of $\bar{I}_C$ and $n = \#I_C$ the number of elements belonging to the initial trial contact domain, $I_C \triangleq I_N \setminus \bar{I}_C$. The set I_C is only a superset of the set I_C^* of actual contact points, since the corrections to the displacements induced by elastic interactions may induce lack of contact in some elements (i, j), i.e., $u_{i,j} > \bar{u}_{i,j}$, where $\bar{u}_{i,j} \triangleq \Delta - \xi_{\max} + \xi_{i,j}$ is the value of the compenetration of the height corresponding to the element (i, j) in the half-plane (see Fig. 4.1).

For a generic $(i, j) \in I_C$ corresponding to an element of the surface which is potentially in contact with the elastic half-plane, we denote by

$$w_{i,j} \triangleq u_{i,j} - \bar{u}_{i,j} \geq 0 \tag{4.10}$$

the corresponding elastic correction to the displacement. Clearly, it must hold that

$$w_{i,j} p_{i,j} = 0, \ \forall (i, j) \in I_C \tag{4.11}$$

since $w_{i,j} > 0$ implies no contact between the surfaces and therefore vanishing contact pressures, while $p_{i,j} > 0$ implies contact, $u_{i,j} = \bar{u}_{i,j}$, or equivalently $w_{i,j} = 0$.

By taking into account that $p_{k,l} = 0$ for all $(k, l) \in \bar{I}_C$, Eq. (4.7) can be recast as the following condition:

$$w_{i,j} + \bar{u}_{i,j} = \sum_{(k,l) \in I_C} H_{i-k,j-l}\, p_{k,l}, \ \forall (i, j) \in I_C, \tag{4.12}$$

which is limited to the nodes belonging to the initial trial contact domain I_C, whose number of elements is in general significantly smaller than those of I_N. The relations (4.9)–(4.12) can be recast in matrix form as the following Linear Complementarity Problem (LCP) (Cottle et al. 1992):

$$\mathbf{w} = \mathbf{H}\mathbf{p} - \bar{\mathbf{u}} \tag{4.13a}$$

$$\mathbf{w} \geq \mathbf{0}, \ \mathbf{p} \geq \mathbf{0}, \ \mathbf{w}^{\mathrm{T}}\mathbf{p} = 0, \tag{4.13b}$$

where $\mathbf{w} \in \mathbb{R}^n$ is the vector of unknown elastic corrections $w_{i,j}$, $(i, j) \in \bar{I}_C$, $\mathbf{w}^{\mathrm{T}}$ denotes its transpose, $\mathbf{p} \in \mathbb{R}^n$ is the vector of unknown boundary element contact forces $p_{i,j}$, $(i, j) \in I_C$, $\bar{\mathbf{u}} \in \mathbb{R}^n$ is the vector of compenetrations $\bar{u}_{i,j}$, $(i, j) \in I_C$, and $\mathbf{H} = \mathbf{H}^{\mathrm{T}}$ is the matrix obtained by collecting the compliance coefficients $H_{i-k,j-l}$, for $(i, j), (k, l) \in I_C$. Due to the properties of linear elasticity (Johnson 1985, p. 144), we have that

$$\mathbf{H} = \mathbf{H}^{\mathrm{T}} \succ 0, \tag{4.14}$$

that is, $\mathbf{H}$ is a symmetric positive definite matrix (with the additional property deriving from (4.8) of having all its entries positive). After solving (4.13), the vector $\mathbf{u} \in \mathbb{R}^n$ of normal displacements $u_{i,j}$, $(i, j) \in I_C$, is simply retrieved as $\mathbf{u} = \bar{\mathbf{u}} + \mathbf{w}$.

By the positive definiteness property (4.14) of $\mathbf{H}$, we inherit immediately the following important property (Cottle et al. 1992, Theorem 3.3.7): the discretized version (4.6), (4.9)–(4.12) of the contact problem admits a unique solution **p**, **u**, for all $\Delta \geq 0$.

The LCP problem (4.13) corresponds to the Karush–Kuhn–Tucker (KKT) conditions for optimality of the following convex quadratic program (QP):

$$\min_p \frac{1}{2}\mathbf{p}^{\mathrm{T}}\mathbf{H}\mathbf{p} - \bar{\mathbf{u}}^{\mathrm{T}}\mathbf{p} \tag{4.15a}$$

$$\text{s.t. } \mathbf{p} \geq \mathbf{0} \tag{4.15b}$$

in that the solution **p** of (4.15) and its corresponding optimal dual solution **w** solve (4.13), and vice versa.

The QP problem is consistent with former considerations by Kalker and van Randen (1972) and also summarized in Johnson (Johnson 1985, pp. 151–152). In fact, the contact pressures solving the unilateral contact problem can be obtained by minimizing the total complementary energy W of the linear elastic system, subject to the constraint $p(\mathbf{x}) \geq 0, \forall \mathbf{x} \in \mathcal{S}$. For a continuous system, the total complementary energy is

$$W = U - \int_{\mathcal{S}} p(\mathbf{x})\bar{u}(\mathbf{x}, \Delta)\, d\mathbf{x}, \tag{4.16}$$

where U is the internal complementary energy of the deformed half-plane in contact. For linear elastic materials, we have

$$U = \frac{1}{2}\int_{\mathcal{S}} p(\mathbf{x})u(\mathbf{x})\, d\mathbf{x}. \tag{4.17}$$

Although such an energy-based approach can be used to derive finite element formulations, it is also possible to remain within the boundary element method and introduce a surface discretization as before. By invoking the averaged Green functions in (4.8), the discretized version of W, say $\widetilde{W}$, reads

$$\widetilde{W} = \frac{1}{2}\sum_{(i,j)\in I_C}\sum_{(k,l)\in I_C} H_{i-k,j-l}\, p_{k,l}p_{i,j} - \sum_{(i,j)\in I_C} p_{i,j}\bar{u}_{i,j}, \tag{4.18}$$

which represents a quadratic function of **p** to be minimized, under the constraints $p_{i,j} \geq 0, \forall (i,j) \in I_C$, as in (4.15). Since it is unlikely that the contact area is known a priori, the active set of nodes in contact results only after solving problem (4.13) or equivalently (4.15).

A large variety of solvers for LCP and QP problems were developed in the last 60 years (Beale 1955; Fletcher 1971; Goldfarb and Idnani 1983; Cottle et al. 1992; Schmid and Biegler 1994; Patrinos and Bemporad 2014), and is still an active area of research in the optimization and control communities. Historically, in the mechanics community, Kalker and van Randen (1972) proposed the simplex method, although it was found to be practical only for relatively small N. More recent contributions adopt algorithms to solve the unconstrained linear system of equations and then correct the solution by eliminating the boundary elements bearing tensile tractions (Francis 1983; Borri-Brunetto et al. 1999, 2001), or use a constrained version of the conjugate gradient (CG) algorithm (Polonsky and Keer 1999). These methods are simply initialized by considering arbitrary nonnegative entries in $\mathbf{p}$, without taking advantage of the monotonic increase (or decrease) of pressures by increasing (or decreasing) the far-field displacement, an important property guaranteed by rigorous elasticity theorems (Barber 1974). The history of pressures can be saved during a contact simulation and it is easy to access and use and it can be beneficial to save computation time, as proved by Bemporad and Paggi (2015).

Optimization algorithms. Since now on, we use the subscript i to denote the ith component of a vector or the ith row of a matrix, the subscript $\mathcal{I}$ to denote the subvector obtained by collecting all the components $i \in \mathcal{I}$ of a vector (or all the rows i of a matrix), and the double subscript $\mathcal{I}, \mathcal{I}_1$ to denote the submatrix obtained by collecting the ith row and jth column, for all $i \in \mathcal{I}$, $j \in \mathcal{I}_1$.

In the sequel, following the content in Bemporad and Paggi (2015), a brief overview of algorithms to solve the constrained contact problem is provided, starting first with those for the solution of the LCP, namely, the greedy method and the constrained conjugate gradient algorithm. Next, optimization algorithms for the solution of the corresponding QP are discussed, such as the nonnegative least squares and the alternative direction method of multipliers. Finally, a comparison of the algorithms above in terms of computation performance is presented. Further acceleration strategies and advanced methods are also discussed.

A greedy method corresponds to solve problem (4.15) by iteratively solving the unconstrained linear system of equations $\mathbf{w} = \mathbf{H}\mathbf{p} - \bar{\mathbf{u}} = \mathbf{0}$ with respect to $\mathbf{p}$ and increasingly zeroing negative elements of $\mathbf{p}$ until the condition $\mathbf{p} \geq \mathbf{0}$ is satisfied. By construction we obtain $\mathbf{w}^{\mathrm{T}}\mathbf{p} = 0$. The method is described in Algorithm 1, in which a standard conjugate gradient (CG) is employed to solve the unconstrained linear system of equations. Steps 2.1–2.4 can be replaced by any other algorithm for solving the linear system of equations, like the Gauss–Seidel iterative scheme as in Borri-Brunetto et al. (1999, 2001), the MATLAB's `mldivide` solver, or even the FFT algorithm as in Karpenko and Akay (2001), Batrouni et al. (2002), Vollebregt (2014).

Assuming that the prescribed initial $\mathbf{p}$ and $\mathcal{I}$ are such that $p_j = 0$ for all $j \in \{1, \ldots, n\} \setminus \mathcal{I}$, and $K_{\max}$ is sufficiently large, the output of the greedy algorithm leads to a contact pressure vector $\mathbf{p}^*$ and a normal displacement vector $\mathbf{u}^*$ satisfying $\mathbf{u}^* = \mathbf{H}\mathbf{p}^*$, $\mathbf{p}^* \geq \mathbf{0}$, $(\mathbf{u}^* - \bar{\mathbf{u}})^{\mathrm{T}}\mathbf{p}^* = 0$. In fact, condition $\mathbf{p}^* \geq \mathbf{0}$ is guaranteed by the condition in Step 2 up to ϵ precision. By letting $\mathbf{w}^* \triangleq \mathbf{u}^* - \bar{\mathbf{u}}$, at termination of the algorithm we have $\mathbf{w}^*_{\mathcal{I}} = \mathbf{H}_{\mathcal{I},\mathcal{I}}\mathbf{p}^*_{\mathcal{I}} - \bar{\mathbf{u}}_{\mathcal{I}} = \mathbf{0}$ because of the solution of the CG method (Step 2.4), or equivalently $\mathbf{u}^*_{\mathcal{I}} = \bar{\mathbf{u}}_{\mathcal{I}}$ (cf. Step 4). By setting $\mathbf{u}^*_{\bar{\mathcal{I}}} \triangleq \mathbf{H}_{\bar{\mathcal{I}},\mathcal{I}}\mathbf{p}_{\mathcal{I}}$ in Step 4, and recalling that $\mathbf{p}^*_{\bar{\mathcal{I}}} = \mathbf{0}$, we have

Input: Matrix $\mathbf{H} = \mathbf{H}^{\mathrm{T}} \succ 0$, vector $\bar{\mathbf{u}}$; initial guess $\mathbf{p}$ and initial active set $\mathcal{I} \subseteq \{1, \ldots, n\}$ such that $\mathbf{p}_{\{1,\ldots,n\}\setminus\mathcal{I}} = \mathbf{0}$; maximum number $K_{\max}$ of iterations, tolerance $\epsilon > 0$.

---

1. $i \leftarrow 0; \bar{\mathcal{I}} \leftarrow \{1, \ldots, n\} \setminus \mathcal{I}$;
2. **while** $(i \leq K_{\max}$ **and** $\min(\mathbf{p}) < -\epsilon)$ **or** $i = 0$ **do**:
   - (2.1) $\mathbf{w}_{\mathcal{I}} \leftarrow \mathbf{H}_{\mathcal{I},\mathcal{I}}\mathbf{p}_{\mathcal{I}} - \bar{\mathbf{u}}_{\mathcal{I}}$;
   - (2.2) $n_w \leftarrow \|\mathbf{w}_{\mathcal{I}}\|_2$;
   - (2.3) $\mathbf{b}_{\mathcal{I}} \leftarrow -\mathbf{w}_{\mathcal{I}}$
   - (2.4) **while** $n_w > \epsilon$ **and** $i \leq K_{\max}$ **do**:
     - (2.4.1) $\mathbf{s}_{\mathcal{I}} \leftarrow \mathbf{H}_{\mathcal{I},\mathcal{I}}\mathbf{b}_{\mathcal{I}}$;
     - (2.4.2) $\mathbf{p}_{\mathcal{I}} \leftarrow \mathbf{p}_{\mathcal{I}} - \dfrac{\mathbf{w}_{\mathcal{I}}^{\mathrm{T}}\mathbf{b}_{\mathcal{I}}}{\mathbf{b}_{\mathcal{I}}^{\mathrm{T}}\mathbf{s}_{\mathcal{I}}}\mathbf{b}_{\mathcal{I}}$;
     - (2.4.3) $\bar{\mathbf{w}}_{\mathcal{I}} \leftarrow \mathbf{H}_{\mathcal{I},\mathcal{I}}\mathbf{p}_{\mathcal{I}} - \bar{\mathbf{u}}_{\mathcal{I}}$;
     - (2.4.4) $\mathbf{b}_{\mathcal{I}} \leftarrow -\bar{\mathbf{w}}_{\mathcal{I}} + \dfrac{\bar{\mathbf{w}}_{\mathcal{I}}^{\mathrm{T}}\mathbf{s}_{\mathcal{I}}}{\mathbf{b}_{\mathcal{I}}^{\mathrm{T}}\mathbf{s}_{\mathcal{I}}}\mathbf{b}_{\mathcal{I}}$;
     - (2.4.5) $\mathbf{w}_{\mathcal{I}} \leftarrow \bar{\mathbf{w}}_{\mathcal{I}}$;
     - (2.4.6) $n_w \leftarrow \|\mathbf{w}_{\mathcal{I}}\|_2$;
     - (2.4.7) $i \leftarrow i + 1$;
   - (2.5) **for** $j \in \mathcal{I}$ **do**:
     - (2.5.1) **if** $\mathbf{p}_j < -\epsilon$ **then** $\mathbf{p}_j \leftarrow 0; \mathcal{I} \leftarrow \mathcal{I} \setminus \{j\}; \bar{\mathcal{I}} \leftarrow \bar{\mathcal{I}} \cup \{j\}$;
3. $\mathbf{p}^* \leftarrow \mathbf{p}$;
4. $\mathbf{u}_{\mathcal{I}}^* = \bar{\mathbf{u}}_{\mathcal{I}}, \mathbf{u}_{\bar{\mathcal{I}}}^* \leftarrow \mathbf{H}_{\bar{\mathcal{I}},\mathcal{I}}\mathbf{p}_{\mathcal{I}}$;
5. **end**.

Output: Contact force vector $\mathbf{p}^*$ and normal displacement vector $\mathbf{u}^*$.

Algorithm 1: Greedy method with Conjugate Gradient (greedy CG)

$$\begin{bmatrix} \mathbf{w}_{\mathcal{I}}^* \\ \mathbf{w}_{\bar{\mathcal{I}}}^* \end{bmatrix} = \begin{bmatrix} \mathbf{0} & \mathbf{0} \\ \mathbf{H}_{\bar{\mathcal{I}},\mathcal{I}} & \mathbf{0} \end{bmatrix} \begin{bmatrix} \mathbf{p}_{\mathcal{I}}^* \\ \mathbf{0} \end{bmatrix} + \begin{bmatrix} \mathbf{0} \\ -\bar{\mathbf{u}}_{\bar{\mathcal{I}}} \end{bmatrix} = \begin{bmatrix} \mathbf{H}_{\mathcal{I},\mathcal{I}} & \mathbf{H}_{\mathcal{I},\bar{\mathcal{I}}} \\ \mathbf{H}_{\bar{\mathcal{I}},\mathcal{I}} & \mathbf{H}_{\bar{\mathcal{I}},\bar{\mathcal{I}}} \end{bmatrix} \begin{bmatrix} \mathbf{p}_{\mathcal{I}}^* \\ \mathbf{p}_{\bar{\mathcal{I}}}^* \end{bmatrix} + \begin{bmatrix} -\bar{\mathbf{u}}_{\mathcal{I}} \\ -\bar{\mathbf{u}}_{\bar{\mathcal{I}}} \end{bmatrix}$$

and hence $\mathbf{u}^* = \mathbf{w}^* + \bar{\mathbf{u}} = \mathbf{H}\mathbf{p}^*$. The complementarity condition $(\mathbf{u}^* - \bar{\mathbf{u}})^{\mathrm{T}}\mathbf{p}^* = (\mathbf{w}^*)^{\mathrm{T}}\mathbf{p}^* = 0$ follows by construction, as Step 2.4 zeroes all the components of $\mathbf{w}_j^*$ that correspond to nonnegative $\mathbf{p}_j^*$, $\forall j \in \mathcal{I}$, and zeroes all the components $\mathbf{p}_j^*$ that correspond to possible nonzero components $\mathbf{w}_j^*$, $\forall j \in \bar{\mathcal{I}}$.

However, Bemporad and Paggi (2015) demonstrated that there is no formal proof that the condition $\mathbf{w}_{\bar{\mathcal{I}}}^* \geq \mathbf{0}$ (i.e., that $\mathbf{u}^* \geq \bar{\mathbf{u}}$) is satisfied after the algorithm terminates. If the algorithm is applied to randomly generated $\bar{\mathbf{u}}$ vectors and $\mathbf{H}$ positive definite matrices with positive coefficients, Bemporad and Paggi (2015) found that in many cases the LCP is not solved exactly. This problem was noticed especially when the contact domain is densely packed, with many boundary elements close to each other and all in contact. A MATLAB routine of the counterexample is available for download at http://musam.imtlucca.it/counterexample.m.

Therefore, as a word of caution, the reliability of the greedy method should be carefully checked in case of applications of the boundary element method to contact problems governed by other forms of $\mathbf{H}$, as in the case of contact with an anisotropic or an inhomogeneous half-plane.

Another drawback of the algorithm is the difficulty to warm start the method with a proper choice of the initial active set $\mathcal{I}$. Since at Step 2.5.1 the number of elements in the sequence $\mathcal{I}$ is decreased by removing negative enough components $\mathbf{p}_j$ of the current solution vector, i.e., eliminating the points bearing tensile (negative) forces, in a monotonic way (no index j that has been removed from $\mathcal{I}$ can be added back), a safe cold start is to set $\mathcal{I} = \{1, \ldots, n\}$ and pick up a vector $\mathbf{p} \geq \mathbf{0}$, usually a vector with arbitrary nonnegative numbers. The history of boundary element contact forces obtained during the solution of a sequence of imposed displacements is not taken into account by the method to accelerate its convergence, although we know that contact forces are monotonically increasing functions of the far-field displacement. In any case, for a complex sequence of loading with an increased or decreased far-field displacement, any warm starting on forces cannot be implemented in the method, since the elimination of contact points is irreversible.

A constrained conjugate gradient (CG) algorithm was proposed by Polonsky and Keer (1999) based on the theory by Hestenes (1980, Chaps. 2, 3), to solve the linear system of equations and rigorously impose the satisfaction of the contact constraints. Algorithm 2 has been applied by Polonsky and Keer (1999) to simulations under load control. However, it can be used also for displacement control. The condition for convergence set by Polonsky and Keer (1999) in terms of relative variation in the local contact forces from an iteration to the next has been recast in terms of the error in the local contact displacements. The two criteria are completely equivalent.

This constrained CG algorithm does not remove the points bearing tensile forces from the active set, as the Greedy algorithms do. Therefore, the size of the linear system of equations is not reduced during the iterations, increasing the computation time for its solution. On the other hand, the method assures the satisfaction of the LCP conditions (4.13) and it was found in Bemporad and Paggi (2015) to convergence with a reduced number of iterations as compared to the Greedy CG algorithm. Although not investigated in Polonsky and Keer (1999), it can be warm-started in case of a sequence of loading steps by considering both an initial trial contact domain and a set of contact pressures derived from the previous converged solution. The FFT method can be used to accelerate step (3.8) of Algorithm 2, as in Polonsky and Keer (2000a).

The QP problem with positive definite Hessian matrix having the special form (4.15) and corresponding to the LCP can be effectively solved as a nonnegative least squares problem, as proposed by Bemporad and Paggi (2015).

Thanks to property (4.14), matrix $\mathbf{H}$ admits a Cholesky factorization $\mathbf{H} = \mathbf{C}^{\mathrm{T}}\mathbf{C}$. Hence, we can *theoretically* recast problem (4.15) as the nonnegative least squares (NNLS) problem:

$$\min_p \frac{1}{2}\|\mathbf{C}\mathbf{p} - \mathbf{C}^{-\mathrm{T}}\bar{\mathbf{u}}\|_2^2 \tag{4.19a}$$

$$\text{s.t. } \mathbf{p} \geq \mathbf{0}. \tag{4.19b}$$

Input: Matrix $\mathbf{H} = \mathbf{H}^{\mathrm{T}} \succ 0$, vector $\bar{\mathbf{u}}$, initial guess $\mathbf{p} \geq \mathbf{0}$, initial active set $\mathcal{I} = \{1, \ldots, n\}$; maximum number $K_{\max}$ of iterations, tolerance $\epsilon > 0$.

---

1. $i \leftarrow 0, n_{w,\text{old}} = 1, d = 0, err = +\infty$;
2. $\mathbf{w} \leftarrow \mathbf{H}\mathbf{p} - \bar{\mathbf{u}}$;
3. **while** ($i \leq K_{\max}$ **and** $err > \epsilon$):

   (3.1) **if** $i = 0$;
   then $\mathbf{t} \leftarrow \mathbf{w}$;
   else: $\mathbf{t} \leftarrow \mathbf{w} + d\dfrac{n_w}{n_{w,\text{old}}}\mathbf{t}_{\text{old}}$;

   (3.2) $\tau = \dfrac{\mathbf{w}^{\mathrm{T}}\mathbf{t}}{\mathbf{t}^{\mathrm{T}}\mathbf{H}\mathbf{t}}$;

   (3.3) $\mathbf{p} \leftarrow \mathbf{p} - \tau\mathbf{t}$;

   (3.4) $\forall j \in \mathcal{I} : p_j \leftarrow \max\{p_j, 0\}$;

   (3.5) Find $I_{ol} = \{j \in \mathcal{I} : p_j = 0, w_j < 0\}$;
   if $I_{ol} = \emptyset$ **then** $d = 1$ **else** $d = 0$; $p_j \leftarrow p_j - \tau w_j, \forall j \in I_{ol}$;

   (3.6) $\mathcal{I} \leftarrow \{j : p_j > 0\} \cup I_{ol}$;

   (3.7) $\mathbf{t}_{\text{old}} \leftarrow \mathbf{t}, n_{w,\text{old}} \leftarrow n_w$;

   (3.8) $\mathbf{w} \leftarrow \mathbf{H}\mathbf{p} - \bar{\mathbf{u}}$;

   (3.9) $n_w = \|\mathbf{w}\|_2$;

   (3.10) $err \leftarrow |n_w - n_{w,\text{old}}|/n_{w,\text{old}}$;

   (3.11) $i \leftarrow i + 1$;

4. $\mathbf{p}^* \leftarrow \mathbf{p}$; $\mathbf{u}^* = \mathbf{H}\mathbf{p}^*$;
5. **end**.

Output: Contact force vector $\mathbf{p}^*$ and normal displacement vector $\mathbf{u}^*$.

Algorithm 2: Constrained Conjugate Gradient

A simple and effective active-set method for solving the NNLS problem (4.19) is the one in Lawson and Hanson (1974, p. 161) that was extended by Bemporad and Paggi (2015) in Algorithm 3 to directly solve (4.15) without explicitly computing the Cholesky factor $\mathbf{C}$ and its inverse $\mathbf{C}^{-1}$, and to handle warm starts. After a finite number of steps, Algorithm 3 converges to the optimal contact force vector $\mathbf{p}^*$ and returns the normal displacement vector $\mathbf{u}^*$ whose components $p_{i,j}$, $u_{i,j}$ satisfy $p_{i,j} \geq 0$, $u_{i,j} \geq \bar{u}_{i,j}$, $(u_{i,j} - \bar{u}_{i,j})p_{i,j} = 0$, and (4.12), $\forall(i, j) \in I_C$.

The method is easy to warm start in case of a loading scenario consisting of an alternating sequence of increasing or decreasing far-field displacements. The contact forces determined for a given imposed displacement are used to initialize vector $\mathbf{p}$. Due to the monotonicity of the contact solution, this initialization is certainly much closer to the optimal solution $\mathbf{p}^*$ than a zero vector. This usually significantly reduces the iterations of the method to convergence. Such a warm start has a fast implementation requiring a projection of the forces of the points belonging to $I_C^*(\Delta_k)$ to the same points of the trial domain $I_C^*(\Delta_{k+1})$ for a new imposed far-field displacement Δ_{k+1}. For an increasing far-field displacement, i.e., $\Delta_{k+1} > \Delta_k$ the forces in the elements belonging to $I_C^*(\Delta_{k+1}) - I_C^*(\Delta_k)$ are simply initialized equal to zero.

Input: Matrix $\mathbf{H} = \mathbf{H}^{\mathrm{T}} \succ 0$, vector $\bar{\mathbf{u}}$, initial guess $\mathbf{p}$; maximum number $K_{\max}$ of iterations, tolerance $\epsilon > 0$.

1. $\mathcal{I} \leftarrow \{i \in \{1, \ldots, n\} : \mathbf{p}_i > \mathbf{0}\}$; $init \leftarrow$ FALSE; $k \leftarrow 0$;
2. **if** $\mathcal{I} = \emptyset$ **then** $init \leftarrow$ TRUE;
3. $\mathbf{w} \leftarrow \mathbf{Hp} - \bar{\mathbf{u}}$;
4. **if** (($\mathbf{w} \geq -\epsilon$ **or** $\mathcal{I} = \{1, \ldots, n\}$) **and** $init =$ TRUE) **or** $k \geq K_{\max}$ **then go to** Step 13;
5. **if** $init =$ TRUE **then** $i \leftarrow \arg\min_{i \in \{1,\ldots,n\} \setminus \mathcal{I}} \mathbf{w}_i$; $\mathcal{I} \leftarrow \mathcal{I} \cup \{i\}$; **else** $init \leftarrow$ TRUE;
6. $\mathbf{s}_I \leftarrow$ solution of the linear system $\mathbf{H}_I \mathbf{s}_I = \bar{\mathbf{u}}_I$
7. **if** $\mathbf{s}_{\mathcal{I}} \geq -\epsilon$ **then** $\mathbf{p} \leftarrow \mathbf{s}$ and **go to** Step 3;
8. $j \leftarrow \arg\min_{h \in \mathcal{I}:\, \mathbf{s}_h \leq \mathbf{0}} \left\{ \frac{\mathbf{p}_h}{\mathbf{p}_h - \mathbf{s}_h} \right\}$;
9. $\mathbf{p} \leftarrow \mathbf{p} + \frac{\mathbf{p}_j}{\mathbf{p}_j - \mathbf{s}_j} (\mathbf{s} - \mathbf{p})$;
10. $\mathcal{I}_0 \leftarrow \{h \in \mathcal{I} : \mathbf{p}_h = \mathbf{0}\}$;
11. $\mathcal{I} \leftarrow \mathcal{I} \setminus \mathcal{I}_0$; $k \leftarrow k + 1$;
12. **go to** Step 6;
13. $\mathbf{p}^* \leftarrow \mathbf{p}$;
14. $\mathbf{u}^* \leftarrow \mathbf{w} + \bar{\mathbf{u}}$;
15. **end**.

Output: Contact force vector $\mathbf{p}^*$ and normal displacement vector $\mathbf{u}^*$ satisfying $\mathbf{u}^* = \mathbf{Hp}$, $\mathbf{u}^* \geq \bar{\mathbf{u}}$, $\mathbf{p}^* \geq \mathbf{0}$, $(\mathbf{u}^* - \bar{\mathbf{u}})^{\mathrm{T}} \mathbf{p} = 0$.

Algorithm 3: Non-Negative Least Squares (NNLS)

Note that Step 6 of Algorithm 3 is equivalent to Step 2.4 of Algorithm 3 and it has been performed in Algorithm 1 by using the MATLAB's `mldivide` solver. This step can be accelerated by the use of an approach based on the FFT (for its implementation, see, e.g., Batrouni et al. (2002)). Alternatively, since the set $\mathcal{I}_0$ changes incrementally during the iterations of the algorithm, more efficient iterative QR (Lawson and Hanson, 1974, Chap. 24) or $\mathrm{LDL}^{\mathrm{T}}$ Bemporad (2014) factorization methods can be employed.

An alternative method to solve the QP problem (4.15) is to use an accelerated gradient projection (GP) method for QP (Nesterov 1983; Patrinos and Bemporad 2014). Because of the simple nonnegative constraints in (4.15), rather than going to the dual QP formulation as in Patrinos and Bemporad (2014), the GP problem was formulated directly for the primal QP problem (4.15). Numerical experiments have shown slow convergence of a pure accelerated GP method to solve (4.15). However, the method can be used to *warm start* Algorithm 3, as described in Algorithm 4 proposed in Bemporad and Paggi (2015). If Algorithm 4 is executed ($K > 0$), it returns a vector $\mathbf{p}$ that is immediately used as an input to Algorithm 3, otherwise one can simply set $\mathbf{p} = \mathbf{0}$ (cold start). As shown in the algorithms' comparison in the sequel, GP iterations provide large benefits in warm starting the NNLS solver, therefore allowing taking the best advantages of the two methods: quickly getting in the neighborhood of the optimal solution (GP iterations of Algorithm 4) and getting solutions up to machine precision after a finite number of iterations (the active-set NNLS Algorithm 3).

Input: Matrix $\mathbf{H} = \mathbf{H}^{\mathrm{T}} \succ 0$ and its Frobenius norm L, vector $\bar{\mathbf{u}}$, initial guess $\mathbf{p}$, number K of iterations.

---

1. $\bar{\mathbf{p}} \leftarrow \mathbf{p}$;
2. **for** $i = 0, \ldots, K-1$ **do**:

   (2.1) $\beta = \max\{\frac{i-1}{i+2}, 0\}$;
   (2.2) $\mathbf{s} = \mathbf{p} + \beta(\mathbf{p} - \bar{\mathbf{p}})$;
   (2.3) $\mathbf{w} = \mathbf{H}\mathbf{s} - \bar{\mathbf{u}}$;
   (2.4) $\bar{\mathbf{p}} \leftarrow \mathbf{p}$;
   (2.5) $\mathbf{p} \leftarrow \max\{\mathbf{s} - \frac{1}{L}\mathbf{w}, \mathbf{0}\}$;

3. **end**.

Output: Warm start for contact force vector $\mathbf{p}$ and elastic correction vector $\mathbf{w}$.

Algorithm 4: Accelerated Gradient Projection (GP) to be used to warm start the NNLS algorithm

The QP problem (4.15) can also be solved by the alternating direction method of multipliers (ADMM), which belongs to the class of augmented Lagrangian methods. The reader is referred to Boyd et al. (2011) for mathematical details, while its application to contact problems was proposed in Bemporad and Paggi (2015). The method treats the QP (4.15) as the following problem:

$$\begin{array}{c}\min_{\mathbf{p},\mathbf{s}} \frac{1}{2}\mathbf{p}^{\mathrm{T}}\mathbf{H}\mathbf{p} - \bar{\mathbf{u}}^{\mathrm{T}}\mathbf{p} + g(\mathbf{s}) \\ \text{s.t. } \mathbf{p} = \mathbf{s},\end{array} \tag{4.20}$$

where

$$g(\mathbf{s}) = \begin{cases} 0 \text{ if } \mathbf{s} \geq \mathbf{0} \\ +\infty \text{ if } \mathbf{s} < \mathbf{0} \end{cases}.$$

Then, the augmented Lagrangian function

$$L_{\rho}(\mathbf{p}, \mathbf{s}, \mathbf{w}) = \frac{1}{2}\mathbf{p}^{\mathrm{T}}\mathbf{H}\mathbf{p} - \bar{\mathbf{u}}^{\mathrm{T}}\mathbf{p} + g(\mathbf{s}) + \mathbf{w}^{\mathrm{T}}(\mathbf{p} - \mathbf{s}) + \frac{\rho}{2}\|\mathbf{p} - \mathbf{s}\|_2^2$$

is considered, where $\rho > 0$ is a parameter of the algorithm. The basic ADMM algorithm consists of the following iterations:

$$\begin{aligned} \mathbf{p}^{k+1} &= \arg\min_{\mathbf{p}} L_{\rho}(\mathbf{p}, \mathbf{s}^k, \mathbf{w}^k) \\ \mathbf{s}^{k+1} &= \arg\min_{\mathbf{s}} L_{\rho}(\mathbf{p}^{k+1}, \mathbf{s}, \mathbf{w}^k) \\ \mathbf{w}^{k+1} &= \mathbf{w}^k + \rho(\mathbf{p}^{k+1} - \mathbf{s}^{k+1}). \end{aligned} \tag{4.21}$$

A scaled form with over-relaxation of the ADMM iterations (4.21) is summarized in Algorithm 5. The algorithm is guaranteed to converge asymptotically to the solution $\mathbf{p}^*$, $\mathbf{u}^*$ of the problem. The over-relaxation parameter $\alpha > 1$ is introduced to improve convergence. Typical values for α suggested in Boyd et al. (2011) are $\alpha \in [1.5, 1.8]$.

A warm start of the algorithm that takes into account the loading history is possible in a way analogous to that described for the NNLS approach. However, as an additional complexity, also an initialization for the dual variable vector $\mathbf{w}$ must be provided, possibly obtained by projecting the solution obtained for a certain Δ_k to that for Δ_{k+1}.

Input: Matrix $\mathbf{H} = \mathbf{H}^{\mathrm{T}} \succ 0$, vector $\bar{\mathbf{u}}$, initial guesses $\mathbf{p}$, $\mathbf{w}$, parameter $\rho > 0$, over-relaxation parameter $\alpha > 1$, maximum number $K_{\max}$ of iterations, tolerance $\epsilon > 0$.

---

1. $\mathbf{M} \leftarrow (\frac{1}{\rho}\mathbf{H} + \mathbf{I})^{-1}$;
2. $\mathbf{w}_\rho \leftarrow -\frac{1}{\rho}\mathbf{w}$;
3. $\mathbf{s} \leftarrow \mathbf{p}$;
4. $i \leftarrow 0$;
5. **while** ($i \leq K_{\max}$ **and** $\|\mathbf{p} - \mathbf{s}\|_\infty > \epsilon$) **or** $i = 0$ **do**:
   - (5.1) $\mathbf{s} \leftarrow \mathbf{M}(\mathbf{p} - \mathbf{w}_\rho - \frac{1}{\rho}\bar{\mathbf{u}})$;
   - (5.2) $\bar{\mathbf{s}} \leftarrow \alpha\mathbf{s} + (1 - \alpha)\mathbf{p}$;
   - (5.3) $\mathbf{p} \leftarrow \max\{\bar{\mathbf{s}} + \mathbf{w}_\rho, \mathbf{0}\}$;
   - (5.4) $\mathbf{w}_\rho \leftarrow \mathbf{w}_\rho + \bar{\mathbf{s}} - \mathbf{p}$;
   - (5.5) $i \leftarrow i + 1$;
6. $\mathbf{p}^* \leftarrow \mathbf{p}$;
7. $\mathbf{u}^* \leftarrow \bar{\mathbf{u}} - \rho\mathbf{w}_\rho$;
8. **end**.

Output: Contact force vector $\mathbf{p}^*$ and normal displacement vector $\mathbf{u}^*$ satisfying $\mathbf{u}^* = \mathbf{H}\mathbf{p}$, $\mathbf{u}^* \geq \bar{\mathbf{u}}$, $\mathbf{p}^* \geq \mathbf{0}$, $(\mathbf{u}^* - \bar{\mathbf{u}})^{\mathrm{T}}\mathbf{p} = 0$.

Algorithm 5: Alternative Direction Method of Multipliers (ADMM)

Comparison of the algorithms' performance. To assess the computation efficiency and performance in terms of number of iterations required to achieve convergence, the optimization algorithms reviewed in the previous section can be compared in relation to a benchmark frictionless normal contact problem involving a numerically generated fractal rough surface and a half-plane.

To this aim, the random midpoint displacement algorithm (Peitgen and Saupe 1988) can be used to generate the synthetic height field of surfaces with multiscale fractal roughness, i.e., with a power spectral density (PSD) function of the height field of power-law type. The surface with a given resolution (pre-fractal) is realized by a successive refinement of an initial coarse representation by adding a sequence of

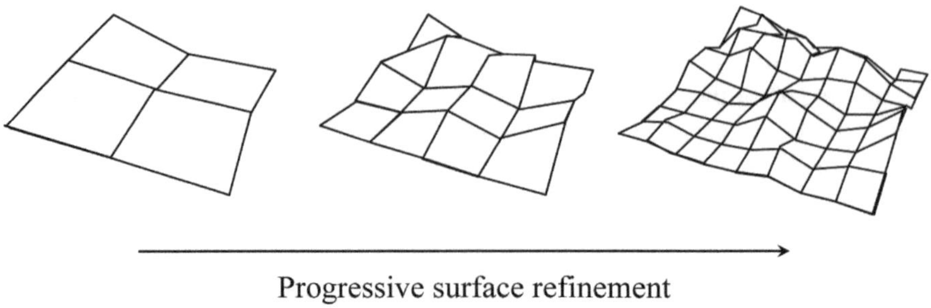

Fig. 4.2 Rough surfaces with multiscale roughness and different resolutions, numerically generated by the random midpoint displacement algorithm. From Bemporad and Paggi (2015)

intermediate heights whose elevation is extracted from a Gaussian distribution with a suitable rescaled variance, see a qualitative sketch in Fig. 4.2. Several applications of the method to model rough surfaces for contact mechanics simulations are available in Zavarise et al. (2004a, b), Paggi and Ciavarella (2010).

In particular, let us consider a test problem consisting of a surface with Hurst exponent $H = 0.7$, lateral size $L = 100\ \mu\text{m}$, and 512 heights per side. The surface is brought into contact with an elastic half-plane under displacement control. Ten displacement steps are imposed to reach a maximum far-field displacement which is set equal to $(\xi_{\max} - \xi_{\text{ave}})/2$, where $\xi_{\max}$ and ξ_{ave} are the maximum and the average elevations of the rough surface, respectively. All the simulations were carried out with the server 653745-421 Proliant DL585R07 from Hewlett Packard with 128 GB Ram, 4 processors AMD Opteron 6282 SE 2.60 GHz with 16 cores running MATLAB R2014b.

The parameters for the Greedy CG method are the maximum number of iterations $K_{\max} = 1 \times 10^5$ and the convergence tolerance $\epsilon = 1 \times 10^{-8}$. The contact forces are initialized at zero (cold start). The constrained CG method also considers $K_{\max} = 1 \times 10^5$ and the same tolerance $\epsilon = 1 \times 10^{-8}$. Both the original version by Polonsky and Keer (1999) (labeled P&K1999 in Fig. 4.3) and its warm-started variant (labeled P&K1999 + warm start in Fig. 4.3) are considered.

For the NNLS algorithm (Algorithm 3), the warm start strategy based on the projection of contact forces from the solution corresponding to a previous displacement step is adopted. Alternatively, warm starting using gradient projections (denoted as NNLS+GP) is examined, using 100 gradient projections to initialize vector $\mathbf{p}$. The parameters for the ADMM method are $\alpha = 1.5$, $\rho = 1$, $K_{\max} = 3 \times 10^3$, and $\epsilon = 10^{-8}$. The total number n of optimization variables is varying with the amount of imposed displacement Δ and therefore with the force level. For the highest indentation level of the present test, $n = 35{,}555$. Warm starting the algorithm is achieved by projecting primal variables as for the NNLS and dual variables $\mathbf{w}$ as well. The

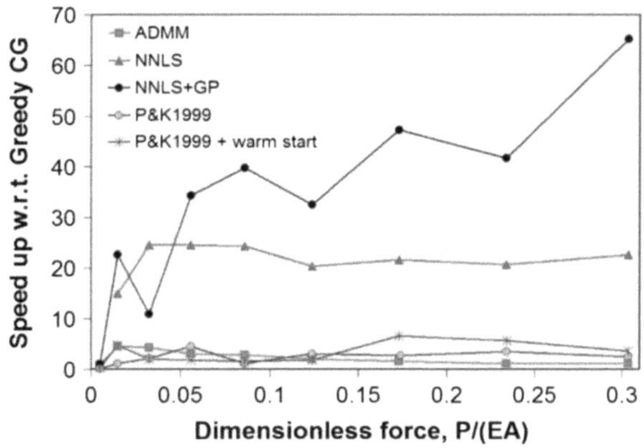

Fig. 4.3 Comparison between the optimization algorithms in terms of speedup of computation time

projection simply consists of assigning the values of $p^*_{i,j}$ and w^* of the boundary elements in contact for the step Δ_k to the same boundary elements belonging to the trial contact domain I_C corresponding to the higher indentation Δ_{k+1}.

Once convergence is achieved for each imposed far-field displacement, the optimization algorithms provide the same normal force P and contact domains, with small roundoff errors due to finite machine precision. The ratio between the CPU time required by each method to achieve convergence and the CPU time employed by the Greedy CG algorithm, which is the slowest, is considered as a measure of speedup. This ratio is plotted vs. the dimensionless normal force $P/(EA)$ in Fig. 4.3, where $A = L^2$ is the nominal contact area. The best performance is achieved by the application of the NNLS method with 100 gradient projections (GP), which is 25 times faster than the original constrained CG method by Polonsky and Keer (1999) and about two orders of magnitude faster than the ADMM and the Greedy CG algorithms, with an increasing efficiency for high loads. The NNLS with warm start is also very well performing, with a stable speedup of about 25 times for any load level.

As outlined in the introduction, the Greedy method can be used in conjunction with other algorithms for solving the unconstrained linear system of equations (Step 2.4) than the CG algorithm. For instance, the CG Step 2.4 in the Greedy algorithm can be replaced with the optimized built-in `mldivide` function of MATLAB, or with the Gauss–Seidel algorithm, as proposed in Borri-Brunetto et al. (1999, 2001).

The MATLAB's `mldivide` solver (which employs the Cholesky factorization) leads to a reduction of computation time of 30–40% with respect to the CG method, almost regardless of the size of the system n, see Fig. 4.4. Even with this gain in computation speed, the overall performance is still quite far from that of the NNLS Algorithm 3 on the platform used for the tests. Moreover, the MATLAB solver leads to an error of lack of memory for $n >$20,000, a serious problem for large systems that are not suffered by the CG solver described in Step 2.4 of Algorithm 1. The Gauss–Seidel algorithm does not suffer for the lack of memory but it is about 3 times slower than the CG method.

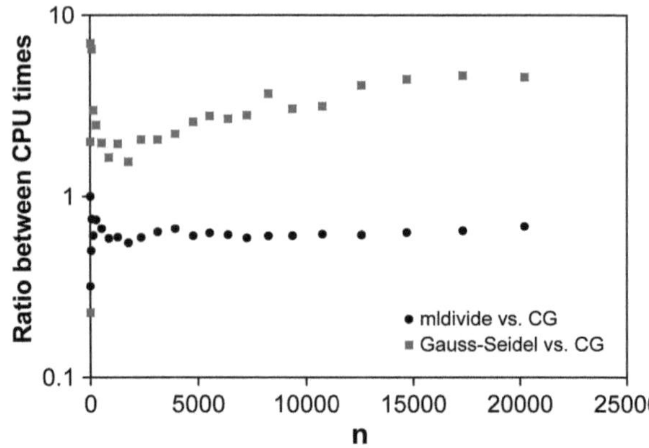

Fig. 4.4 Ratio between computation times for the Greedy method using different solvers (MATLAB's `mldivide` solver or Gauss–Seidel algorithm) as compared to the conjugate gradient (CG) algorithm, for different sizes n of the contact superset I_C. Adapted from Bemporad and Paggi (2015)

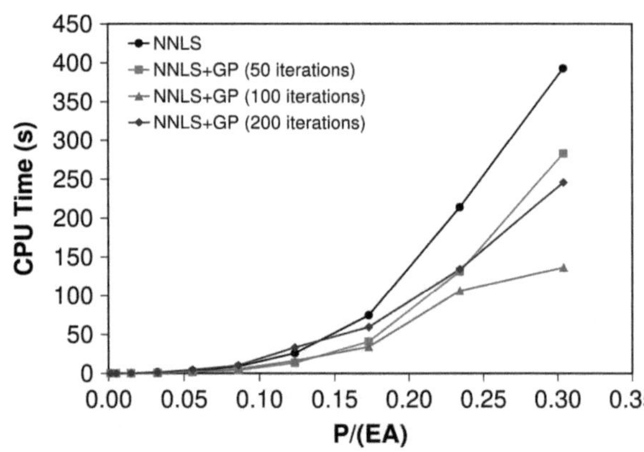

Fig. 4.5 Computation times of the NNLS algorithm depending on the number K of gradient projection (GP) iterations. Adapted from Bemporad and Paggi (2015)

The effect of the number K of GP iterations applied before the NNLS algorithm was also investigated in Bemporad and Paggi (2015). Figure 4.5 shows, for the same test problem whose results are shown in Fig. 4.3, the effect of K on the total computation time. For K from 0 to 100, we observe a reduction in the total computation time due to a decrease in the number of iterations requested by the NNLS algorithm to achieve convergence, thanks to a better initial guess of $\mathbf{p}$. However, a further increase in K (see, e.g., the curve in Fig. 4.5 corresponding to $K = 200$ iterations) does not correspond to further savings of CPU time. This is due to the fact that the number of NNLS iterations was already reduced to its minimum for $K = 100$ GP iterations, so that the application of further gradient projections is just leading to additional CPU time without further benefit.

Further acceleration strategies and variants of the boundary element method. A further speedup of computation time, as compared to the NNLS method, can be achieved by improving the criterion for the guess of the initial set I_C of points in contact. The standard criterion based on checking the interpenetration of the surface

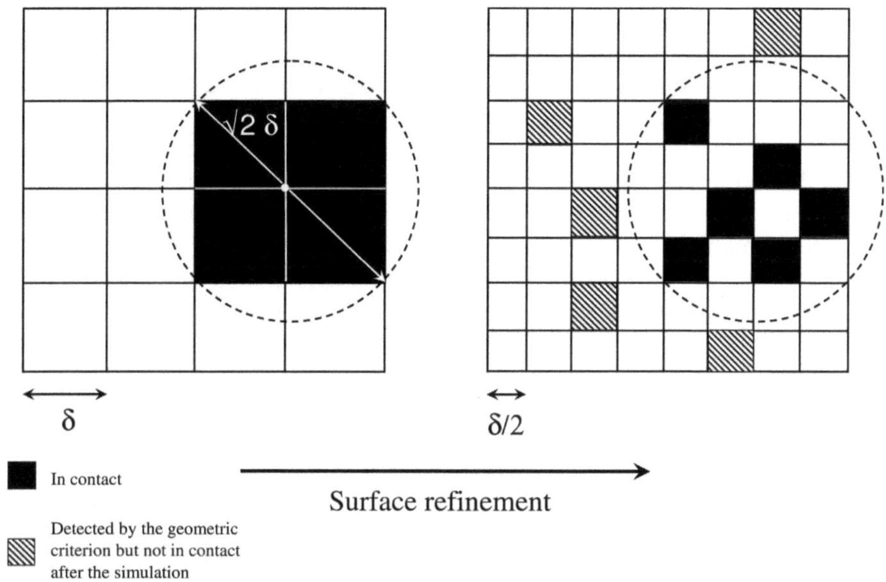

Fig. 4.6 A sketch illustrating the property of lacunarity of the contact domain: the real contact area progressively diminishes by refining the surface, until vanishes in the fractal limit of $\delta \to 0$. This implies that some boundary elements detected by the rigid body interpenetration criterion (dashed gray elements) can be neglected a priori since they are outside the real contact area corresponding to the coarse scale contact solution. From Bemporad and Paggi (2015)

heights into the half-plane in case of a rigid body motion is the most conservative one. However, at convergence, we know that only a small subset I_C^* of that initial set is actually in contact. Therefore, a better choice of the initial trial contact domain would reduce the size of the system of linear equations with an expected benefit in terms of computation time.

As shown in Borri-Brunetto et al. (1999) via numerical simulations on pre-fractal surfaces with Hurst exponent $H > 0.5$ and different resolutions by refining the surface height field via a recursive application of the random midpoint displacement algorithm, the real contact area of each surface representation decreases by reducing the grid spacing δ, as illustrated in the sketch in Fig. 4.6. In the fractal limit of $\delta \to 0$, the real contact area vanishes. Therefore, this property of lacunarity implies that the heights that are not in contact for a coarser surface representation are not expected to come into contact by a successive refining of the height field, for the same imposed far-field displacement.

Therefore, a possible better criterion was proposed by Bemporad and Paggi (2015) and was called cascade multiresolution (CMR) algorithm. In the method, the initial trial contact domain is selected by retaining, among all the heights selected by the rigid body interpenetration check, only those located within the areas of influence of the nodes belonging to the contact domain of a coarser representation of the rough surface for the same imposed displacement Δ. This criterion, inspired by fractal considerations, shares some analogies with multigrid methods, where coarse and fine grids are used to compute the contact solution.

As graphically shown in Fig. 4.6, an area of influence of a given node in contact can be defined by the radius $\sqrt{2}\delta$, where δ is the grid size of the coarser surface representation. Since the criterion is not exact, Bemporad and Paggi (2015) suggested to consider a multiplicative factor h larger than unity for the radius defining the nodal area of influence. It is remarkable to note that this numerical scheme can be applied recursively to a cascade of coarser representations of the same rough surface. As a general trend, computation time is expected to drastically diminish by increasing the number of cascade projections. However, the propagation of errors due to the wrong exclusion of heights that would actually make contact cannot be controlled by the algorithm and it is expected to increase with the number of projections as well. The advantage of the method is represented by the fact that, in addition to saving computation time with respect to that required by the NNLS algorithm to solve just one contact problem for the finest surface, all the contact predictions for the coarser scale representations of the same surface are provided without additional costs, which is a useful result for the multiscale characterization of contact problems. Moreover, the CMR method can be used in conjunction with any of the optimization algorithms reviewed in the previous sections. The algorithm is illustrated in Algorithm 6.

Input: $s = 1, \ldots, l$ surface representations with different resolution or grid spacing $\delta(s)$; area of influence parameter $h \geq 1$.

---

1. **for** $s = 1, \ldots, l$ **do**:

   (1.1) Determine $I_C(s) = \{(i, j) \in I_N(s) : \xi_{i,j} \geq \xi_{\max}(s) - \Delta\}$;
   (1.2) **if** $s = 1$ **then** $I_{C,p}(s) = I_C(s)$
   else
   $I_{C,p}(s) = \{(i, j) \in I_C(s) : r_{i-k,j-l} = \|\mathbf{x}_{i,j} - \mathbf{x}_{k,l}\| \leq h\delta(s-1)\}, \forall (k, l) \in I_C^*(s-1)$
   end

2. Construct **H** based on the projected trial contact domain $I_{C,p}(s)$;
3. Apply optimization algorithms (e.g., NNLS) and determine $\mathbf{p}^*$, $\mathbf{u}^*$, $I_C^*(s)$;
4. **end**.

Algorithm 6: Cascade multi-resolution (CMR) algorithm

To assess the computational performance of the method, the CMD algorithm was applied in conjunction with the NNLS algorithm to pre-fractal surfaces with different H numerically generated by the RMD method (Bemporad and Paggi 2015). As an example, the lateral size was set equal to 100 μm for all the surfaces and the finest resolution whose contact response has to be sought corresponded to 256 heights per side. The method requires the storage of the coarser representations of such surfaces that are in any case available by the RMD algorithm during its various steps of random addition.

The cascade of projections was applied starting with a coarser representation of the surfaces with only 16 heights per side and then considering 32, 64, 128, and finally 256 heights per side. A parameter $h = 2$ was used for the definition of the area of influence. The solution of the contact problem for the surface with 16 heights per side was obtained in an exact form, since it is the starting point of the cascade projections, whereas the contact predictions for the finer surface representations can be affected by an error intrinsic in the criterion. The approximate predictions for the surface with 256 heights per side were compared with the reference solution corresponding to the application of the NNLS algorithm with warm start directly to the finest representation of the rough surface.

The computation time of the CMR+NNLS solution is the sum of the CPU time required to solve all the coarser surface representations and it is found to be much less than the CPU time required by the NNLS algorithm to solve just one single surface with the finest resolution, see Fig. 4.7, where we observe a reduction of about 50% in CPU time almost regardless of H. The relative error in the computation of the maximum normal force between the predicted solution and the reference one is a

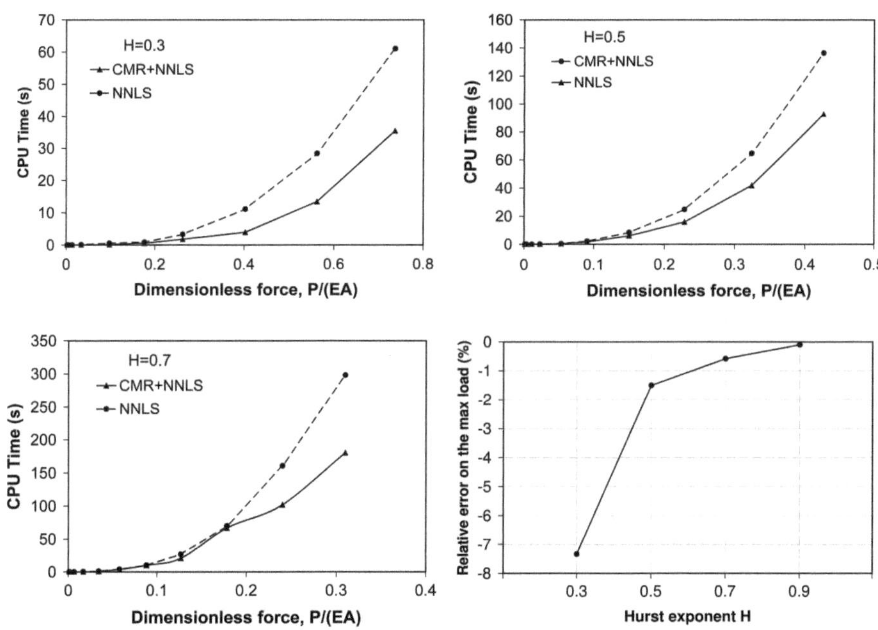

Fig. 4.7 Performance of the CMR+NNLS method applied to numerically generated fractal surfaces with a different Hurst exponent H and $h = 2$. Adapted from Bemporad and Paggi (2015)

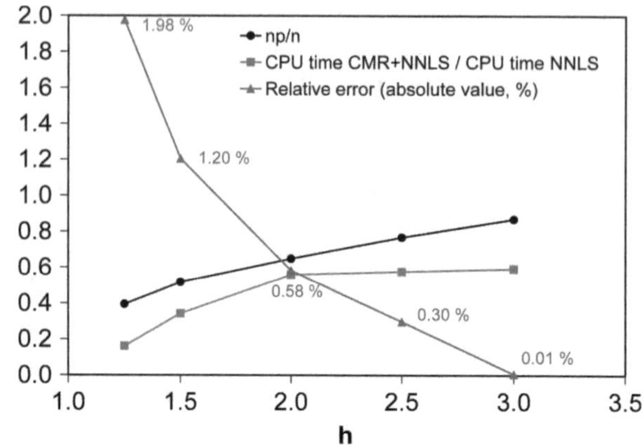

Fig. 4.8 Performance of the CMR+NNLS method with respect to NNLS for a numerically generated fractal surface with $H = 0.7$, depending on the parameter h. Adapted from Bemporad and Paggi (2015)

rapidly decreasing function of H, as shown in Fig. 4.7. Considering that real surfaces have often a Hurst exponent $H > 0.5$, the method is very promising.

A synthetic diagram illustrating the effect of the parameter h for the surface with $H = 0.7$ and for a single imposed displacement corresponding to the maximum load is shown in Fig. 4.8. The relative error is rapidly decreasing to values less than 1% by increasing h. The ratio between the number of points expected to be in contact after the application of the CMR projection criterion, n_p, and the number of points that would be included by using the classic rigid body interpenetration check, n, is ranging from 0.4 to 0.8 by increasing h from 1.25 to 3.0. The ratio between CPU times, on the other hand, tends to an asymptotic value of 0.6, which implies a saving of 40% of computation time as compared to the exact solution, with less than 0.01% of relative error.

Among the variants of the boundary element method published in the literature, some concerned with the treatment of the key features of roughness and its evolution during contact. Starting from the fundamental assumption of micromechanical contact theories that only the asperities, i.e., the local maxima of the rough surface, make contact (Greenwood and Williamson 1966; Zavarise et al. 2004a; Paggi and Ciavarella 2010), then one could simplify the boundary element approach by treating only the set of asperities as boundary elements. Each asperity can be modeled as a paraboloid whose geometry is defined by its position (x, y) in the plane, its elevation z above a reference plane, and its mean radius of curvature R, as proposed in Greenwood (2006). For such a Hertzian asperity, the theory of elasticity provides the displacement of the half-plane in the location of the asperity itself, as well as the displacement in any other position, see Nowell and Hills (1989). Such an information can be used to define the Green functions for the application of a generalized boundary element method, where a recursive elimination of asperities supporting tensile forces can be implemented till the final active set of asperities in contact and the corresponding normal contact forces are identified, using, for instance, the Greedy algorithm. Alternatively, the problem can be formulated by an iterative correction to the asperity deformation as proposed in Ciavarella et al. (2006), avoiding the inversion of the compliance matrix. Moreover, the scheme allows also studying the effect of short- or long-range elastic interaction effects, as carefully investigated in Paggi and Barber (2011).

The above multi-asperity contact problem overcomes the limitations of the original semi-analytical contact theory by Greenwood and Williamson (1966) which did not include elastic interactions in its original form. Moreover, the correction scheme proposed in Ciavarella et al. (2006) inspired also a way to improve the Greenwood and Williamson contact theory by introducing a yet simple but effective mean pressure elastic interaction effect (Ciavarella et al. 2008b). Moreover, the approach presents several advantages over the standard boundary element method, especially in terms of speedup of computation time which is mainly achieved by the fact that the size of the set of potential asperities in contact is significantly much smaller than the total number of boundary elements. On the other hand, drawbacks regard the fact that the method strongly relies on the geometrical parameters of the asperities, which are resolution-dependent as discussed in Majumdar and Bhushan (1990), Zavarise et al. (2004b). Another limitation as compared to a more general boundary element method regards the fact that the contact problem relies on the undeformed asperity geometry. This limitation has been partially overcome by Afferrante et al. (2012), who proposed an update of the asperity detection and their geometrical parameters during contact, to model the phenomenon of merging of asperities by forming bigger ones with completely different geometrical features as compared to what estimated from the undeformed configuration.

The Contact Problem with Friction

In the most general three-dimensional contact problem with friction, the surface displacement vector projected onto an orthogonal Cartesian frame $Oxyz$ has three components, u_x, u_y, and u_z. The component u_z is the component perpendicular to the mean plane of the nominally flat rough surface and it corresponds to the variable u for the frictionless normal contact problem detailed in the previous section. The other components u_x and u_y correspond, on the other hand, to the in-plane surface displacements. The latter are intimately connected to the corresponding surface tractions q_x, q_y, and p through the Green functions.

For instance, a uniform distributed normal traction $p(\mathbf{y})$ acting over a square surface element S of lateral size δ identified by the position vector $\mathbf{y} = (x', y')^{\mathrm{T}}$ leads to the following surface displacements at another point $\mathbf{x} = (x, y)^{\mathrm{T}}$ at a distance $r = \|\mathbf{x} - \mathbf{y}\|$:

$$u_x = -\frac{1-2\nu}{4\pi G}\int_S \frac{x-x'}{r^2}p(\mathbf{y})\mathrm{d}\mathbf{y}, \tag{4.22a}$$

$$u_y = -\frac{1-2\nu}{4\pi G}\int_S \frac{y-y'}{r^2}p(\mathbf{y})\mathrm{d}\mathbf{y}, \tag{4.22b}$$

$$u_z = \frac{1-\nu}{2\pi G}\int_S \frac{p(\mathbf{y})}{r}\mathrm{d}\mathbf{y}. \tag{4.22c}$$

Similarly, for a uniform distributed tangential traction $q_x(\mathbf{y})$,

$$u_x = \frac{1}{2\pi G}\int_S \left[\frac{1-\nu}{r} + \nu\frac{(x-x')^2}{r^3}\right] q_x(\mathbf{y})\mathrm{d}\mathbf{y}, \tag{4.23a}$$

$$u_y = \frac{1}{2\pi G}\int_S \nu\frac{(x-x')(y-y')}{r^3} q_x(\mathbf{y})\mathrm{d}\mathbf{y}, \tag{4.23b}$$

$$u_z = \frac{1-2\nu}{4\pi G}\int_S \frac{x-x'}{r^2} q_x(\mathbf{y})\mathrm{d}\mathbf{y}, \tag{4.23c}$$

and for a uniform distributed tangential traction $q_y(\mathbf{y})$:

$$u_x = \frac{1}{2\pi G}\int_S \nu\frac{(x-x')(y-y')}{r^3} q_y(\mathbf{y})\mathrm{d}\mathbf{y}, \tag{4.24a}$$

$$u_y = \frac{1}{2\pi G}\int_S \left[\frac{1-\nu}{r} + \nu\frac{(y-y')^2}{r^3}\right] q_y(\mathbf{y})\mathrm{d}\mathbf{y}, \tag{4.24b}$$

$$u_z = \frac{1-2\nu}{4\pi G}\int_S \frac{y-y'}{r^2} q_y(\mathbf{y})\mathrm{d}\mathbf{y}. \tag{4.24c}$$

Therefore, we recognize that the normal contact problem is in general fully coupled with the tangential one, in the sense that a normal pressure induces not only normal displacements but also not-vanishing in-plane deformation. After introducing a boundary element discretization of the nominally rough surface into $N \times N$ elements as for the frictionless normal contact problem, surface displacements in a point defined by the indices i, j are related to uniform surface tractions acting on a square element defined by the indices k, l via a matrix collecting the Green functions, see Love (1999), Pohrt and Li (2014):

$$\begin{Bmatrix} u_x \\ u_y \\ u_z \end{Bmatrix}_{i,j} = \sum_{k=1}^{N}\sum_{l=1}^{N} \begin{bmatrix} H_{xx} & H_{xy} & H_{xz} \\ H_{yx} & H_{yy} & H_{yz} \\ H_{zx} & H_{zy} & H_{zz} \end{bmatrix}_{ijkl} \begin{Bmatrix} q_x \\ q_y \\ p \end{Bmatrix}_{k,l}, \tag{4.25}$$

where $H_{zx} = -H_{xz}$, $H_{zy} = -H_{yz}$, $H_{yx} = H_{xy}$.

The special case $\nu = 1/2$ leads to $H_{xz} = H_{yz} = 0$ and therefore the normal contact problem becomes uncoupled from the tangential one. Another notable case corresponds to $\nu = 0$, where the coefficient H_{xy} vanishes and $H_{xx} = H_{yy} = 2H_{zz}$, leading to coupling between the normal and the tangential contact problems, but uncoupling between the two in-plane directions.

The computation of the surface displacements requires a convolution of the traction effects according to the application of Eq. (4.25), which has a complexity of the order of $O(N^4)$ operations. Although the number of boundary elements $\#I_C$ included in the superset of the possible candidates in contact, I_C, is usually smaller than N^2, a speedup is certainly required. To this aim, a multilevel multi-integration procedure has been proposed in Lubrecht and Ioannides (1991) reducing the complexity to $O(N^2 \log N)$ by carrying out a summation over a coarser grid and then introducing a correction in the vicinity of the point i, j. The same complexity can be reached by doing the convolution in the Fourier space using a Fast Fourier Transform technique, as proposed in Vollebregt (2014), Pohrt and Li (2014).

Suppose now to apply a monotonically increasing displacement in the normal direction, till a given maximum value which is then held constant. Afterward, a monotonically increasing displacement is applied in a tangential (in-plane) direction. While the normal contact problem is ruled by the unilateral contact condition, in the tangential direction it is customary to postulate the existence of two regimes locally valid for any point in contact: *stick*, when the points of the two surfaces are intimately adhering to each other, and *slip*, when the points experience a relative displacement in the tangential direction. The distinction between such two states is ruled by the Coulomb law of friction, which affirms that any point with $q \leq \mu p$ is in the stick condition, being μ the local coefficient of static friction. Therefore, the solution of the tangential contact problem requires finding the boundary elements belonging to the stick or to the slip state, for a given fixed normal displacement and an imposed tangential one.

Due to coupling, tangential tractions determined from the solution of the tangential contact problem lead to an additional contribution to the normal displacements, which would demand the recursive solution of the normal contact problem to identify the corresponding updated normal contact tractions. Since coupling is in general weak, this feedback effect is often neglected also in the case of $\nu \neq 1/2$, simply setting $H_{xz} = Hyz \cong 0$, see, e.g., the implementation in Pohrt and Li (2014). Physically, this approximation implies that the real contact area does not change due to tangential tractions.

Under these assumptions, after solving the frictionless normal contact problem for a given imposed normal displacement Δ, the set I_C^* of boundary elements in contact ($\#I_C^* = N_C^*$) is known. The subsequent application of a tangential displacement Δ_{T} leads to a partition of the contact set into two parts: a set where stick is observed, $I_{C,\mathrm{st}}^*$, and a set where slip is expected, $I_{C,\mathrm{sl}}^*$. Hence, in matrix form, the relation between surface displacements and tangential tractions can be formally partitioned as follows:

$$\begin{Bmatrix} \mathbf{u}_{\rm st} \\ \mathbf{u}_{\rm sl} \end{Bmatrix} = \begin{bmatrix} \mathbf{A}_{\rm st,st} & \mathbf{A}_{\rm st,sl} \\ \mathbf{A}_{\rm sl,st} & \mathbf{A}_{\rm sl,sl} \end{bmatrix} \begin{Bmatrix} \mathbf{q}_{\rm st} \\ \mathbf{q}_{\rm sl} \end{Bmatrix}, \tag{4.26}$$

where $\mathbf{u}_{\rm st}$ collects the value of the imposed far-field tangential displacement $\Delta_{\rm T}$ for all the boundary elements in stick condition. On the other hand, by definition, the vector $\mathbf{q}_{\rm sl}$ collects entries which are given by the product between the friction coefficient μ and the local pressure p acting on the boundary element in slip condition, and therefore it is known.

The solution of the problem requires an iterative algorithm to identify $I^*_{C,\rm st}$, $I^*_{C,\rm sl}$, and all the tractions and the surface displacements. To do so, the set $I^*_{C,\rm st}$ is usually initialized equal to I^*_C, i.e., all the boundary elements in contact are supposed to be in stick condition, while $I^*_{C,\rm sl}$ is empty. Then, the tangential tractions of the boundary elements in stick condition associated to the surface deflections can be computed using the first row of Eq. (4.26):

$$\mathbf{A}_{\rm st,st}\mathbf{q}_{\rm st} = \mathbf{u}_{\rm st} - \mathbf{A}_{\rm st,sl}\mathbf{q}_{\rm sl}, \tag{4.27}$$

which requires the inversion of the matrix $\mathbf{A}_{\rm st,st}$ using an inverse fast convolution (a conjugate gradient algorithm, for instance). In the first iteration, $\mathbf{q}_{\rm sl} = \mathbf{0}$ and $\mathbf{u}_{\rm st}$ is a vector of entries all equal to $\Delta_{\rm T}$.

If all the boundary elements have $q_{\rm st} < \mu p$, then the initial tentative approximation was correct. However, in general, there will be some elements with a tangential traction overcoming the limit value for the stick condition, and therefore they slip and have to be moved to the $I^*_{C,\rm sl}$ set. Correspondingly, the tangential tractions of those boundary elements have to be limited to the maximum admissible value μp. The tangential deflections $\mathbf{u}_{\rm st}$ and $\mathbf{u}_{\rm sl}$ can now be recomputed from the updated distribution of the stick and the slip tractions using again Eq. (4.26). At this point it is still necessary to check if any boundary element in the tentative slip domain presents $u_{\rm sl} \geq \Delta_{\rm T}$. If this is the case, then such boundary elements should be sticking and therefore they have to be transferred back to the stick domain and $u_{\rm sl}$ has to be limited by $\Delta_{\rm T}$. Afterward, for the updated partition of stick and slip domains, another iteration is performed, which requires the computation of $\mathbf{q}_{\rm st}$ from Eq. (4.27) and the surface displacements from Eq. (4.26). The iterative procedure stops when the entries in $I^*_{C,\rm st}$ and $I^*_{C,\rm sl}$ do not change anymore, all the boundary elements belonging to $I^*_{C,\rm st}$ are subject to tangential tractions less than μp, and all the boundary elements belonging to $I_{C,\rm sl}$ have $u_{\rm sl} < \Delta_{\rm T}$.

Applications of this algorithm have been made in Paggi et al. (2014) for the identification of the evolution of the stick and slip contact domains from full stick to full slip for fractal rough surfaces subject to a given normal force and an increasing shearing displacement. Results pinpointed that the last boundary elements entering the contact domain are the first slipping, due to the low normal pressures acting on them. Therefore, the contact domain in stick condition is initially coincident with the normal contact domain and it progressively shrinks to zero, when all the boundary elements slip. From the macroscopical point of view, the shearing force level corresponding to full slip is simply equal to the local friction coefficient multiplied by the applied total normal force, consistently with the Coulomb criterion adopted at the microscopical level.

The Finite Element Method

Variational Formulation

In this section, the variational formulation governing the problem of contact between two bodies across a rough interface is detailed. Since the mathematical formulation leading to the finite element method can easily handle in a consistent manner also adhesive (tensile) interactions at the interface, the most general scenario is herein examined. Therefore, starting from the strong differential form describing the mechanics of the continua and the problem of contact with adhesion along the interface, the corresponding weak form is derived. Afterward, different solution strategies and finite element discretization schemes are discussed, comparing methods based on the explicit discretization of roughness versus a recent method proposed in Paggi and Reinoso (2018) to analytically embed roughness into a special interface finite element.

Governing equations and strong form. Let two deformable bodies occupy the domains $\Omega_i \in \mathbb{R}^2$ $(i = 1, 2)$ in the undeformed configuration defined by the reference system $Oxyz$. The two domains are separated by an interface Γ defined by the opposite boundaries Γ_i $(i = 1, 2)$ of the two bodies, viz., $\Gamma = \bigcup_{i=1,2} \Gamma_i$, where contact or adhesive interactions take place. The whole boundary of the ith body, $\partial\Omega_i$, is therefore split into three parts: (i) a portion where displacements are imposed, i.e., the Dirichlet boundary $\partial\Omega_i^D$; (ii) a portion where tractions are specified, i.e., the Neumann boundary $\partial\Omega_i^N$; (iii) and the interface $\Gamma_i = \Gamma_i^C \bigcup \Gamma_i^A$ where specific boundary conditions have to be imposed to model contact on Γ_i^C or adhesion on Γ_i^A, see Fig. 4.9. The partition of Γ_i in Γ_i^C and Γ_i^A is not known a priori, but it is the result of the solution of the elastic problem.

In the most general case, we postulate the existence of a displacement field for each body, $\mathbf{u}_i = (u_i, v_i, w_i)^{\mathrm{T}}$, that maps the transformation from the underformed configuration to the deformed one, and vice versa. Such functions are thereby assumed to be continuous, invertible, and differentiable functions of the position vector $\mathbf{x} = (x, y, z)^{\mathrm{T}}$ within each body. At the interface Γ^*, on the other hand, the configuration of the system is described by the relative displacement field $\Delta\mathbf{u}$, usually called *gap field* $\mathbf{g}$ across the interface, which is mathematically defined as the projection of the relative displacement $\mathbf{u}_1 - \mathbf{u}_2$ onto the normal and tangential directions of the interface defined by the unit vectors $\mathbf{n}$, $\mathbf{t}_1$ and $\mathbf{t}_2$, respectively. In components, the vector $\Delta\mathbf{u}$ collects the relative tangential displacements, $\Delta u_{t,1}$, $\Delta u_{t,2}$, and the relative normal displacement, Δu_n, i.e., $\Delta\mathbf{u} = (\Delta u_{t1}, \Delta u_{t2}, \Delta u_n)^{\mathrm{T}}$. The total relative

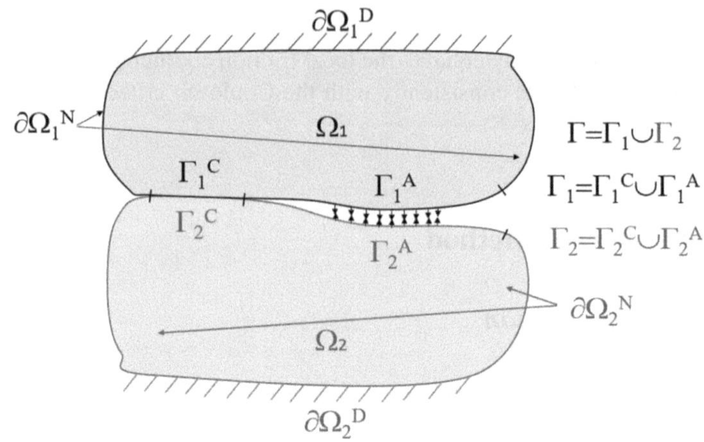

Fig. 4.9 Domains Ω_i $(i = 1, 2)$, their Dirichlet $(\partial\Omega_i^D)$ and Neumann $(\partial\Omega_i^N)$ boundaries, and the interface $\Gamma = \Gamma_1 \bigcup \Gamma_2$ composed of an adhesive part, Γ_i^A, and a contact part, Γ_i^C. Adapted from Paggi and Reinoso (2018)

displacement in the tangential plane is given by $\Delta\mathbf{u}_t = (\Delta u_{t1}, \Delta u_{t2})^{\mathrm{T}}$. In the sequel, we denote $g_{\mathrm{n}} = \Delta u_n$ and $\mathbf{g}_t = \Delta\mathbf{u}_t$, with its modulus $g_t = \Delta u_t = \sqrt{\Delta u_{t1}^2 + \Delta u_{t3}^2}$.

Inside each deformable material, the small deformation strain tensor $\boldsymbol{\epsilon}_i$ $(i = 1, 2)$ is introduced as customary, which is defined as the symmetric part of the displacement gradient: $\boldsymbol{\epsilon}_i = \nabla^s \mathbf{u}_i$. In the sequel, the standard Voigt notation will be used and the strain tensor components will be collected in the vector $\boldsymbol{\epsilon}_i = (\epsilon_{xx}, \epsilon_{yy}, \epsilon_{zz}, \gamma_{xy}, \gamma_{xz}, \gamma_{yz})_i^{\mathrm{T}}$.

In the absence of body forces, the strong (differential) form of equilibrium for each body is provided by the linear momentum equation along with the Dirichlet and the Neumann boundary conditions on $\partial\Omega_i^D$ and $\partial\Omega_i^N$, respectively $(i = 1, 2)$, equipped by unilateral contact conditions in the normal direction on Γ_C^*, Coulomb frictional conditions on the stick and slip partitions $\Gamma_{C,\mathrm{st}}^*$ and $\Gamma_{C,\mathrm{sl}}^*$ of Γ_C^*, and adhesion on Γ_A^*:

$$\nabla \cdot \boldsymbol{\sigma}_i = \mathbf{0} \quad \text{in } \Omega_i, \tag{4.28a}$$

$$\mathbf{u}_i = \bar{\mathbf{u}} \quad \text{on } \partial\Omega_i^D, \tag{4.28b}$$

$$\boldsymbol{\sigma}_i \cdot \mathbf{n} = \mathbf{T} \quad \text{on } \partial\Omega_i^N, \tag{4.28c}$$

$$g_{\mathrm{n}} = 0, \; p_{\mathrm{n}} < 0 \quad \text{on } \Gamma_C^*, \tag{4.28d}$$

$$\mathbf{g}_{\mathrm{t}} = \mathbf{0}, \; \|\mathbf{q}\| < \mu|p_{\mathrm{n}}| \quad \text{on } \Gamma_{C,\mathrm{st}}^*, \tag{4.28e}$$

$$\mathbf{q} = -\mu|p_{\mathrm{n}}|\frac{\dot{\mathbf{g}}_{\mathrm{t}}}{\|\dot{\mathbf{g}}_{\mathrm{t}}\|} \quad \text{on } \Gamma_{C,\mathrm{sl}}^*, \tag{4.28f}$$

$$g_{\mathrm{n}} > 0, \; p_{\mathrm{n}} = p_{\mathrm{A}} > 0 \quad \text{on } \Gamma_{\mathrm{A}}^*, \tag{4.28g}$$

where $\overline{\mathbf{u}}$ denotes the imposed displacement, $\mathbf{T}$ the applied traction vector, $p_A(g_n)$ is a function of the relative displacement $\Delta\mathbf{u}$, and $\mathbf{q}$ is the shearing traction vector. Therefore, the nonlinearity of the problem stems from the fact that the contact and adhesive portions of the interface Γ^* are known only once the displacement field, solution of the problem, is known. As a consequence, the present problem can be ascribed to the family of the so-called moving boundary value problems and it requires an iterative solution scheme.

For its solution, the strong form has to be equipped by the constitutive equations for the bulk and for the interface. For the bulk, recalling standard thermodynamic arguments, general (linear or nonlinear) constitutive stress–strain relations can be postulated without any loss of generality for the ith material domain: $\boldsymbol{\sigma}_i := \partial_{\boldsymbol{\epsilon}_i}\Psi(\boldsymbol{\epsilon}_i)$ and $\mathbb{C}_i := \partial^2_{\boldsymbol{\epsilon}_i\boldsymbol{\epsilon}_i}\Psi(\boldsymbol{\epsilon}_i)$, whereby $\Psi(\boldsymbol{\epsilon}_i)$ is the Helmholtz free-energy function for the body i, whereas its corresponding Cauchy stress tensor and the constitutive operator are, respectively, denoted by $\boldsymbol{\sigma}_i$ and $\mathbb{C}_i$. The two bodies are in general both deformable, but in the present setting it is also possible to consider one of them as rigid. This condition is of paramount interest for contact mechanics in the presence of two dissimilar linear elastic bodies. In such a case, it is possible to simplify the matter by replacing the bi-material system by a rigid body indenting a linear elastic material having composite elastic parameters, function of the Young's moduli E_i, and Poisson's ratios ν_i $(i = 1, 2)$ of the two elastic materials, as previously detailed for the boundary element method, see also Barber (2010, 2018).

Regarding the interface, the constitutive response should be introduced by distinguishing between the normal and the tangential directions. In the normal direction, the contact condition imposes that the displacement field solution leads to a vanishing normal gap $g_\mathrm{n} = 0$ for the points in contact. Correspondingly, contact tractions are negative valued in the finite element method, while it is remarkable to note that the opposite convention was adopted in the boundary element method. For $g_\mathrm{n} > 0$, positive-valued adhesive tractions apply, and they can be, for instance, given by a relation dictated by an adhesion model inspired by the interatomic Lennard-Jones potential:

$$p_A = 24\varepsilon\left[\frac{\kappa^6}{(g_\mathrm{n}+g_{\mathrm{n},0})^7} - 2\frac{\kappa^{12}}{(g_\mathrm{n}+g_{\mathrm{n},0})^{13}}\right], \tag{4.29}$$

where ε and κ are the model parameters and $g_{\mathrm{n},0}$ is the molecular equilibrium distance. The parameter $g_{\mathrm{n},0}$ is such that the condition $g_\mathrm{n} = 0$ leads to vanishing adhesive tractions and it correctly captures the transition from adhesion to contact. Other nonlinear adhesive models can be used in the present approach without any loss of generality, e.g., the surface potential derived from the interatomic Lennard-Jones potential, see Yu and Polycarpou (2004). To treat both contact and adhesive tractions in a unified framework, relaxing at the same time the unilateral contact constraint, a generalized penalty approach can be efficiently exploited (Paggi and Reinoso 2018). Physically speaking, a nonlinear spring model is inserted along the interface between the two bodies, where p_n is given by

$$p_{\rm n}(g_{\rm n}) = \begin{cases} K g_{\rm n}, & \text{if } g_{\rm n} < 0, \\ p_{\rm A}, & \text{if } g_{\rm n} \geq 0. \end{cases} \tag{4.30}$$

This formulation leads to a solution allowing for small compenetration, depending on the value of the penalty stiffness K. Hence, K should be high enough to reduce material penetration between adjacent continua and, at the same time, it should not be too high to cause ill-conditioning of the tangent operator resulting from the computational scheme. Following the pioneering work in Zavarise et al. (2019, 1992), the penalty stiffness K could be related to the normal contact stiffness predicted by semi-analytical micromechanical contact models, giving a physical ground for its estimation.

In the tangential direction, $\mathbf{q}$ is the tangential traction vector which obeys the Coulomb friction law. To simplify its treatment into a computational scheme, a regularized dependency of $\mathbf{q}$ upon $\dot{\mathbf{g}}_t$ is usually put forward, smoothing the sharp transition from the stick to the slip condition, see, e.g., Wriggers (2006, Sect. 5.2.3).

It is remarkable to note here that the treatment of the contact problem with friction in the finite element method is more general than the analogous treatment in the boundary element method, since it allows simulating any loading path in three dimensions. According to Eq. (4.28), the tangential traction vector $\mathbf{q}$ changes sign depending on the velocity of sliding, $\dot{\mathbf{g}}_{\rm t}$, which has to be computed using a time integration routine, usually based on the implicit Euler scheme. Moreover, the stick and slip portions of $\mathbf{q}$ can be computed using a return mapping algorithm, in analogy with elastoplasticity. For small tangential displacements leading to a situation intermediate from full stick to full slip, before the onset of gross sliding, and for simple monotonic loading paths in one given tangential direction, the formulation can be simplified and rewritten in terms of the total relative displacement $\mathbf{g}_{\rm t}$, instead of its time derivative. This leads to a penalty-like formulation similar to that used for Mode II cohesive zone models for fracture, see Paggi et al. (2006), Carpinteri et al. (2008), where tangential tractions are opposing to the relative sliding deformation and are specified as a closed-form equation in terms of $\mathbf{g}_{\rm t}$.

Weak form. According to the principle of virtual work, the weak form associated to the strong form Eq. (4.28) with the penalty regularization in the normal and tangential directions reads

$$\begin{aligned} \Pi = & \int_{\Omega_1} \boldsymbol{\sigma}_1(\mathbf{u}_1)^{\rm T} \boldsymbol{\epsilon}_1(\mathbf{v}_1) {\rm d}\Omega + \int_{\Omega_2} \sigma_2(\mathbf{u}_2)^{\rm T} \boldsymbol{\epsilon}_2(\mathbf{v}_2) {\rm d}\Omega \\ & - \int_{\partial\Omega_1^N} \mathbf{T}^{\rm T}\mathbf{v}_1 {\rm d}\partial\Omega - \int_{\partial\Omega_2^N} \mathbf{T}^{\rm T}\mathbf{v}_2 {\rm d}\partial\Omega \\ & - \int_{\Gamma_C^*} p_{\rm n}(\Delta\mathbf{u}) g_{\rm n}(\Delta\mathbf{v}) {\rm d}\Gamma - \int_{\Gamma_C^*} \mathbf{q}(\Delta\mathbf{u}) \mathbf{g}_{\rm t}(\Delta\mathbf{v}) {\rm d}\Gamma \\ & - \int_{\Gamma_A^*} p_A(\Delta\mathbf{u}) g_{\rm n}(\Delta\mathbf{v}) {\rm d}\Gamma = 0, \end{aligned} \tag{4.31}$$

where $\mathbf{v}_i$ is the test function (virtual displacement field), and $g_n(\Delta\mathbf{v})$ and $\mathbf{g}_t(\Delta\mathbf{v})$ are the virtual normal and tangential relative displacements at the interface Γ^*. The test function in the ith body fulfills the condition $\mathbf{v}_i = \mathbf{0}$ on $\partial\Omega_i^D$ and the adhesive-contact condition on Γ^*. The displacement field $\mathbf{u}_i$ solution of the weak form (4.31) is such that it corresponds to the minimum of Π for any choice of the test functions $\mathbf{v}_i$.

The numerical treatment of the weak form (4.31) within the finite element method requires the introduction of two different types of finite element discretization, one for the bulk, $\Omega_{i,h}$, and another for the interface, Γ_h^*, where the subscript h refers to the respective discretized geometrical feature. For the bulk, standard linear quadrilateral or triangular isoparametric finite elements can be invoked, see classical finite element textbooks (Zienkiewicz and Taylor 2000) for details. For the interface, different strategies can be exploited and they are discussed in the next sections.

Methods Based on the Explicit Discretization of Roughness

Methods based on the explicit discretization of roughness introduce special discretization schemes to model the rough interface topology.

The simplest method was proposed by Hyun et al. (2004) and Pei et al. (2005), who investigated the frictionless normal contact problem between a rough surface and a flat half-plane for elastic (Hyun et al. 2004) or elastoplastic (Pei et al. 2005) continua. In their approach, a three-dimensional mesh for a rough surface was constructed in two stages. First, a flat surface with nodes at each point on the square grid was considered. A local refinement technique was used to achieve a strong mesh gradation with very small elements near the surface and a coarser discretization in the bulk, to reduce the number of finite elements and, therefore, indirectly, the computation cost associated to the solution of the algebraic equations associated to the finite element method. In the second step, all nodes belonging to the surface were displaced to create the desired roughness.

Since only a small fraction of the nodes of the rough surface are in contact after the application of the load, it was convenient to assemble only nonvanishing contact contributions to the weak form. To do that, the conventional master/slave node-to-surface contact search was employed in Hyun et al. (2004), Pei et al. (2005). The contact search computes the value of the normal gap for each node of the rough surface with respect to the opposing master flat surface, considering the projection of the node position in the direction normal to the flat surface, see Wriggers (2006, Sect. 10.1). Depending on the sign of the gap function, nodes experiencing a compenetration are retained in the so-called *active set*. After inserting the finite element discretization, an explicit Newmark time-stepping algorithm was adopted in Hyun et al. (2004), Pei et al. (2005) to further reduce the computation cost. The advantage is that the explicit method leads to a set of uncoupled algebraic equations whose solution can be parallelized. The disadvantage is represented by the fact that an artificial damping has to be introduced to simulate quasi-static contact problems and the algorithm is not unconditionally stable, thus requiring small time steps.

More sophisticate discretization methods available in the literature exploit smooth interpolation schemes based on splines with randomly chosen heights to generate the asperities of the surface (Wriggers and Reinelt 2009), in conjunction with contact search algorithms and implicit Newton–Raphson incremental-iterative solution schemes for the solution of the nonlinear algebraic equations. Such an approach is unconditionally stable, thanks to the computation of the tangent stiffness matrix. However, sparse global stiffness matrices are obtained, which do not allow a straightforward parallelization of the linearized set of algebraic equations as for the explicit approach. Moreover, the introduction of smooth interpolation schemes can be beneficial for modeling wavy surfaces or spheres, as in the case of NURBS used in De Lorenzis and Wriggers (2013) to depict sinusoidal wavy profiles, but it is cumbersome for the description of multiscale roughness features over multiple wavelengths. Due to the complexity of the implicit solution scheme and the explicit discretization of the rough geometry, applications have been confined to the solution of small-scale problems on representative surface elements. This is, for instance, the case of two-scale finite element simulations as proposed in Wriggers and Reinelt (2009), where a reduced fine-scale model of a wavy surface was used for the computation of a microscopically constitutive law to be passed to standard smooth macroscale finite element computations.

Interface Finite Elements with Embedded Profile for Joint Roughness

As an alternative approach to the explicit discretization of roughness, which is computationally demanding and poses severe limitations for the use of the finite element method, a new interface finite element with analytically embedded roughness has been proposed in Paggi and Reinoso (2018).

The method assumes that the boundaries Γ_i in Fig. 4.9 are *nominally flat but microscopically embedding rough profiles*. Γ_1 and Γ_2 can be one the negative of the other, as in the case of an interface originated by fracture, or different from each other, as for two bodies coming into contact, without any restriction.

It is convenient to introduce for the ith rough profile Γ_i its *smoother line* $\overline{h}_i(\xi_i)$ parallel to the average line of the profile and with datum set in correspondence of its deepest valley (see Fig. 4.10). A point along the curve $\overline{h}_i(\xi_i)$ is identified by a value of the curvilinear coordinate $\xi_i = \xi_i(x, y)$, which establishes a one-to-one correspondence with the coordinates of the same point in the global reference system Oxy. It also associates the tangential and normal unit vectors $\mathbf{t}_i(\xi_i)$ and $\mathbf{n}_i(\xi_i)$ to $\overline{h}_i(\xi_i)$, to identify the normal and the tangential directions at any point along the smoothed line $\overline{h}_i(\xi_i)$, with $\mathbf{n}_i$ pointing outward from the domain Ω_i. Due to the assumption that the two non-conformal profiles are microscopically rough but nominally flat, the two smoother lines $\overline{h}_i(\xi_i)$ are parallel to each other and therefore $\mathbf{n}_1(\xi) = -\mathbf{n}_2(\xi)$ and $\mathbf{t}_1(\xi) = -\mathbf{t}_2(\xi)$, $\forall\xi$.

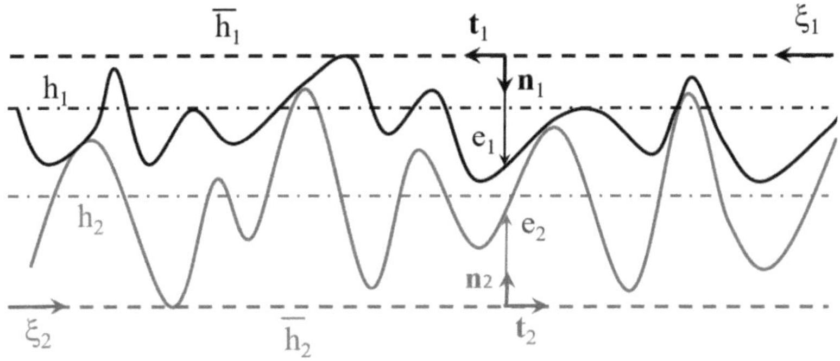

Fig. 4.10 Parametrization of two microscopically rough profiles composing an interface Γ. Adapted from Paggi and Reinoso (2018)

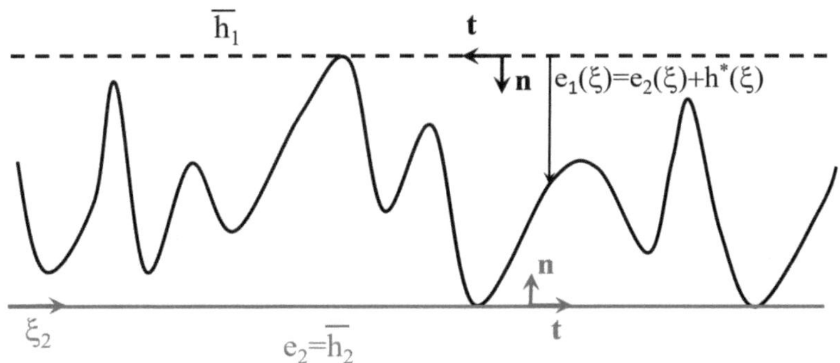

Fig. 4.11 Composite topography of the interface Γ. Adapted from Paggi and Reinoso (2018)

The actual elevation of the rough profile measured from $\overline{h}_i(\xi_i)$ is finally described by the *roughness function* $h_i(\xi_i)$. Therefore, the ith boundary Γ_i is parametrized such that its actual elevation $e_i(\xi_i)$ in the curvilinear setting is given by $e_i(\xi_i) = \overline{h}_i(\xi_i) + h_i(\xi_i)$.

It is in general convenient to exploit the concept of *composite topography* Γ^* of the interface Γ, as also routinely done for the boundary element method. The contact problem between two linear elastic materials with dissimilar rough boundaries is therefore simplified into the contact problem between an infinitely stiff indenter with such a composite topography taken as boundary, and a linear elastic half-plane with composite elastic parameters function of those of the parent elastic bodies. This transformation also allows the study of the contact problem involving a rigid indenter of arbitrary profile (spherical, conical, etc.) and a half-plane.

In the context of the present method, the composite topography is mathematically represented by a flat line, $e_2 = \overline{h}_2(\xi)$, and a profile with elevation $e_1(\xi) = \overline{h}_2(\xi) + h^*(\xi)$, where $h^*(\xi) = \max_\xi[h_1(\xi) + h_2(\xi)] - [h_1(\xi) + h_2(\xi)]$ (see Fig. 4.11).

This transformation does not apply for two elastoplastic or viscoelastic materials, while it does apply for any nonlinear interface constitutive model provided that the two materials are linear elastic. Elastoplastic or viscoelastic contact problems can be

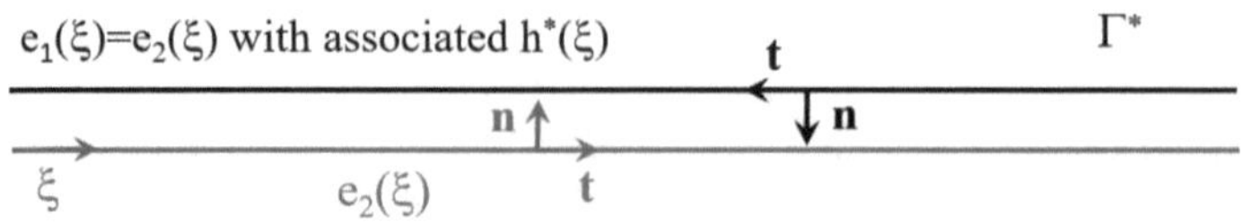

Fig. 4.12 Zero-thickness interface representation Γ^* of the composite topography. Adapted from Paggi and Reinoso (2018)

still simulated using the proposed approach, provide that the attention is restricted to the contact problem between a rigid indenter with rough boundary Γ_1 and a flat half-plane Γ_2 with any prescribed material constitutive relation. In such a case, in fact, the composite topography simply reduces to Γ_1 and the original indenter geometry is kept unchanged.

After this transformation, a *zero-thickness interface model* for Γ^* is introduced and defined by the two initially coincident but distinct (not-joined) flat lines described by the function $e_2(\xi)$, plus the associated function $h^*(\xi)$. This composite topography has also unique tangential and normal unit vectors **t** and **n**, as previously discussed, see Fig. 4.12.

In this framework, the normal gap $g_{\rm n}$ of the composite topography, which represents the actual physical separation between the composite topography and the smooth curve e_2 after deformation, is given by $g_{\rm n} = \Delta u_{\rm n} + h^*$, since material 1 acts as a rigid indenter with a rough profile h^*. Based on the value of $g_{\rm n}$, the portion of the interface in contact, $\Gamma^C = \Gamma_1^C \bigcup \Gamma_2^C$, is identified by the condition $g_{\rm n} = 0$. On the other hand, the portion subject to adhesion, $\Gamma_A = \Gamma_1^A \bigcup \Gamma_2^A$, presents a positive-valued normal gap $g_{\rm n} > 0$. A negative-valued normal gap is not admissible so far by definition, since it would imply compenetration between the bodies.

At the interface, a conforming finite element discretization for the continua can be simply adopted. Consequently, a special interface finite element with embedded profile for joint roughness (MPJR interface finite element) whose kinematics departs from the formulation of interface elements used in nonlinear fracture mechanics for cohesive crack growth (Ortiz and Pandolfi 1999; Paggi and Wriggers 2011, 2012; Reinoso and Paggi 2014; Paggi and Reinoso 2015) has been proposed in Paggi and Reinoso (2018) to be inserted along the interface.

In 2D problems, the interface element is defined by nodes 1 and 2, which belong to $\Gamma_{2,h}^*$, and by nodes 3 and 4, which belong to $\Gamma_{1,h}^*$, see Fig. 4.13.

For frictionless normal contact problems, the contribution of the interface to the weak form is provided by the integral $\int_{\Gamma^*} p(g_{\rm n}) g_{\rm n} {\rm d}\Gamma$ in Eq. (4.31), which can be

Fig. 4.13 Sketch of the interface finite element topology. From Paggi and Reinoso (2018)

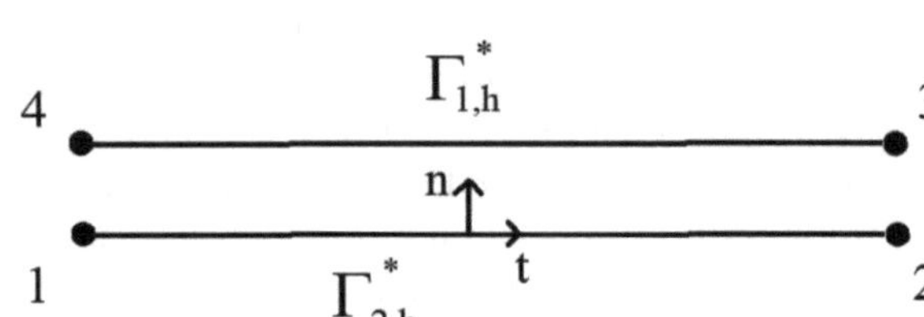

computed as the sum of the contributions of the whole interface elements, invoking the property of compactness of isoparametric shape functions:

$$\int_{\Gamma^*} p(g_n)g_n d\Gamma \cong \int_{\Gamma_h^*} p(g_n)g_n d\Gamma = \mathcal{A}_{e=1}^{n_{el}} \left\{ \int_{\Gamma_e^*} p(g_n)g_n d\Gamma \right\}, \tag{4.32}$$

where the subscript e refers to the eth interface element $e = 1, \ldots, n_{el}$, and $\mathcal{A}$ symbolically denotes an assembly operator.

The interface integral is herein computed exactly, by using the two-point Newton–Cotes quadrature formula which implies the sampling of the integrand at the nodes 1 and 2 (or, equivalently, at nodes 3 and 4):

$$\int_{\Gamma_e^*} p(g_n)g_n d\Gamma = \sum_{j=1,2} p_i(g_n)g_{n,i} \det J, \tag{4.33}$$

where $\det J$ is the standard determinant of the Jacobian of the transformation that maps the geometry of the interface element from its global reference frame to the natural reference system.

To evaluate the normal gap g_n at any point inside the interface element, the nodal displacement vector $\mathbf{d} = (u_1, v_1, \ldots, u_4, v_4)^T$ has to be introduced, which collects the displacements u and v of the four interface finite element nodes. The relative displacement $\Delta\mathbf{u}$ for the nodes 1–4 and 2–3 is then computed by applying a matrix operator $\mathbf{L}$ which makes the difference between the displacements of nodes 1 and 4, and between nodes 2 and 3. The relative displacement within the interface finite element is then given by the linear interpolation of the corresponding nodal values, performed by the multiplication with the matrix $\mathbf{N}$ which collects the shape functions at the element level. Finally, the tangential and the normal gaps are determined by the multiplication with the rotation matrix $\mathbf{R}$ defined by the components of the unit vectors $\mathbf{t}$ and $\mathbf{n}$. In formulae, we have

$$\Delta\mathbf{u} = \mathbf{RNLd}, \tag{4.34}$$

where the operators present the following matrix form:

$$\mathbf{L} = \begin{bmatrix} -1 & 0 & 0 & 0 & 0 & 0 & 1 & 0 \\ 0 & -1 & 0 & 0 & 0 & 0 & 0 & 1 \\ 0 & 0 & -1 & 0 & 1 & 0 & 0 & 0 \\ 0 & 0 & 0 & -1 & 0 & 1 & 0 & 0 \end{bmatrix}, \tag{4.35a}$$

$$\mathbf{N} = \begin{bmatrix} N_1 & 0 & N_2 & 0 \\ 0 & N_1 & 0 & N_2 \end{bmatrix}, \tag{4.35b}$$

$$\mathbf{R} = \begin{bmatrix} t_x & t_y \\ n_x & n_y \end{bmatrix}, \tag{4.35c}$$

where n_x, n_y, t_x, and t_y are the components of the unit vectors **n** and **t** along the x and y directions, respectively.

Once $\Delta\mathbf{u} = (\Delta u_\mathrm{t}, \Delta u_\mathrm{n})^\mathrm{T}$ is determined, the actual normal gap is given by a correction to Δu_n to account for the embedded profile that models the non-planarity of the rigid indenter Γ^*, i.e., $g_\mathrm{n} = \Delta u_\mathrm{n} + h^*$. The normal gap is used to compute the normal traction p_n according to Eq. (4.30). Similarly, for further extensions to adhesive-contact problems with friction in the tangential direction, a relationship between the shearing traction p_t and the relative sliding displacement $\mathbf{g}_\mathrm{t}$, or its velocity, should be introduced, in analogy with the normal problem.

Due to the intrinsic nonlinearity, a full Newton–Raphson iterative and incremental scheme was adopted in Paggi and Reinoso (2018) to solve the implicit nonlinear algebraic system of equations resulting from the finite element discretization:

$$\mathbf{K}^{(k)}\Delta\mathbf{d}^{(k)} = -\mathbf{R}^{(k)}, \tag{4.36a}$$

$$\mathbf{d}^{(k+1)} = \mathbf{d}^{(k)} + \Delta\mathbf{d}^{(k)}, \tag{4.36b}$$

where the superscript k denotes the iteration inside the Newton–Raphson loop. The residual vector $\mathbf{R}_e^{(k)}$ and the tangent stiffness matrix $\mathbf{K}_e^{(k)}$ associated with the eth interface finite element, to be assembled to the global residual vector **R** and to the global stiffness matrix **K**, are

$$\mathbf{R}_e^{(k)} = \int_{\Gamma_e^*} \mathbf{L}^\mathrm{T}\mathbf{N}^\mathrm{T}\mathbf{R}^\mathrm{T}\mathbf{p}\,\mathrm{d}\Gamma, \tag{4.37a}$$

$$\mathbf{K}_e^{(k)} = \int_{\Gamma_e^*} \mathbf{L}^\mathrm{T}\mathbf{N}^\mathrm{T}\mathbf{R}^\mathrm{T}\mathbb{C}\mathbf{R}\mathbf{N}\mathbf{L}\,\mathrm{d}\Gamma, \tag{4.37b}$$

where $\mathbf{p} = (p_\mathrm{t}, p_\mathrm{n})^\mathrm{T} = (0, p_\mathrm{n})^\mathrm{T}$ for frictionless normal contact problems, and $\mathbb{C}$ is the linearized interface constitutive matrix:

$$\mathbb{C} = \begin{bmatrix} \dfrac{\partial p_\mathrm{t}}{\partial g_\mathrm{t}} & \dfrac{\partial p_\mathrm{t}}{\partial g_\mathrm{n}} \\ \dfrac{\partial p_\mathrm{n}}{\partial g_\mathrm{t}} & \dfrac{\partial p_\mathrm{n}}{\partial g_\mathrm{n}} \end{bmatrix}, \tag{4.38}$$

where, again for the frictionless normal contact problem, one needs to specify only $\partial p_\mathrm{n}/\partial g_\mathrm{n}$ depending on the sign of the normal gap, distinguishing between the penalty relation in compression or the adhesive relation in tension.

In principle, the MPJR interface element contributions to the stiffness matrix could be added only for the elements in contact, using a conventional contact search algorithm. However, in case of adhesive contact with long-range adhesive effects, all the interface elements can contribute to tension or compression and have to be assembled in any case. Due to the simplicity in modeling roughness according to this approach, which is embedded in the computation of the normal gap, the cost of assembling the whole set of interface finite elements is much less important than for the methods relying on an explicit discretization of roughness. Therefore, contact search algorithms can be skipped, further simplifying the numerical implementation and the robustness of the computational method.

Comparison Between Different Approaches

As a benchmark problem to compare the different approaches presented in Sects. 4.3.2 and 4.3.3, we simulate the bidimensional frictionless normal contact problem without adhesion between a rigid cylinder indenting a half-plane. For comparison purposes, we recall the Hertzian analytical solution, which is available to assess the model accuracy.

The standard procedure for solving this problem within the finite element method requires modeling of the circular cross section of the cylinder and the use of a contact formulation to enforce the unilateral contact constraint along the interface between the cylinder and the half-plane. For that, among the possible numerical strategies, the penalty approach, the Lagrange multiplier method, and the mortar method are among the most popular formulations, see Wriggers (2006). In spite of the simplicity of this nonconforming contact problem, it is well known that all such methods require very fine meshes to resolve the contact area and the contact traction distribution, especially near the edges of the contact strip. This is primarily due to the fact that a C^1 linear finite element interpolation scheme is not sufficiently accurate to describe the circular shape of the cylinder. To overcome this drawback and increase the accuracy in the boundary element method and in the finite element method relying on the explicit discretization of the interface geometry, adaptive mesh refinement was proposed by Oysu (2007), see Fig. 4.14.

Alternatively, the NURBS finite element technique, which adopts shape functions with a very high regularity and smoothness to approximate curvilinear shapes, can be adopted to explicitly discretize the interface geometry, since it has been demonstrated in Dimitri et al. (2014) to provide the best accuracy over other discretization techniques. In spite of that, NURBS still presents problems in capturing the analytical Hertzian contact solution for the frictionless normal contact problem between a sphere and a half-plane, though a very fine mesh was used in Dimitri et al. (2014), see Fig. 4.15.

In the MPJR interface finite element, instead of modeling the geometry of the circular cross section, the non-planarity of the interface is simply embedded in the interface finite element with its exact analytical function. The actual circular shape of the boundary Γ_1 is therefore given by the composite topography of the interface profile: $e_1(x) = \overline{h}_2 + h^*(x)$, where $\overline{h}_2 = x_l$ and $h^*(x) = R - \sqrt{R^2 - x^2}$. This strategy resolves the issues related to the accuracy of finite element interpolation schemes for the bulk and the interface, which can now have low-order linear shape functions. In this context, the geometry of the cylinder of radius R occupying the domain Ω_1 can be simply replaced by a rectangular block of lateral size x_l and thickness $x_l/20$, while

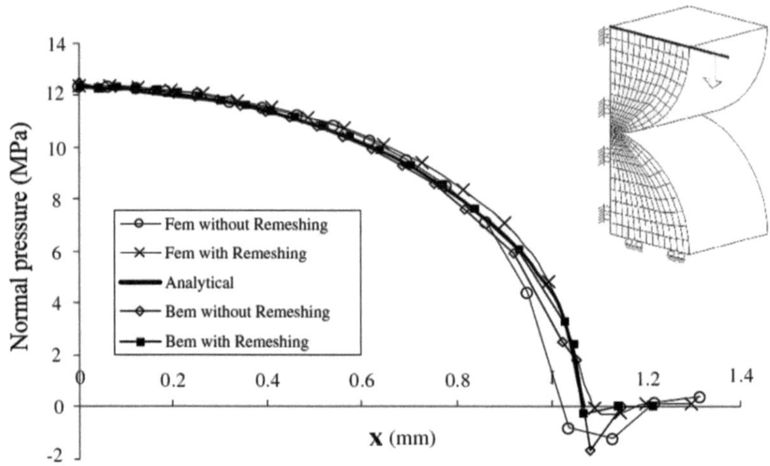

Fig. 4.14 Problems in resolving contact tractions in Hertzian normal contact problems arising from the boundary element method and the finite element method with or without mesh refinement. Adapted from Oysu (2007)

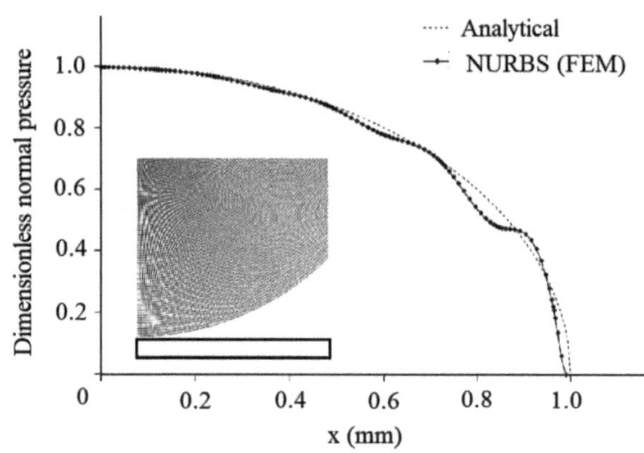

Fig. 4.15 Problems in resolving contact tractions in Hertzian normal contact problems arising from the finite element method with NURBS discretization of the interface geometry. Adapted from Dimitri et al. (2014)

the half-plane occupying the domain Ω_2 can be modeled as a plane strain domain with size x_l, see Fig. 4.16. Since the indenter is rigid, the equivalent model is exact, because there is no error resulting from the deformation associated to the different geometries of body 1. The low-order interpolation scheme used for the interface, on the other hand, is enhanced by the exact analytical representation of the circular indenter profile.

To achieve the condition of a rigid cylinder pressed onto an elastically deformable half-plane, E_1 can be simply set equal to $1000E_2$, where the subscripts 1 and 2 identify the rigid (indenter) and deformable (half-plane) bodies, respectively. Regarding the ratio between the cylinder radius and the lateral size of the half-plane, R/x_l, two cases were examined in Paggi and Reinoso (2018): (i) $R/x_l = 100$, which corresponds

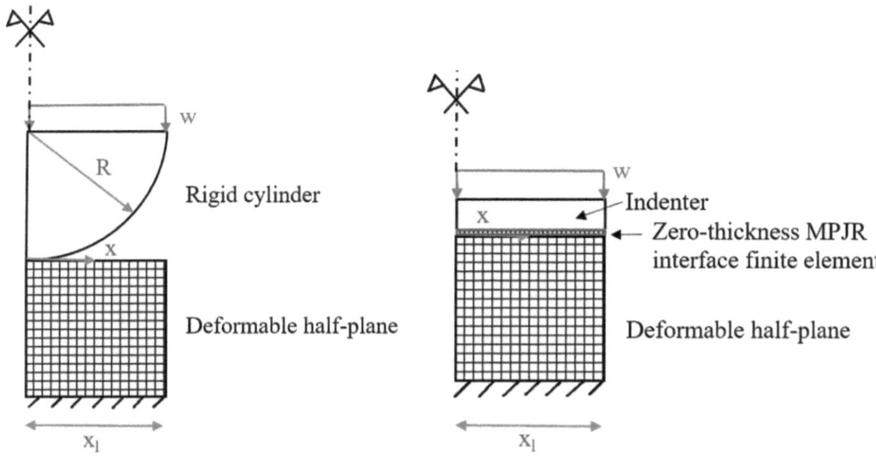

Fig. 4.16 The actual geometry of the Hertz contact problem (left), and its finite element model based on the present variational approach with embedded roughness (right), which incorporates the analytical expression of the curved interface profile into the MPJR interface finite elements instead of explicitly discretizing the interface geometry

to a slightly nonplanar interface; (ii) $R/x_l = 1$, which corresponds to a significant deviation of the interface from the non-planarity. For both cases, uniform meshes for the domains Ω_1 and Ω_2 were used, employing four nodes linear finite elements for the bulk and the proposed MPJR interface elements with embedded roughness for the interface. The whole interface was discretized in the horizontal direction by only $n_{el} = 100$ finite elements, which is much less than what was used in Fig. 4.15 for NURBS, and without adopting any mesh refinement.

Dirichlet boundary conditions are represented in this test by imposed downward vertical displacements w on the topmost side of the domain Ω_1, monotonically increasing with a pseudo-time variable to simulate the quasi-static normal contact problem; a fully restrained lower side of the domain Ω_2; and a symmetry condition on the vertical size of domains Ω_1 and Ω_2 to account for the symmetry in the geometry and in the loading (Fig. 4.16).

Numerical predictions are provided in terms of the dimensionless normal contact pressure, p/E, versus the dimensionless position along the interface, x/R. The contact pressure p is given by $p = -p_\mathrm{n}$, and therefore it is positive valued on the portion of the interface Γ_C in contact and it must be zero elsewhere, since adhesion is not considered here. The penalty stiffness K is set $K = 10E_1/x_l$, to model a very stiff interface and avoid material penetration.

Results from the current simulations are shown in Fig. 4.17 for the case $R/x_l = 100$ and $R/x_l = 1$, considering nine increasing values for the imposed far-field displacement w. Analytical Hertzian results, corresponding to the same contact radii, are also superimposed by circles. As can be observed in these graphs, the agreement between the present model predictions and theory is excellent, also for the case

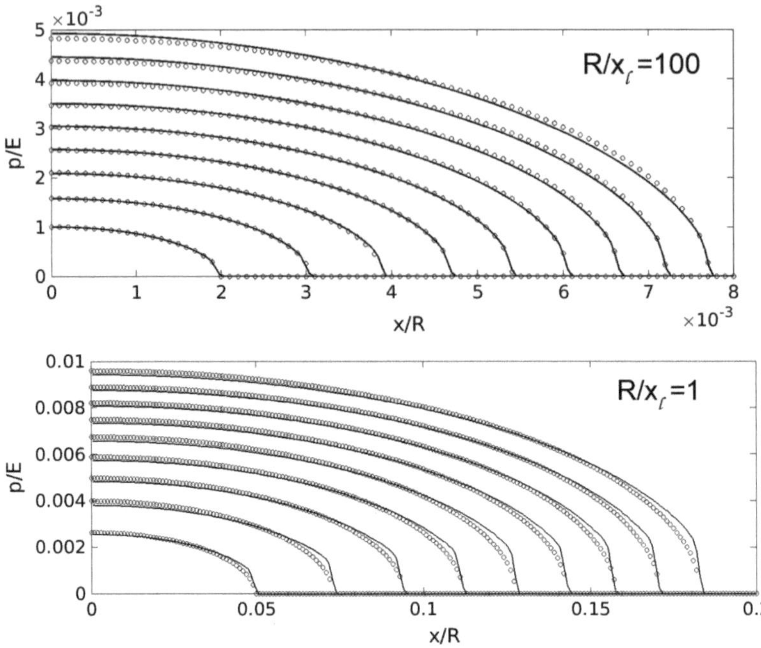

Fig. 4.17 Dimensionless contact pressure along the interface for different imposed far-field displacements and two different values of R/x_l. E, R, and x_l denote, respectively, the composite Young's modulus, the cylinder radius, and the lateral size of the domain, respectively. The analytical Hertzian solution is superimposed with circles. Adapted from Paggi and Reinoso (2018)

$R/x_l = 1$, which is indeed very challenging from the computational point of view due to the significant non-planarity of the interface.

Conclusive Remarks

The solution of the contact problem between bodies separated by rough boundaries is very important in many tribological applications and it is a challenging research topic due to the multiscale features of roughness that span over multiple length scales. So far, especially in relation to the linear elastic frictionless normal contact problem, the boundary element method has been proved to be very efficient from the computational point of view and preferable over the finite element method. It has been used to validate semi-analytical approaches based on asperities or other prominent contact theories, see, e.g., Mueser et al. (2017), inspiring also further developments in the framework of multi-asperity contact models including elastic interaction effects. Moreover, it has been applied to identify valuable trends on the effect of roughness on the emerging contact response, such as the real contact area

Table 4.1 Synopsis of the computational methods reviewed in this chapter

Approach	Roughness discretization	Continua	Interface	Pros	Cons
BEM	Exact representation	Linear elasticity or elastoplasticity	Recently extended to adhesion	Fast and accurate	Nonlinearities and multi-field problems are difficult to be addressed
Multi-asperity BEM	Only asperities	Linear elasticity	Resolution-dependent discretization	Very fast	As BEM, plus the limits of defining the asperities
FEM with explicit discretization of the interface geometry (NURBS, splines, etc.)	Modeling multiscale roughness is problematic	No restrictions	No restrictions	Ideal for nonlinear and multi-field problems	Limits in the discretization of roughness; computationally demanding
FEM-MPJR with embedded roughness	Exact	No restrictions	No restrictions	Ideal for nonlinear and multi-field problems	Faster and simpler than FEM, but still computationally demanding as compared to BEM

versus load relation (Paggi and Ciavarella 2010), the normal contact stiffness versus load dependency (Paggi and Barber 2011), as well as the evolution of the free volume between rough surfaces during the progress of contact (Paggi and He 2015).

If the research on the frictionless normal contact problem between linear elastic materials has seen a significant impulse during the last decades, the contact problem in the presence of friction still presents open issues and aspects deserving investigation. So far, the evolution of the stick and slip contact domains in the case of a monotonically increasing tangential displacement applied by keeping constant the normal load has been investigated using the boundary element method in Paggi et al. (2014). Another important topic regards the effect of roughness on hysteretic energy dissipation caused by cyclic tangential loading paths, see, e.g., Borri-Brunetto et al. (2006), Barber et al. (2011).

Finally, there is a range of contact problems involving nonlinear constitutive relations, nonlinear geometric (finite elasticity) effects, and coupled multi-field problems that have been only marginally challenged so far. For instance, the problem of electromechanical interaction between excitable deformable cells in finite elasticity investigated in Lenarda et al. (2018) is an exemplary problem requiring the extension of the methods of contact mechanics to soft biological matter. For this class of problems, the constitutive nonlinearities for the interface and the continua suggest passing to the finite element method. In this regard, further advancements in the interface finite element discretization are indeed required for an efficient treatment of contact problems with roughness features covering multiple length scales. The novel approach based on the interface finite element with embedded roughness proposed in Paggi and Reinoso (2018) opens new perspectives in this direction.

As a guideline for new researchers entering the field, and also for experienced researchers willing to explore future directions of research in contact mechanics between rough surfaces, a synopsis of the computational methods reviewed in this chapter is provided in Table 4.1, emphasizing the advantages and the disadvantages of each approach.

Acknowledgements This chapter is derived in part from: an article published in the International Journal of Solids and Structures (Elsevier), available online June 16, 2015, doi: http://doi.org/10.1016/j.ijsolstr.2015.06.005; an article published in Mechanics of Advanced Materials and Structures (Taylor & Francis), available online November 5, 2018, doi: http://doi.org/10.1080/15376494.2018.1525454.

References

Afferrante, L., Carbone, G., & Demelio, G. (2012). Interacting and coalescing Hertzian asperities: A new multiasperity contact model. *Wear*, *278–279*, 28–33.

Andersson, T. (1981). The boundary element method applied to two-dimensional contact problems with friction. *Boundary Element Methods*, *3*, 239–258.

Barber, J. R. (1974). Determining the contact area in elastic-indentation problems. *Journal of Strain Analysis*, *9*, 230–232.

Barber, J. R. (2003). Bounds on the electrical resistance between contacting elastic rough bodies. *Proceedings of the Royal Society of London, Series A*, *459*, 53–66.

Barber, J. R. (2010). *Elasticity* (3rd ed.). Dordrecht: Springer.

Barber, J. R. (2018). *Contact mechanics*. Springer International Publishing.

Barber, J. R., Davies, M., & Hills, D. A. (2011). Frictional elastic contact with periodic loading. *International Journal of Solids and Structures*, *48*, 2041–2047.

Batrouni, G. G., Hansen, A., & Schmittbuhl, J. (2002). Elastic response of rough surfaces in partial contact. *Europhysics Letters*, *60*, 724–730.

Beale, E. M. L. (1955). On minimizing a convex function subject to linear inequalities. *Journal of the Royal Statistical Society, Series B*, 173–184.

Bemporad, A. (2014). A quadratic programming algorithm based on nonnegative least squares with applications to embedded model predictive control. *IEEE Transactions on Automatic Control*.

Bemporad, A., & Paggi, M. (2015). Optimization algorithms for the solution of the frictionless normal contact between rough surfaces. *International Journal of Solids and Structures*, *69–70*, 94–105.

Borri, C., & Paggi, M. (2015). Topological characterization of antireflective and hydrophobic rough surfaces: Are random process theory and fractal modeling applicable? *Journal of Physics D: Applied Physics*, *48*, 045301.

Borri, C., & Paggi, M. (2016). Topology simulation and contact mechanics of bifractal rough surfaces. *Proceedings of the Institution of Mechanical Engineers, Part J: Journal of Engineering Tribology*, *230*, 1345–1358.

Borri-Brunetto, M., Carpinteri, A., & Chiaia, B. (1999). Scaling phenomena due to fractal contact in concrete and rock fractures. *International Journal of Fracture*, *95*, 221–238.
Borri-Brunetto, M., Chiaia, B., & Ciavarella, M. (2001). Incipient sliding of rough surfaces in contact: A multiscale numerical analysis. *Computer Methods in Applied Mechanics and Engineering*, *190*, 6053–6073.
Borri-Brunetto, M., Carpinteri, A., Invernizzi, S., & Paggi, M. (2006). Micro-slip of rough surfaces under cyclic tangential loading. In P. Wriggers & U. Nackenhorst (Eds.), *Analysis and simulation of contact problems*. Lecture notes in applied and computational mechanics (Vol. 27, pp. 191–200). Berlin, Heidelberg: Springer.
Boyd, S., Parikh, N., Chu, E., Peleato, B., & Eckstein, J. (2011). Distributed optimization and statistical learning via the alternating direction method of multipliers. *Foundations and Trends in Machine Learning*, *3*, 1–122.
Campaña, C., Persson, B. N. J., & Mueser, M. H. (2001). Transverse and normal interfacial stiffness of solids with randomly rough surfaces. *Journal of Physics: Condensed Matter*, *23*, 085001.
Carbone, G., & Bottiglione, F. (2008). Asperity contact theories: Do they predict linearity between contact area and load? *Journal of the Mechanics and Physics of Solids*, *56*, 2555–2572.
Carbone, G., & Putignano, C. (2013). A novel methodology to predict sliding and rolling friction of viscoelastic materials: Theory and experiments. *Journal of the Mechanics and Physics of Solids*, *61*, 1822–1834.
Carpinteri, A., & Paggi, M. (2005). Size-scale effects on the friction coefficient. *International Journal of Solids and Structures*, *42*, 2901–2910.
Carpinteri, A., & Paggi, M. (2009). A fractal interpretation of size-scale effects on strength, friction and fracture energy of faults. *Chaos, Solitons & Fractals*, *39*, 540–546.
Carpinteri, A., Paggi, M., & Zavarise, G. (2008). The effect of contact on the decohesion of laminated beams with multiple microcracks. *International Journal of Solids and Structures*, *45*, 129–143.
Chang, W. R., Etsion, I., & Bogy, D. B. (1987). An elastic-plastic model for the contact of rough surfaces. *Journal of Tribology*, *109*, 257–263.
Chen, W. W., Wang, Q. J., Wang, F., Keer, L. M., & Cao, J. (2008). Three-dimensional repeated elasto-plastic point contacts, rolling, and sliding. *Journal of Applied Mechanics*, *75*, 021021.
Ciavarella, M., Demelio, G., Barber, J. R., & Jang, Y. H. (2000). Linear elastic contact of the Weierstrass profile. *Proceedings of the Royal Society of London, Series A*, *456*, 387–405.
Ciavarella, M., Murolo, G., Demelio, G., & Barber, J. R. (2004). Elastic contact stiffness and contact resistance for the Weierstrass profile. *Journal of the Mechanics and Physics of Solids*, *52*, 1247–1265.
Ciavarella, M., Delfine, V., & Demelio, G. (2006). A "re-vitalized" Greenwood and Williamson model of elastic contact between fractal surfaces. *Journal of the Mechanics and Physics of Solids*, *54*, 2569–2591.
Ciavarella, M., Dibello, S., & Demelio, G. (2008a). Conductance of rough random profiles. *International Journal of Solids and Structures*, *45*, 879–893.
Ciavarella, M., Greenwood, J. A., & Paggi, M. (2008b). Inclusion of "interaction" in the Greenwood and Williamson contact theory. *Wear*, *265*, 729–734.
Cottle, R. W., Pang, J.-S., & Stone, R. E. (1992). *The linear complementarity problem*. Academic Press.
Dimitri, R., De Lorenzis, L., Scott, M. A., Wriggers, P., Taylor, R. L., & Zavarise, G. (2014). Isogeometric large deformation frictionless contact using T-splines. *Computer Methods in Applied Mechanics and Engineering*, *269*, 394–414.
Fletcher, R. (1971). A general quadratic programming algorithm. *IMA Journal of Applied Mathematics*, *7*, 76–91.
Francis, H. A. (1983). The accuracy of plane strain models for the elastic contact of three-dimensional rough surfaces. *Wear*, *85*, 239–256.
Goldfarb, D., & Idnani, A. (1983). A numerically stable dual method for solving strictly convex quadratic programs. *Mathematical Programming*, *27*, 1–33.
Greenwood, J. A. (2006). A simplified elliptic model of rough surface contact. *Wear*, *261*, 191–200.

Greenwood, J. A., & Williamson, J. B. P. (1966). Contact of nominally flat surfaces. *Proceedings of the Royal Society of London, Series A, 295*, 300–319.
Greenwood, J. A., & Wu, J. J. (2001). Surface roughness and contact: An apology. *Meccanica, 36*, 617–630.
Hendriks, C. P., & Visscher, M. (1995). Accurate real area of contact measurements on polyuretane. *ASME Journal of Tribology, 117*, 607–611.
Hestenes, M. R. (1980). *Conjugate direction methods in optimization* (Chaps. 2 and 3). New York: Springer.
Hyun, S., & Robbins, M. O. (2007). Elastic contact between rough surfaces: Effect of roughness at large and small wavelengths. *Tribology International, 40*, 1413.
Hyun, S., Pei, L., Molinari, J.-F., & Robbins, M. O. (2004). Finite-element analysis of contact between elastic self-affine surfaces. *Physical Review E, 70*, 026117.
Johnson, K. L. (1985). *Contact mechanics*. Cambridge, UK: Cambridge University Press.
Kalker, J. J., & van Randen, Y. A. (1972). A minimum principle for frictionless elastic contact with application to non hertzian problems. *Journal of Engineering Mathematics, 6*, 193–206.
Karpenko, Y. A., & Akay, A. (2001). A numerical model of friction between rough surfaces. *Tribology International, 34*, 531–545.
Kogut, L., & Etsion, I. (2002). Elastic-plastic contact analysis of a sphere and a rigid flat. *Journal of Applied Mechanics, 69*, 657–662.
Kubo, A., Okamoto, T., & Kurokawa, N. (1981). Contact stress between rollers with surface irregularity. *Journal of Tribology, 116*, 492–498.
Lawson, C., & Hanson, R. (1974). *Solving least squares problems* (Vol. 161, Chap. 24). SIAM.
Lenarda, P., Gizzi, A., & Paggi, M. (2018). A modeling framework for electro-mechanical interaction between excitable deformable cells. *European Journal of Mechanics—A/Solids, 72*, 374–392.
Leroux, J., Fulleringer, B., & Nélias, D. (2010). Contact analysis in presence of spherical inhomogeneities within a half-space. *International Journal of Solids and Structures, 47*, 3034–3049.
Li, J., & Berger, E. J. (2003). A semi-analytical approach to three-dimensional normal contact problems with friction. *Computational Mechanics, 30*, 310–322.
Li, Q., Argatov, I., & Popov, V. (2018). Onset of detachment in adhesive contact of an elastic half-space and flat-ended punches with non-circular shape: Analytic estimates and comparison with numeric analysis. *Journal of Physics D: Applied Physics*.
De Lorenzis, L., & Wriggers, P. (2013). Computational homogenization of rubber friction on rough rigid surfaces. *Computational Materials Science, 77*, 264–280.
Love, H. A. E. (1999). The stress produced in a semi-infinite solid by pressure on part of the boundary. *Phylosophical Transactions of the Royal Society of London, 228*, 377–420.
Lubrecht, A. A., & Ioannides, E. (1991). A fast solution of the dry contact problem and the associated sub-surface stress field, using multilevel techniques. *ASME Journal of Tribology, 113*, 128–133.
Majumdar, A., & Bhushan, B. (1990). Role of fractal geometry in roughness characterization and contact mechanics of surfaces. *ASME Journal of Tribology, 112*, 205–216.
Man, K. W. (1994). *Contact mechanics using boundary elements, topics in engineering* (Vol. 22). Boston: Southampton.
McCool, J. I. (1986). Comparison of models for the contact of rough surfaces. *Wear, 107*, 37–60.
Mueser, M. H., Dapp, W. B., Bugnicourt, R., Sainsot, P., Lesaffre, N., Lubrecht, T. A., et al. (2017). Meeting the contact-mechanics challenge. *Tribology Letters, 65*, 118.
Nelias, D., Boucly, V., & Brunet, M. (2006). Elastic-plastic contact between rough surfaces: Proposal for a wear or running-in model. *Journal of Tribology, 128*, 236–244.
Nesterov, Y. (1983). A method of solving a convex programming problem with convergence rate $O(1/k^2)$. *Soviet Mathematics Doklady, 27*, 372–376.
Nogi, T., & Kato, T. (1997). Influence of a hard surface layer on the limit of elastic contact—Part I: Analysis using a real surface model. *Journal of Tribology, 110*, 376–493.
Nosonovsky, M., & Bhushan, B. (2005). Roughness optimization for biomimetic superhydrophobic surfaces. *Microsystem Technologies, 11*, 535–376.

Nowell, D. D., & Hills, D. A. (1989). Hertzian contact of ground surfaces. *ASME Journal of Tribology*, *111*, 175–179.

O'Callaghan, P. W., & Probert, S. D. (1970). Real area of contact between a rough surface and a softer optically flat surface. *Journal of Mechanical Engineering Science*, *11*, 259–267.

Ortiz, M., & Pandolfi, A. (1999). Finite deformation irreversible cohesive elements for three-dimensional crack-propagation analysis. *International Journal for Numerical Methods in Engineering*, *44*, 1267–1282.

Oysu, C. (2007). Finite element and boundary element contact stress analysis with remeshing technique. *Applied Mathematical Modelling*, *31*, 2744–2753.

Paggi, M., & Barber, J. R. (2011). Contact conductance of rough surfaces composed of modified rmd patches. *International Journal of Heat and Mass Transfer*, *54*, 4664–4672.

Paggi, M., & Ciavarella, M. (2010). The coefficient of proportionality κ between real contact area and load, with new asperity models. *Wear*, *268*, 1020–1029.

Paggi, M., & He, Q.-C. (2015). Evolution of the free volume between rough surfaces in contact. *Wear*, *336–337*, 86–95.

Paggi, M., & Reinoso, J. (2015). An anisotropic large displacement cohesive zone model for fibrillar and crazing interfaces. *International Journal of Solids and Structures*, *69*, 106–120.

Paggi, M., & Reinoso, J. (2018). A variational approach with embedded roughness for adhesive contact problems. *Mechanics of Advanced Materials and Structures*. https://doi.org/10.1080/15376494.2018.1525454.

Paggi, M., & Wriggers, P. (2011). A nonlocal cohesive zone model for finite thickness interfaces—Part II: FE implementation and application to polycrystalline materials. *Computational Materials Science*, *50*(5), 1634–1643.

Paggi, M., & Wriggers, P. (2012). Stiffness and strength of hierarchical polycrystalline materials with imperfect interfaces. *Journal of the Mechanics and Physics of Solids*, *60*(4), 557–572.

Paggi, M., & Zavarise, G. (2011). Contact mechanics of microscopically rough surfaces with graded elasticity. *European Journal of Mechanics—A/Solids*, *30*, 696–704.

Paggi, M., Carpinteri, A., & Zavarise, G. (2006). A unified interface constitutive law for the study of fracture and contact problems in heterogeneous materials. In *Analysis and simulation of contact problems. Lecture notes in applied and computational mechanics* (Vol. 27, pp. 297–304). Springer.

Paggi, M., Pohrt, R., & Popov, V. L. (2014). Partial-slip frictional response of rough surfaces. *Scientific Reports*, *4*, 5178.

Patrinos, P., & Bemporad, A. (2014). An accelerated dual gradient-projection algorithm for embedded linear model predictive control. *IEEE Transactions and Automatic Control*, *59*, 18–33.

Pei, L., Hyun, S., Molinari, J. F., & Robbins, M. O. (2005). Finite element modeling of elasto-plastic contact between rough surfaces. *Journal of the Mechanics and Physics of Solids*, *53*, 2385–2409.

Peitgen, H. O., & Saupe, D. (1988). *The science of fractal images*. New York: Springer.

Persson, B. N. J. (2006). Contact mechanics for randomly rough surfaces. *Surface Science Reports*, *261*, 201–227.

Persson, B. N. J., Albohr, O., Tartaglino, U., Volokitin, A. I., & Tosatti, E. (2005). On the nature of surface roughness with application to contact mechanics, sealing, rubber friction and adhesion. *Journal of Physics: Condensed Matter*, *17*, R1.

Pohrt, R., & Li, Q. (2014). Complete boundary element formulation for normal and tangential contact problems. *Physical Mesomechanics*, *17*, 334–340.

Polonsky, I. A., & Keer, L. M. (1999). A numerical method for solving rough contact problems based on the multi-level multi-summation and conjugate gradient techniques. *Wear*, *231*, 206–219.

Polonsky, I. A., & Keer, L. M. (2000a). A fast and accurate method for numerical analysis of elastic layered contacts. *Journal of Tribology*, *122*, 30–35.

Polonsky, I. A., & Keer, L. M. (2000b). Fast methods for solving rough contact problems: A comparative study. *Journal of Tribology*, *122*, 36–41.

Popov, V. L., Pohrt, R., & Li, Q. (2017). Strength of adhesive contacts: Influence of contact geometry and material gradients. *Friction*, *5*, 308–325.

Prodanov, N., Dapp, W. B., & Mueser, M. H. (2014). On the contact area and mean gap of rough, elastic contacts: Dimensional analysis, numerical corrections and reference data. *Tribology Letters*, *53*, 433–448.

Putignano, C., Carbone, G., & Dini, D. (2015). Mechanics of rough contacts in elastic and viscoelastic thin layers. *International Journal of Solids and Structures*, *69–70*, 507–517.

Raous, M. (1999). Quasistatic signorini problem with coulomb friction and coupling to adhesion. *New developments in contact problems* (Vol. 384, pp. 101–178). CISM Series: Springer.

Reinoso, J., & Paggi, M. (2014). A consistent interface element formulation for geometrical and material nonlinearities. *Computational Mechanics*, *54*, 1569–1581.

Rey, V., Anciaux, G., & Molinari, J.-F. (2017). Normal adhesive contact on rough surfaces: Efficient algorithm for FFT-based BEM resolution. *Computational Mechanics*, *60*, 69–81.

Sapora, A., & Paggi, M. (2014). A coupled cohesive zone model for transient analysis of thermoelastic interface debonding. *Computational Mechanics*, *53*, 845–857.

Scaraggi, M., Putignano, C., & Carbone, G. (2013). Elastic contact of rough surfaces: A simple criterion to make 2D isotropic roughness equivalent to 1D one. *Wear*, *297*, 811–817.

Schmid, C., & Biegler, L. T. (1994). Quadratic programming methods for reduced hessian SQP. *Computers & Chemical Engineering*, *18*, 817–832.

Sridhar, M. R., & Yovanovich, M. M. (1994). Review of elastic and plastic contact conductance models: Comparison with experiments. *Journal of Thermophysics Heat Transfer*, *8*, 633–640.

Vakis, A. I., Yastrebov, V. A., Scheibert, J., Nicola, L., Dini, D., Minfray, C., et al. (2018). Modeling and simulation in tribology across scales: An overview. *Tribology International*, *125*, 169–199.

Vollebregt, E. A. H. (2014). A new solver for the elastic normal contact problem using conjugate gradients, deflation, and an FFT-based preconditioner. *Journal of Computational Physics*, *257*, 333–351.

Woo, K. L., & Thomas, T. R. (1980). Contact of rough surfaces: A review of experimental work. *Wear*, *58*, 331–340.

Wriggers, P. (2006). *Computational contact mechanics*. Berlin Heidelberg: Springer.

Wriggers, P., & Reinelt, J. (2009). Multi-scale approach for frictional contact of elastomers on rough rigid surfaces. *Computer Methods in Applied Mechanics and Engineering*, *198*, 1996–2008.

Yastrebov, V. A., Anciaux, G., & Molinari, J.-F. (2015). From infinitesimal to full contact between rough surfaces: Evolution of the contact area. *International Journal of Solids and Structures*, *52*, 83–102.

Yu, N., & Polycarpou, A. A. (2004). Adhesive contact based on the Lennard-Jones potential: A correction to the value of the equilibrium distance as used in the potential. *Journal of Colloid and Interface Science*, *278*, 428–435.

Zavarise, G., Wriggers, P., Stein, E., & Schrefler, B. A. (2019). Real contact mechanisms and finite element formulation—A coupled thermomechanical approach. *International Journal for Numerical Methods in Engineering*, *35*, 767–785.

Zavarise, G., Wriggers, P., Stein, E., & Schrefler, B. A. (1992). A numerical model for thermomechanical contact based on microscopic interface laws. *Mechanics Research Communications*, *19*(3), 173–182.

Zavarise, G., Borri-Brunetto, M., & Paggi, M. (2004a). On the reliability of microscopical contact models. *Wear*, *257*, 229–245.

Zavarise, G., Borri-Brunetto, M., & Paggi, M. (2004b). On the resolution dependence of micromechanical contact models. *Wear*, *262*, 42–54.

Zienkiewicz, O. C., & Taylor, R. L. (2000). *The finite element method: Solid mechanics* (Vol. 2). Butterworth-Heinemann.

Chapter 5
Emergent Properties from Contact Between Rough Interfaces

Marco Paggi

Abstract Interface phenomena at the micro- and nanoscales are of paramount importance in nature and technology. Real surfaces present roughness over multiple scales, and understanding the role of roughness in surface physics (heat and electric transfer, hydrophobic properties), surface chemistry (chemical reactions) and tribology (stress transfer, adhesion, lubrication) is a very active research topic. This chapter focuses on the key research question of how nonlinear interactions between contact patches induced by roughness across different length scales influence the emergent physico-mechanical properties of an interface. Special attention is given to the scaling of the real area of contact with the applied normal load, the dependency of the thermal and electric contact conductance on the normal pressure, the evolution of the free volume network between rough surfaces in contact, the role of adhesion and also the evolution of partial slip in frictional contacts.

Introduction

Due to the technological trend of pushing the design and the production of structures down to the micro- and nanoscales (see, e.g. micro- and nanoelectromechanical systems), surface-related phenomena become predominant over bulk properties (Luan and Robbins 2005). Therefore, local imperfections and deviation from the ideal flatness of surfaces (Raja et al. 2002), and especially waviness, roughness and texturing, have a fundamental effect on surface physics (heat and electric transfer, optical properties, fluid–solid interactions including hydrophobic properties, etc.), surface chemistry (adhesion, chemical reactions, diffusion, etc.), and tribology (stress transfer between interacting surfaces in relative motion, friction, wear, lubrication, etc.), see (Paggi and Hills 2016a, b; Vakis et al. 2018) for a wide overview. Moreover, these problems are a neat example of a highly interdisciplinary research area which draws on many academic fields including physics, chemistry, materials science, mathemat-

M. Paggi (✉)
IMT School for Advanced Studies Lucca, Piazza San Francesco 19, 55100 Lucca, Italy
e-mail: marco.paggi@imtlucca.it

M. Paggi and D. Hills (eds.), *Modeling and Simulation of Tribological Problems in Technology*, CISM International Centre for Mechanical Sciences 593,
https://doi.org/10.1007/978-3-030-20377-1_5

ics, biology and engineering. In this regard, the role of mechanics is essential for understanding, modelling and simulating the stress and the deformation fields experienced by rough surfaces in contact, as well as for the description of their evolution over time (Rabinowicz 1965; Johnson 1985; Goryacheva 1998; Persson 2000; Popov 2010; Popov and Hess 2015; Barber 2018).

A closer look at the scientific literature shows that a significant effort has been devoted to understanding and predicting the emergent physico-mechanical properties resulting from surfaces in contact, towards the identification of possible universal trends and scaling laws. Even in the simplest case of linear elastic continua, the presence of roughness highly complicates the solution of the contact problem since the real area of contact is changing with the applied load level, and it is therefore a source of nonlinearity. Therefore, understanding how the geometrical/topological features of roughness influence the relation between the real area of contact and the normal force, the relation between the thermal/electrical contact resistance and the contact pressure, or the apparent value of the friction coefficient, just as few exemplary problems, is an intriguing research question with also practical technological implications.

The relation between the real area of contact and the normal load has been for the first time investigated by micromechanical contact theories (Greenwood and Williamson 1966; Onions and Archard 1973; Nayak 1973; Bush et al. 1976; Greenwood and Wu 2001; Greenwood 2006; Ciavarella et al. 2008b; Paggi and Ciavarella 2010) relying on random process theory (Cartwright and Longuet-Higgins 1956; Longuet-Higgins 1957a, b; Whitehouse and Archard 1970; Nayak 1971; Greenwood 1984, 2006) for the description of the geometrical features of asperities—the 3D maxima of the rough surface—modelled as Hertzian paraboloids with joint statistical distributions of heights and curvatures. Further achievements in the identification of synthetic relations between the topology of the contact domain and the scale-invariant features of roughness were made in (Majumdar and Bhushan 1990, 1991; Bhushan and Majumdar 1992; Borodich and Mosolov 1992; Borodich 1997; Borri-Brunetto et al. 1999; Ciavarella and Demelio 2000; Persson 2001a, 2006; Persson et al. 2002; Carbone and Bottiglione 2008; Yastrebov et al. 2015) by exploiting fractal geometry concepts, without hinging on the asperity definition. Of particular interest was the study of the linearity in the relation between the real contact area and the applied load. If the linearity is proved, then this has the important implication that the Amontons' second law of friction holds true, i.e. the force of friction is independent of the apparent area of contact (Bowden and Tabor 1964).

The relation between the thermal contact conductance and the applied pressure has been pioneeringly investigated in a series of combined experimental and theoretical studies by Yovanovich and his group (Cooper et al. 1968; Mikic 1974; Blahey et al. 1980; Sridhar and Yovanovich 1996a, b, 1994; Milanez et al. 2003b). This topic is of major importance in the design of electronic packaging, where the conductance between electronic devices and dissipators should be increased as much as possible to maintain low operating temperatures and safety conditions. This need would suggest the use of large contact areas to enhance dissipation, which is however a trend opposite to miniaturization. Therefore, a careful analysis of the effect of roughness on the heat flux at asperities is required, see Holm (1958) and the references therein given. In this field, the theoretical proof of the mathematical analogy between the solution of the normal contact problem and the thermal or the electric contact conduction established by Barber (2003) paved the way to a series of theoretical investigations on the scaling relation between the normal contact stiffness and the applied pressure (Ciavarella et al. 2004a, b, 2008a; Campaña et al. 2001; Paggi and Barber 2011). In particular, Paggi and Barber (2011) revisited the whole problem by applying dimensional analysis considerations and incomplete self-similarity concepts, with the important result of reconciling a wide range of empirically identified scaling laws within a unified theoretical framework. Moreover, Paggi and Barber (2011) clearly elucidated on the effect of the long wavelength cut-off of rough surfaces on the normal contact stiffness, pinpointing the limitations of previous numerical results published in the literature.

Frictional phenomena, and in particular sliding as a result of an avalanche process, where asperities progressively change their state from full stick to full slip, were also subject of investigation for their importance in controlling sliding friction of tyres and engineering components (Persson 2001b; Persson et al. 2005; Carbone and Mangialardi 2004; Carbone and Putignano 2013; Paggi et al. 2014), and also to assess the amount of energy dissipated during cyclic loading (Borri-Brunetto et al. 2006; Barber et al. 2011). The understanding of the amount of friction involved in the sliding of natural faults (Carpinteri and Paggi 2005, 2008, 2009) led to the discovery of important size-scale effects on the nominal friction coefficient caused by the complex fractal pattern of the contact domain. They were predicted by theory and confirmed by the experimental evidence in geophysics and rock mechanics (Bandis et al. 1981), across several length scales ranging from the laboratory one up to the planetary size-scale.

Adhesion effects in the presence of wavy (Guduru 2007; Guduru and Bull 2007; Waters et al. 2009) and rough interfaces (Persson 2002a, b; Yu and Polycarpou 2004) have also been subject of experimental and theoretical investigation, with recent developments towards the proposal of novel numerical techniques for studying adhesive contact problems in the presence of roughness based on the boundary element method (Popov et al. 2017; Rey et al. 2017; Li et al. 2018) and on the finite element method (Paggi and Reinoso 2018).

The topology of surfaces is also governing other physical properties of engineering interest, such as the ability of a surface to repel water, which inspires the design of surfaces possessing a self-cleaning behaviour. The relation between surface texture and hydrophobic properties has been studied in (Sherge and Gorb 2001; Nosonovsky 2007; Nosonovsky and Bhushan 2008), also in reference to bio-inspired solutions. In this regard, Lotus and Ginkgo Biloba leaves possess such features through a complex texture involving channels and roughness over multiple length scales, as experimentally characterized by Borri and Paggi (2015, 2016) using a confocal profilometer with different magnifications and a scanning electron microscope (Fig. 5.1).

Other physical surface properties governed by roughness are the optical ones and in particular the ability of reflecting or absorbing the incident light. This feature is

Fig. 5.1 Multi-scale texture and roughness of a Ginkgo Biloba leaf observed at different magnifications within a scanning electron microscope

of technological importance in silicon solar cells, whose surface has a rough texture designed to enhance trapping of the light between the asperities and increase in its turn the solar energy conversion efficiency (Fig. 5.2).

Another important topic regards the transport properties of rough surfaces in contact. A free volume between bodies in contact separated by roughness is present in conditions far from full contact. Such a free volume constitutes a fractal network whose properties are important for flow and transport of hydrothermal fluids, water and contaminants in groundwater systems, but also for oil and gas in petroleum reservoirs (Berkowitz 2002; Scaraggi and Persson 2012). For instance, the transport properties of proppant through fracture networks are relevant for hydraulic fracturing

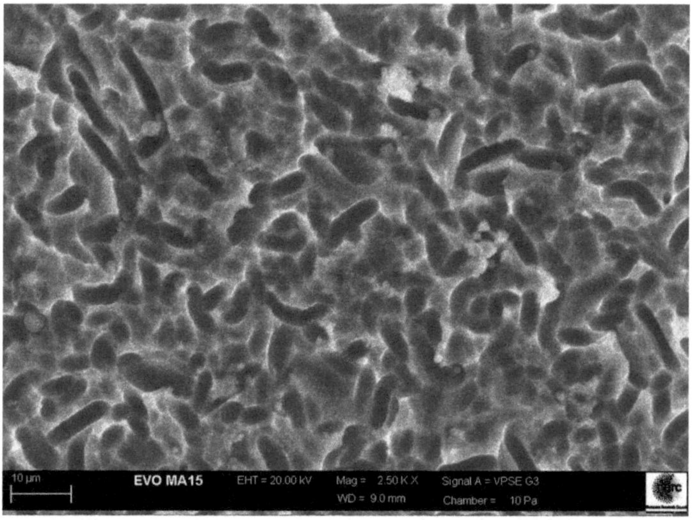

Fig. 5.2 Rough texture of the surface of silicon solar cells observed with a scanning electron microscope

(Tarabay 2014). At a much smaller scale, welded surfaces in microelectromechanical systems may present a free volume forming channels and capillaries of random distribution. Such channels are critical for gas leakage that may penetrate the soldered joint and affect the reliability of the system itself (Han 2012). These problems are also relevant in materials for energy applications, such as in solid oxide fuel cells (Green 2007) and in photovoltaic modules (Lenarda and Paggi 2016; Gagliardi et al. 2017) where moisture can diffuse along the interface between the textured surface of solar cells and the encapsulating polymer, promoting chemical oxidation of electric contacts. The topological features of roughness in seal contacts are also very important for the onset of wear (Leachman et al. 2014).

The presence of a network of microscopical channels is a key issue also for lubrication (Almqvist and Dasht 2006; Almqvist et al. 2014). It is in fact technologically well established that surface texturing, such as micro-dimples, can enhance lubrication by trapping oil between surfaces in relative motion (Scaraggi 2012).

Numerical Methods for the Simulation of Rough Surfaces

The generation of synthetic random rough surfaces to be used in contact mechanics simulations has received a beneficial input from fractal geometry in the 1980s and 1990s, with the pioneering article by Mandelbrot et al. (1984) investigating the fractal character of fracture surfaces of metals, which was stimulating the discussion between mathematicians (Feder 1988; Russ 1994) and engineers (Carpinteri 1994; Carpinteri and Chiaia 1995). Fractals have been largely explored for the simulation of realistic natural landscapes and forms, especially in connection with computer graphics applications, see Peitgen and Saupe (1988) for a comprehensive overview of algorithms. At the same time, the scale-invariant properties of roughness generated using these methods were opening new perspectives for the quantification of the effect of the resolution of roughness measurement systems, dramatically increased with the advent of non-contact profilometers, confocal and interferometric techniques, as well as atomic force microscopy (Majumdar and Bhushan 1990, 1991; Bhushan and Majumdar 1992; Borri-Brunetto et al. 1999; Zavarise et al. 2004b). Moreover, at the same time, the fractal generation algorithms were opening a long dispute on the use of random process theory and in particular Gaussian models for the statistical distribution of surface elevations and asperity curvatures (Zavarise et al. 2004a), which is still debated today in the case of textured surfaces (Borri and Paggi 2015).

In this section, the Weierstrass–Mandelbrot (WM) function, the Random Midpoint Displacement (RMD) algorithm and the Spectral Synthesis Method (SSM) are considered and reviewed for their popularity in contact mechanics. The algorithms are also provided in a pseudo-MATLAB language. All such methods attempt at generating a profile in 2D whose power spectral density function Φ is a power-law of the spatial frequency ω:

$$\Phi_p(\omega) = K_p \omega^{-(5-2D)}, \quad 1 < D < 2, \tag{5.1}$$

where D stands for the fractal dimension of the rough profile, which is a scale-invariant geometrical property. The same methods can be generalized to generate 3D surfaces with a power spectral density function scaling with ω as

$$\Phi_s(\omega) = K_s \omega^{-(7-2D)}, \quad 2 < D < 3, \tag{5.2}$$

where now the fractal dimension D is augmented by unity. Roughly speaking, the generated rough profile has a topological dimension intermediate between that of a Euclidean (smooth) line ($D = 1$) and that of a Euclidean (flat) plane ($D = 2$), while a rough surface has a topological dimension ranging from that of a Euclidean plane ($D = 2$) and that of a Euclidean volume ($D = 3$). These topological properties correspond to an *invasive* character of roughness.

Another measure of the topological dimension is the Hurst exponent, see (Feder 1988; Russ 1994) for its rigorous mathematical definition. In the present applications, D and H are often related to each other, being H given by the largest Euclidean dimension (2 for a profile or 3 for a surface), minus the corresponding fractal dimension D. The only exception concerns bi-fractal randomly rough surfaces, having their power spectral density function separated into two distinct power-law regimes, see Borri and Paggi (2016). In the first range, for low spatial frequencies, the Hurst exponent H is adopted for the topology characterization, while in the second range, for higher spatial frequencies above a crossover value, the fractal dimension D is used.

The Weierstrass–Mandelbrot Function

One interesting model of random roughness based on the Weierstrass–Mandelbrot (WM) function was first explored by Majumdar (1989) to generate rough profiles and rough surfaces. The method was exploited by many authors and in particularly by Ciavarella in a series of fundamental articles (see Ciavarella et al. (2000), Ciavarella et al. (2006), Paggi and Ciavarella (2010) for a selection). In this method, a simple and closed-form analytic equation is providing the pointwise elevation $z(x)$ for a rough profile (Peitgen and Saupe 1988, Sect. 1.4.5):

$$z(x) = A_0 \sum_{n=0}^{\infty} \gamma^{(D-2)n} \cos\left(2\pi \frac{\gamma^n x}{\lambda}\right), \tag{5.3}$$

where A_0, $\gamma(\gamma > 1)$ and $D(1 < D < 2)$ are model parameters, λ is the longest wavelength of the profile and $n \in \mathbb{N}$ denotes the number of length scales of roughness involved. The sum is usually truncated at a given $n_{\max}$, obtaining a pre-fractal rough profile. Its extension to 3D is straightforward and a possible expression is that used in Paggi and Ciavarella (2010):

$$\begin{aligned} z(x, y) =& A_0 \sum_{m=1}^{M} \sum_{n=0}^{n_{\max}} \gamma^{(D-3)n} \left\{\cos \phi_{m,n} \right. \\ & \left. - \cos\left[2\pi \gamma^n \frac{\sqrt{x^2 + y^2}}{\lambda} \cos\left(\tan^{-1}\left(\frac{y}{x}\right) - \frac{\pi m}{M}\right) + \phi_{m,n}\right]\right\}, \end{aligned} \tag{5.4}$$

where m is an additional integer summation index ranging from 1 to M, and $\phi_{m,n}$ are randomly generated phases for each value of m and n.

The advantage of using WM profiles and surfaces is given by the fact that they are constructed by adding cosinusoidal functions with random phases and rescaled amplitudes. However, for a given m and n pair, the additional contribution to the interface topology is a smooth and differentiable function whose geometrical properties and scaling can be quantified in closed form. This advantage, in conjunction with contact models based on asperities, led to the development of theoretical models for the prediction of the dependency of contact quantities, such as the real area of contact and the normal contact stiffness, on the surface resolution (Ciavarella et al. 2000).

Just as an illustrative example, examining a rough profile, a single cosinusoidal function of x/λ, with a λ-periodicity in the horizontal direction, is obtained for $n = 0$. This function has a peak (maximum) at $x/\lambda = 0$ and at $x/\lambda = 1$, and a valley

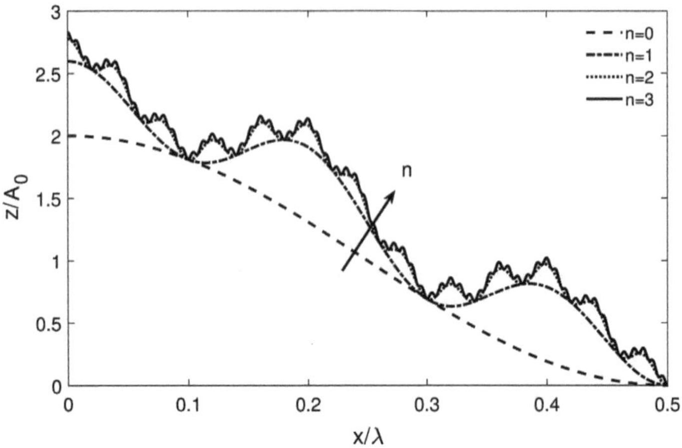

Fig. 5.3 Weierstrass–Mandelbrot profiles depending on the resolution parameter n ranging from 0 up to 3, fractal dimension $D = 1.25$ and scaling parameter $\gamma = 5$

(minimum) at $x/\lambda = 0.5$ (see the dashed curve shown in Fig. 5.3 in the range $0 \leq x/\lambda \leq 0.5$, symmetric in the range $0.5 \leq x/\lambda \leq 1$). The profile with $n = 1$ obtained by superimposing to the profile with $n = 0$ a finer cosinusoidal function with a shorter wavelength λ/γ and a rescaled amplitude $A_0\gamma^{(D-2)}$ is also shown in Fig. 5.3 with a dashed-dotted curve. The addition of further length scales proceeds in a similar manner by adding finer roughness with shorter wavelengths up to λ/γ^n and rescaled amplitudes $A_0\gamma^{(D-2)n}$. For $n = 2$ and $n = 3$, this procedure leads to the profiles shown in Fig. 5.3 with dotted line and continuous line, respectively.

The Random Midpoint Displacement Algorithm

Rough surfaces with fractal properties can also be numerically generated according to the Random Midpoint Displacement (RMD) algorithm (Peitgen and Saupe 1988, Sect. 2.2.4). This method allows generating rough surfaces with a power spectral density function approximating a power-law dependency of the spatial frequency. The method was proposed by Borri-Brunetto et al. (1999) for the generation of synthetic randomly rough surfaces to be used as input for the boundary element method for contact mechanics simulations.

In the algorithm, square surfaces with different resolutions can be generated by successively refining an initial mesh by a successive addition of a series of intermediate heights. The number of successive refinements is defined by the parameter n, which is related to the number of heights per side of the squared generated grid, $2^n + 1$. Given L the lateral size of the surface, the grid spacing is therefore $\delta = L/2^n$, and the resolution is defined as $s = 1/\delta$. The method generates surfaces with higher

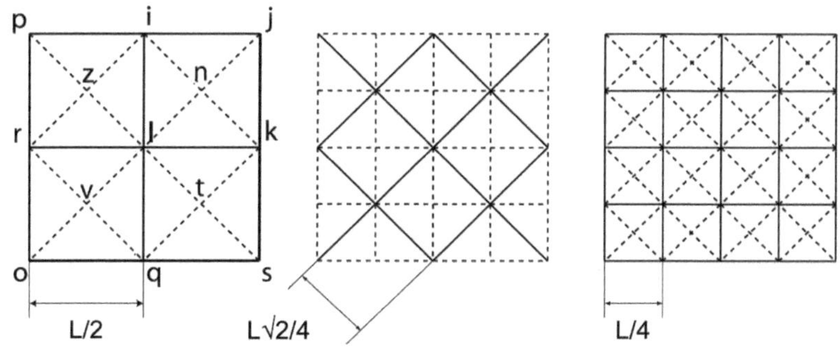

Fig. 5.4 Recursive steps for the generation of rough surfaces using the RMD algorithm (from Paggi and He (2015))

n that are finer representations of the coarser ones, i.e. the height field of a surface with $n = i, i \in \mathbb{N}$, contains the height field of the coarser realizations with $n < i$.

A sketch showing how the RMD algorithm operates is available in Fig. 5.4. Starting with $n = 1$, the elevation of the four corner nodes of the grid, nodes o, p, j, s in Fig. 5.4, are set equal to zero. Afterward, the elevation of the central point of the grid, l, is determined by the average value of the elevations of the corner nodes, plus a random number extracted from a Gaussian distribution with zero mean and variance $\sigma_1^2 = \sigma_0^2/2^{(3-D)/2}$, where σ_0^2 is a free parameter, often set equal to $1/\sqrt{0.09}$ to reproduce surfaces consistent with real ones. The elevations of the nodes i, k, q, r are then assigned by averaging over three elevations, those of the two corner nodes and that of the central node, plus a random number extracted from a Gaussian distribution with zero mean and reduced variance $\sigma_2^2 = \sigma_1^2/2^{(3-D)/2}$. This procedure is further iterated at the next refinement, $n = 2$. This version differs from the original RMD algorithm detailed in Peitgen and Saupe (1988) by the single fact that the elevations of the four initial corner nodes are set equal to zero rather than randomly assigned. The reason for that is to avoid topologies dominated by these initial values, which might constitute a bias especially at low resolution. The Algorithm is detailed in Algorithm 1.

An increase in the fractal dimension D from 2.1 to 2.5 leads to an increase in the amount of roughness, as shown in Fig. 5.5. Although any fractal dimension between 2 and 3 could be theoretically explored, measurements on real surfaces suggest limiting D up to 2.2 or 2.3 (Bouchaud 1997; Jones et al. 2016). The effect of the surface resolution can be visualized by varying the generation parameter n, see Fig. 5.6 by comparing the same surface generated with $n = 5$ or with $n = 8$. It has to be remarked that the RMD algorithm allows generating surfaces whose spectral density function has an approximated power-law dependency on ω with a power-law exponent related to the expected one, based on the input fractal dimension D, which is however retrieved only for large values of n.

Function: z=RMD(σ_0^2,n,D); **Input**: generation parameter σ_0^2, resolution parameter n, fractal dimension D.

(1) `rand`('state',0), rnd=`rand`(1)*100, `randn`('state',rnd);
(2) $N = 2^n$;
(3) $x = 1 : 1 : 2^n + 1$;
(4) $z(1, 1) = 0, z(1, N + 1) = 0, z(N + 1, 1) = 0, z(N + 1, N + 1) = 0$;
(5) $\alpha = \sigma_0^2$;
(6) $M = N, m = N/2$;
(7) **for** $i = 1 : n$;

  (7.1) $\alpha = \alpha/2^{(3-D)/2}$;
  (7.2) **for** $j = m + 1 : M : N - m + 1$
    (7.2.1) **for** $k = m + 1 : M : N - m + 1$
    (7.2.2) $z(j, k) = (z(j + m, k + m) + z(j + m, k - m) + z(j - m, k + m) + z(j - m, k - m))/4 + \alpha \times$ `randn`;
    (7.2.3) **end**
  (7.3) **end**

(8) $\alpha = \alpha/2^{(3-D)/2}$;
(9) **for** $j = m + 1 : M : N - m + 1$

  (9.1) $z(j, 1) = (z(j + m, 1) + z(j - m, 1) + z(j, m + 1))/3 + \alpha \times$ `randn`;
  (9.2) $z(j, N + 1) = (z(j + m, N + 1) + z(j - m, N + 1) + z(j, N - m + 1))/3 + \alpha \times$ `randn`;
  (9.3) $z(1, j) = (z(1, j + m) + z(1, j - m) + z(m + 1, j))/3 + \alpha \times$ `randn`;
  (9.4) $z(N + 1, j) = (z(N + 1, j + m) + z(N + 1, j - m) + z(N - m + 1, j))/3 + \alpha \times$ `randn`;

(10) **end**
(11) **for** $j = m + 1 : M : N - m + 1$

  (11.1) **for** $k = m + 1 : M : N - m + 1$
  (11.2) $z(j, k) = (z(j, k + m) + z(j, k - m) + z(j + m, k) + z(j - m, k))/4 + \alpha \times$ `randn`;
  (11.3) **end**

(12) **end**
(13) **for** $j = M + 1 : M : N - m + 1$

  (13.1) **for** $k = m + 1 : M : N - m + 1$
  (13.2) $z(j, k) = (z(j, k + m) + z(j, k - m) + z(j + m, k) + z(j - m, k))/4 + \alpha \times$ `randn`;
  (13.3) **end**

(14) **end**
(15) $M = M/2, m = m/2$;
(16) **end**

Output: matrix z containing the generated height field.

Algorithm 1: Random Midpoint Displacement algorithm

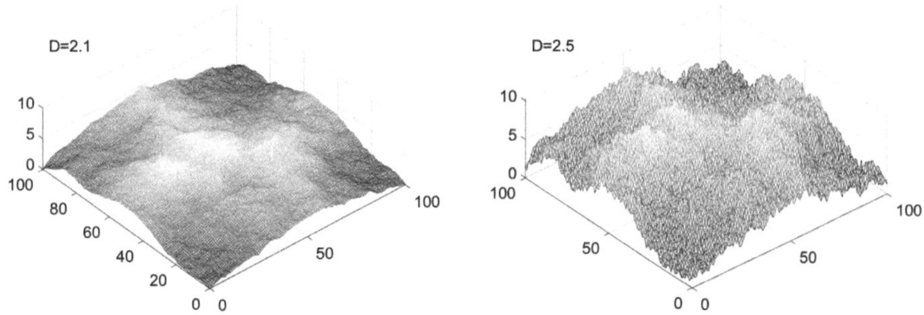

Fig. 5.5 The effect of the fractal dimension D on the roughness of numerically generated RMD surfaces ($n = 7$)

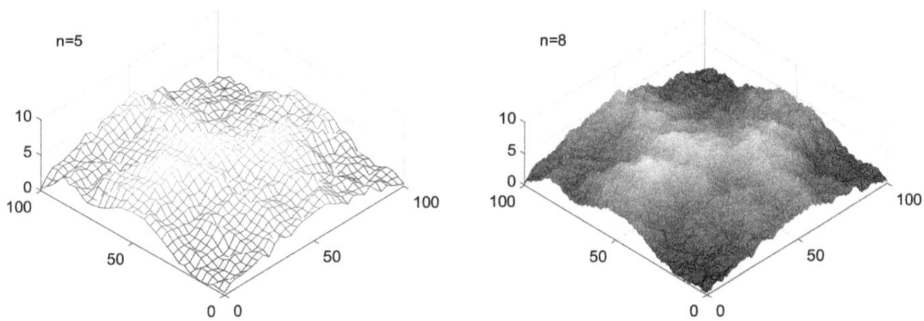

Fig. 5.6 The effect of the generation parameter n on the resolution of numerically generated RMD surfaces ($D = 2.3$)

The Generalized Spectral Synthesis Method for Fractal or Bi-fractal Surfaces

Synthetic rough surfaces can also be generated based on the Spectral Synthesis Method (SSM) (Peitgen and Saupe 1988, Sect. 2.5.3). The basic algorithm allows generating rough surfaces directly from an imposed power-law spectral density function with a given slope related to the desired surface fractal dimension D. Therefore, it is more accurate than RMD in reproducing randomly rough surfaces with a prescribed power-law spectral density function.

Hence, in this algorithm, the starting point is the power spectral density function $\Phi(\omega)$. In its generalized form proposed in Borri and Paggi (2016) for isotropic bi-fractal surfaces, Φ is defined in terms of a Hurst exponent $H(0 < H < 1)$ in the low-frequency domain, and in terms of a fractal dimension $D(2 < D < 3)$ in the high-frequency range, see Fig. 5.7:

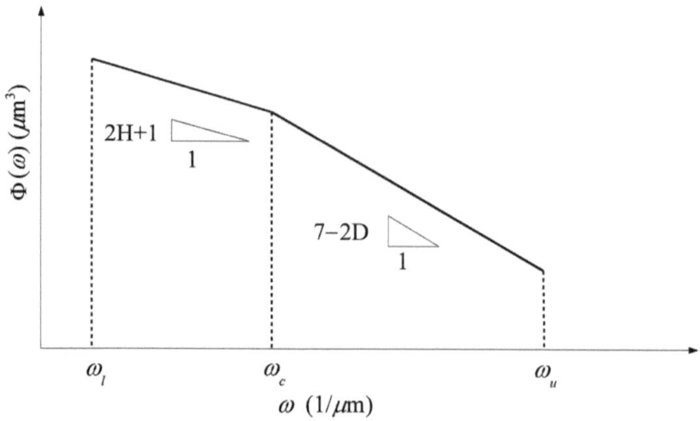

Fig. 5.7 The power spectral density function in a bi-logarithmic diagram used in input for the SSM algorithm, generalized to deal with bi-fractal surfaces

$$\Phi(\omega) = \begin{cases} \dfrac{G_1^{2(1-H)}}{\omega^{1+2H}}, & \text{for}\, \omega_l < \omega \le \omega_c \\ \dfrac{G_2^{2(D-2)}}{\omega^{7-2D}}, & \text{for}\, \omega_c \le \omega < \omega_u \end{cases} \tag{5.5}$$

where $\omega_l = 2\pi/L$ is the lowest frequency related to the sample lateral size L, $\omega_u = 2\pi/\delta$ is the highest frequency related to the sampling interval δ and ω_c is a crossover frequency which denotes the transition between the two power-law regimes.

The algorithm is detailed in all its steps in Algorithms 2 and 3, and it basically employs a discrete inverse fast Fourier transform to generate the height field from the expression of the power spectral density function given in input.

As remarked, the modified SSM algorithm can generate surfaces with a bi-fractal behaviour that shows a kink in the power spectral density function for low frequencies. This kind of surface is typically the result of machining and refining processes, like grinding, where the threshold frequency ω_c is related to physical parameters like the dimension of the grit particles.

As an example, bi-fractal surfaces have been generated in Borri and Paggi (2016) with a lateral size of 850 μm and 512 heights per side. A crossover frequency ω_c was set equal to 15 μm^{-1}. The fractal dimension in the high-frequency range was kept constant and equal to $D = 2.05$, while it was varied in the low-frequency range by acting on the Hurst exponent H. The simulated surfaces have the power spectral density function computed a posteriori on the generated height fields with a clear change of slope for $\omega = \omega_c$ as expected, see the bi-logarithmic plot in Fig. 5.8a. The 3D topographies of two cases corresponding to $H = 0.05$ and $H = 0.95$ are shown in Fig. 5.8b and 5.8c, respectively.

Bi-fractal rough surfaces obtained via a modification of the fractal dimension D at high frequencies is also a case of practical interest, since it can be the outcome of a

---

Function: z=SSM(n, ω_c, H, D) **Input**: generation parameter n, crossover frequency ω_c, Hurst exponent H, fractal dimension D.

---

(1) `rand`('state',0); $N = 2^n$;
(2) **for** $i = 0 : N/2$
  (2.1) **for** $j = 0 : N/2$
    $phase = 2\pi \times \texttt{rand}$; $\omega = \sqrt{i^2 + j^2}$; $radm = \omega_l^{-(H+1)}$; $radp = \omega_l^{-(4-D)}$;
    if$(\omega < \omega_l)$
      if$(i \neq 0 | j \neq 0)$
      $rad = (radp/radm) \times \omega^{-(H+1)} \times \texttt{random('norm'}, 0, 1)$;
      else
      $rad = 0$;
      end
    else
      if$(i \neq 0 | j \neq 0)$
      $rad = \omega^{-(4-D)} \times \texttt{random('norm'}, 0, 1)$;
      else
      $rad = 0$;
      end
    end
  (2.2) $a(i+1, j+1) = \texttt{complex}(rad \times \cos(phase), rad \times \sin(phase))$;
  (2.3) **if**$(i == 0)$
    $i_0 = 0$;
  (2.4) **else**
    $i_0 = N - i$;
  (2.5) **end**
  (2.6) **if**$(j == 0)$
    $j_0 = 0$;
  else;
    $j_0 = N - j$;
  end
  (2.7) $a(i_0+1, j_0+1) = \texttt{complex}(rad \times \cos(phase), -rad \times \sin(phase))$;
  (2.8) **end**
(3) **end**

Algorithm 2: First part of the algorithm of the modified Spectral Synthesis Method for bi-fractal surface generation.

microscopical texturing. As an example, Fig. 5.9a shows the power spectral density function of SSM surfaces generated with a Hurst exponent in the low-frequency range equal to 0.95, which is typical of a very smooth surface at the mesoscale. The fractal dimension in the high-frequency range is varied from $D = 2.05$ to $D = 2.95$. The topography of fractal surfaces corresponding to $D = 2.05$ and $D = 2.95$ is shown in Fig. 5.9b and 5.9c, respectively. As expected, the long wavelength features of roughness are not influenced by changing D, while the overall roughness is augmented by increasing the fractal dimension, as well as the depth of peaks and valleys.

(4) $a(N/2+1,1) = \texttt{complex}(\texttt{real}(a(N/2+1,1)),0);$
(5) $a(1,N/2+1) = \texttt{complex}(\texttt{real}(a(1,N/2+1)),0);$
(6) $a(N/2+1,N/2+1) = \texttt{complex}(\texttt{real}(a(N/2+1,N/2+1)),0);$
(7) **for** $i = 1 : N/2-1$

for $j = 1 : N/2-1$
$phase = 2\pi \times \texttt{rand}; \omega = \sqrt{i^2+j^2};$
$radm = \omega_l^{-(H+1)}; radp = (\omega_l)^{-(4-D)};$
if $(\omega < \omega_l)$
$rad = (radp/radm) \times \omega^{-(H+1)} \times \texttt{random}('\texttt{norm}', 0, 1);$
else
$rad = \omega^{-(4-D)} \times \texttt{random}('\texttt{norm}', 0, 1);$
end
$a(i+1, N-j+1) = \texttt{complex}(rad \times \cos(phase), rad \times \sin(phase));$
$a(N-i+1, j+1) = \texttt{complex}(rad \times \cos(phase), -rad \times \sin(phase));$
end

(8) **end**
(9) $z = \texttt{ifft2}(a);$

Output: matrix z containing the generated height field.

Algorithm 3: Second part of the algorithm of the modified Spectral Synthesis Method for bi-fractal surface generation.

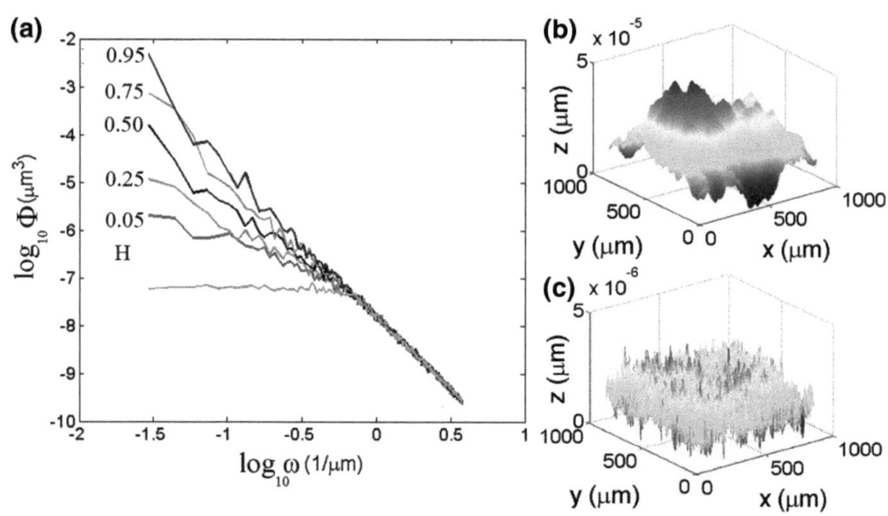

Fig. 5.8 Power spectral density functions of bi-fractal rough surfaces generated with the modified SSM algorithm with $D = 2.05$ and different values of H (**a**). Surface with $H = 0.05$ (**b**). Surface with $H = 0.95$ (**c**)

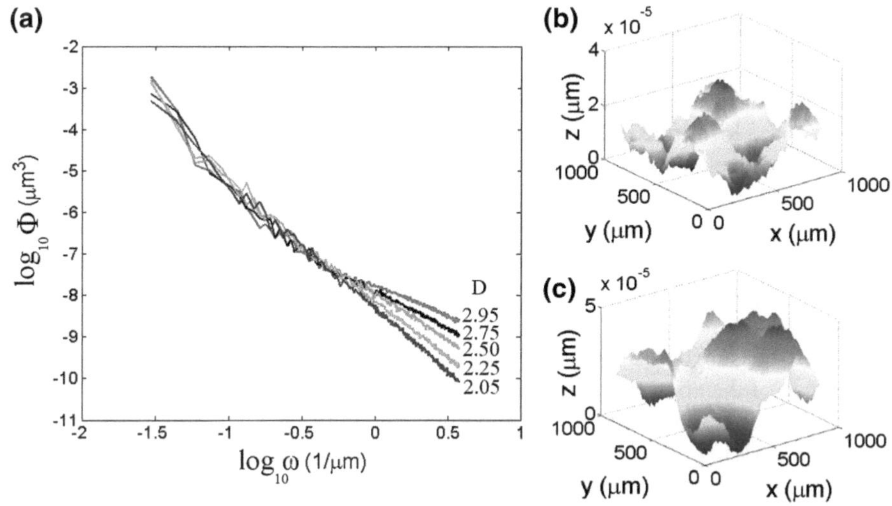

Fig. 5.9 Power spectral density functions of bi-fractal rough surfaces generated with the modified SSM algorithm with $H = 0.95$ and different values of D (**a**). Surface with $D = 2.05$ (**b**). Surface with $D = 2.95$ (**c**)

Therefore, the SSM algorithm gives more flexibility than the WM and the RMD algorithms in generating complex forms of roughness. On the other hand, the main difference from the other algorithms, which has some relevance for multi-scale analyses performed to assess the role of the surface resolution, regards the fact the height field of a surface generated with a given n is not included in the height field of a finer representation of the same surface obtained for $n' > n$. In doing that, the SSM algorithm is reproducing exactly what happens in reality when a surface is sampled using a confocal or an interferometric profilometer at different magnifications by using different lenses. Lower magnification measures allow sampling a portion of the surface around a given point (x_c, y_c) with a lateral dimension L and a discretization consisting in N heights per side (usually $N = 256$ or 512). The repetition of the same measure at the point (x_c, y_c) with a higher magnification provides again a sample of N heights per side, but this time with a lateral dimension $L' < L$, depending on the ratio between the magnifications used. Therefore, the finer surface has a height field which has few points in common with the coarser one, while it provides a zoom into the surface microstructure at point (x_c, y_c). The WM and RMD algorithms, on the other hand, keep the lateral dimension L the same regardless of the resolution used, and simply refine the height field over the whole surface.

Scaling of the Real Area of Contact

When a rough surface is pressed against a half-plane, for low loads only the asperities, which are the local maxima of the interface elevations, come into contact forming isolated contact patches. By increasing the normal load, the number of contact patches progressively increases. Therefore, if one examines the contact domain, which is given by the collection of patches in contact as seen from a top view perpendicular to the half-plane, a progressive transition from a very rarefied contact domain up to a dense one takes place, till the limit of full contact when the real contact area becomes equal to the nominal one. However, in many applications, the real area of contact, which is a measure of the size of the contact domain, remains confined within a small percentage of the nominal one, which is the square of the lateral specimen size.

The surface resolution plays an important role in the predicted real contact area, and the underlying effect was intensively investigated in (Borri-Brunetto et al. 1999; Ciavarella and Demelio 2000; Ciavarella et al. 2000).

To fix ideas, let solve the frictionless normal contact problem without adhesion between a rigid indenter whose shape is defined by the WM profile described by Eq. (5.3) ($D = 1.25$, $\gamma = 5$, $A_0/\lambda = 0.0025$, see Fig. 5.3) and a deformable half-plane, using the finite element method with interface finite elements with eMbedded Profile for Joint Roughness (MPJR interface finite elements) to discretize the rough interface, as detailed in Chap. 4, section "Interface Finite Elements with Embedded Profile for Joint Roughness". Therefore, in this setting, roughness associated to the indenter is analytically defined as $e_1(x) = \overline{h}_2 + h^*(x)$, where $h^*(x) = z(0) - z(x)$. Different resolution parameters n can be examined, with n ranging, for instance, from zero up to three.

To solve the present problem numerically, it is convenient to take into account the λ-periodicity of the profile by imposing periodic boundary conditions at $x/\lambda = 0$ and at $x/\lambda = 1$. Then, due to the symmetry in geometry and loading, which is given by an imposed far-field displacement w to the rigid indenter, the contact predictions can be shown only in the range $0 \leq x/\lambda \leq 0.5$, since the solution is symmetric elsewhere. The finite element discretization is chosen to have $h/\lambda_{\min} = 0.125$, where $\lambda_{\min} = \lambda/\gamma^n$, in order to properly resolve the contact traction distribution for any rough profile.

Contact pressures $p = -p_n$ along the interface are shown in Fig. 5.10 by varying n from 0 (topmost figure) to 3 (bottommost figure). The set of curves shown in the subfigures correspond to the same values of the various imposed far-field displacements. The solution for $n = 0$ shows that the size of the contact domain where tractions are not vanishing is a continuous increasing function of w and it spreads all over the interface, achieving full contact. A refinement of the profile by adding another term in the series ($n = 1$) leads to an increase in the contact pressures for the same imposed displacement w. The full contact condition is attained in this case, but it requires higher values of w. The addition of further length scales ($n = 2$ and $n = 3$) drastically increases the value of the contact pressures and it reduces the real contact area which localizes near the asperities, inhibiting the achievement of full

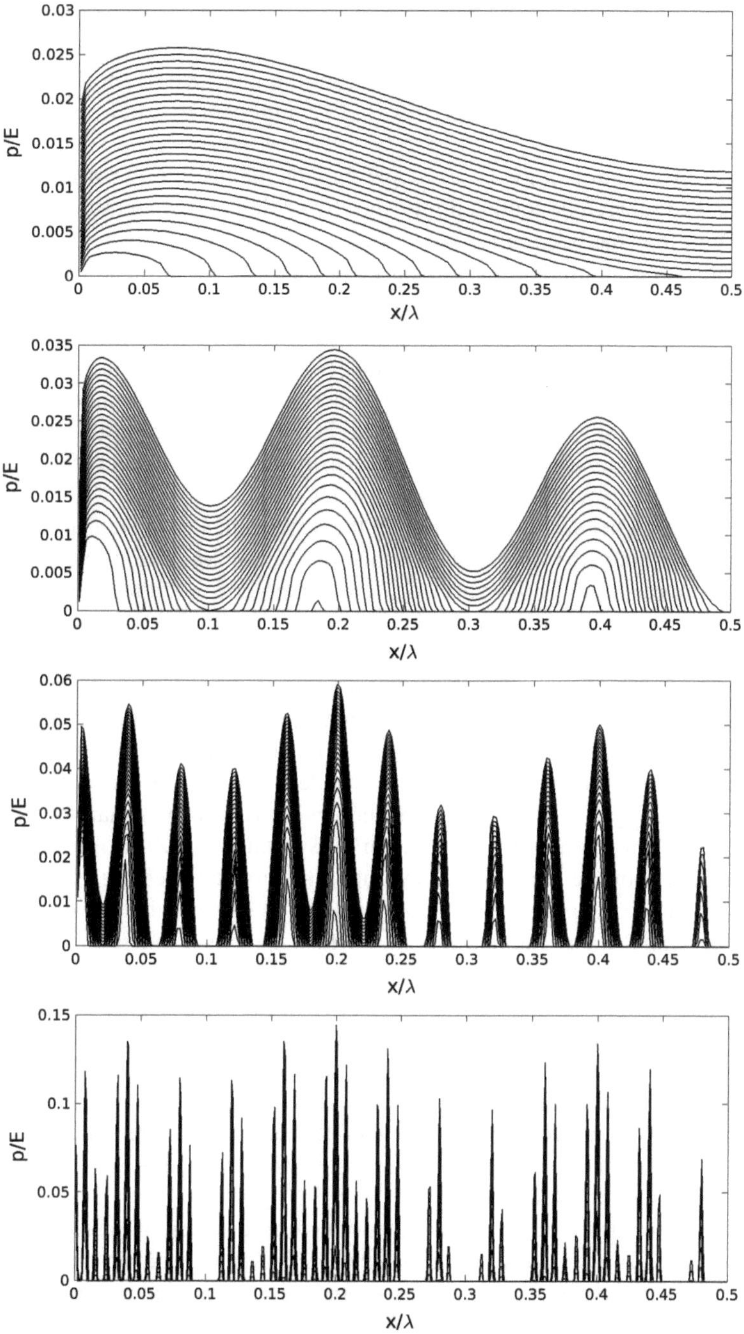

Fig. 5.10 Dimensionless contact pressure along the WM rough profiles shown in Fig. 5.3 in contact with a half-plane, depending on the resolution parameter n from 0 to 3, from the top to the bottom. E, λ denote, respectively, the composite Young's modulus and the longest wavelength of the profile (from Paggi and Reinoso (2018))

contact. The contour plot of the dimensionless stress field component σ_{yy}/E in the bulk in correspondence to the maximum imposed displacement is shown in Fig. 5.11, quantitatively showing up to which depth from the interface the non-uniformity in the contact tractions influences the stress field component σ_{yy}, before smearing out its effect.

A closer observation at the trends depending on the resolution parameter n highlights that the real area of contact diminishes by adding more and more details to the profile. This effect has been called *lacunarity* of the contact domain by Borri-Brunetto et al. (1999). A quantitative assessment of lacunarity can be made by plotting the dimensionless real contact area, $A/(\lambda L)$, versus the total dimensionless compressive normal force, $F/(E\lambda L)$, where L denotes the out-of-plane thickness of the model that is unity in the present plane strain setting (see Fig. 5.12). The value $A/(\lambda L) = 1$ corresponds to full contact. The addition of finer length scales of roughness leads to a progressive reduction of the real contact area in correspondence of the same applied force, with a trend consistent with analytical estimates in Ciavarella et al. (2000) for WM profiles, strictly valid for very large values of n.

A more comprehensive investigation of the dependency of the real area of contact on the fractal dimension and of the surface resolution was made in Paggi and He (2015), where randomly rough surfaces were numerically simulated using the RMD algorithm. Their contact with an elastic half-plane was simulated using the boundary element method, see Chap. 4, section "The Boundary Element Method" for details on the computational approach. Numerical predictions in terms of dimensionless real area of contact versus dimensionless contact pressure corresponding to the same applied far-field normal displacement, related to surfaces with resolution parameter $n = 8$ and different fractal dimensions D, considering ten random repetitions for each set, is shown in Fig. 5.13. In that plot, σ denotes the root mean square of the surface heights, which is a parameter dependent on D as discussed in Zavarise et al. (2004b). The dependency of the real contact area upon the contact pressure is almost linear and, by increasing the surface fractal dimension, the slope of the curves diminishes. This is physically due to the fact that the surface appears rougher and rougher, and less asperities come into contact for the same level of imposed far-field displacement.

The dependency of the real contact area upon the surface resolution is analysed in Fig. 5.14, where the contact response of a RMD surface with $D = 2.3$ and different values of the generation parameter n is depicted. Considering a linear approximation for the relation between real contact area and pressure, a power-law dependency of the type $A/p \sim \delta^{0.37}$ is noticed, which is in good agreement with theoretical predictions provided by Persson theory of contact (Persson 2001b), suggesting $A/p \sim \delta^{D-2} = \delta^{0.3}$ for the present problem.

It has to be remarked that a careful quantitative assessment of the slope of the real contact area–load curve was leading to a wide debate in the contact mechanics community, directed towards a comparison of predictions provided by different contact models. The point of departure was a seminal article by Bush et al. (1975), where an asperity contact model was proposed within the statistical framework in Greenwood and Williamson (1966) but introducing a joint probability density function for the asperity radii of curvature and their elevations, rather than adopting an average

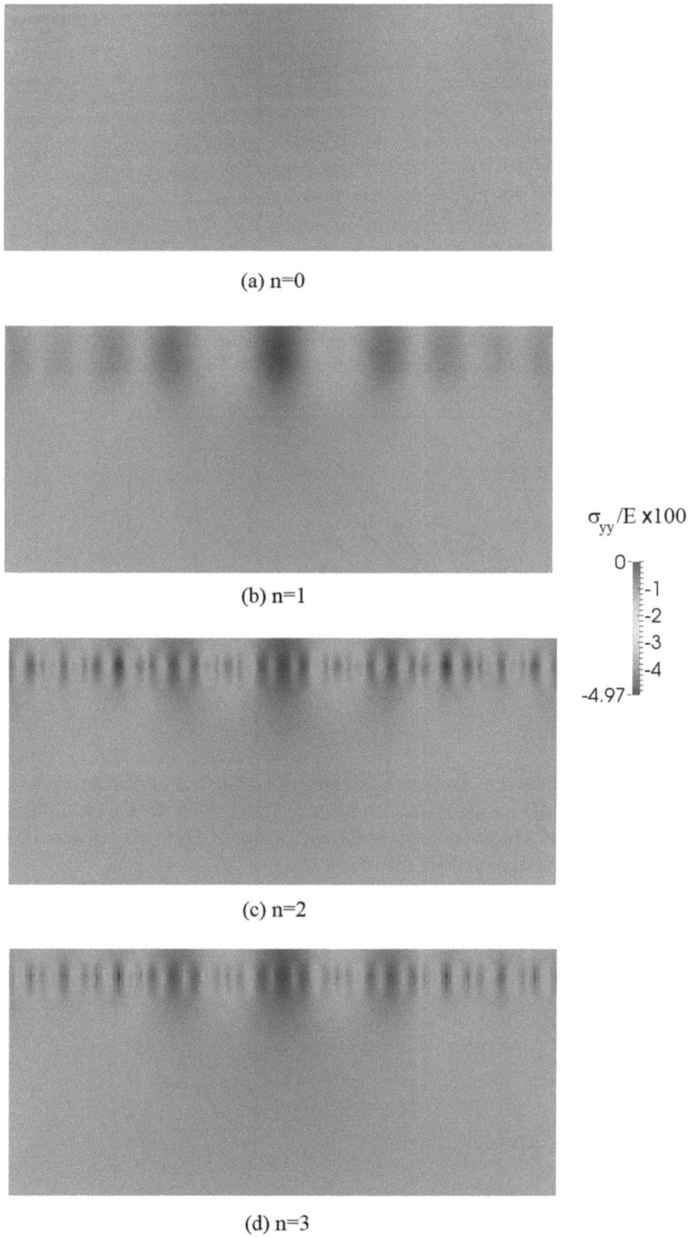

Fig. 5.11 Dimensionless vertical contact stress σ_{yy} in the bulk corresponding to the maximum imposed far-field displacement leading to the contact pressures in Fig. 5.10, depending on the resolution parameter n. Adapted from Paggi and Reinoso (2018)

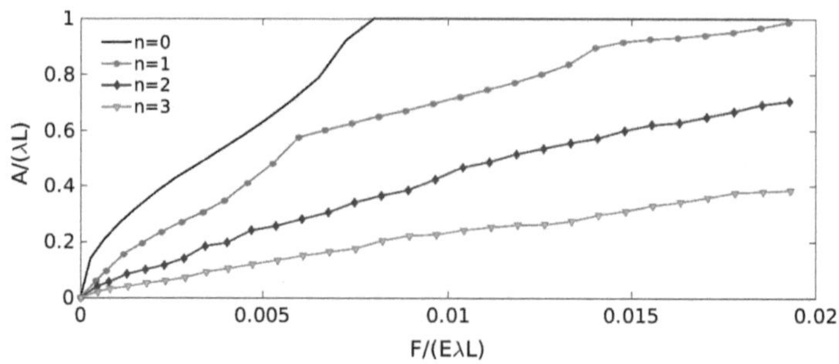

Fig. 5.12 Dimensionless real contact area versus dimensionless contact force for WM rough profiles, depending on the resolution parameter n. Adapted from Paggi and Reinoso (2018)

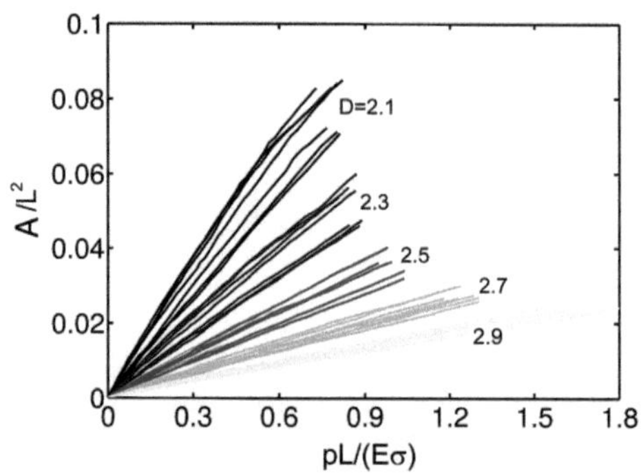

Fig. 5.13 Dependency of the dimensionless real area of contact versus dimensionless pressure for numerically generated RMD rough surfaces with $n = 8$ and different fractal dimensions D. Ten randomly generated surfaces have been tested for each value of D, to assess also the scatter in the predictions. Adapted from Paggi and He (2015)

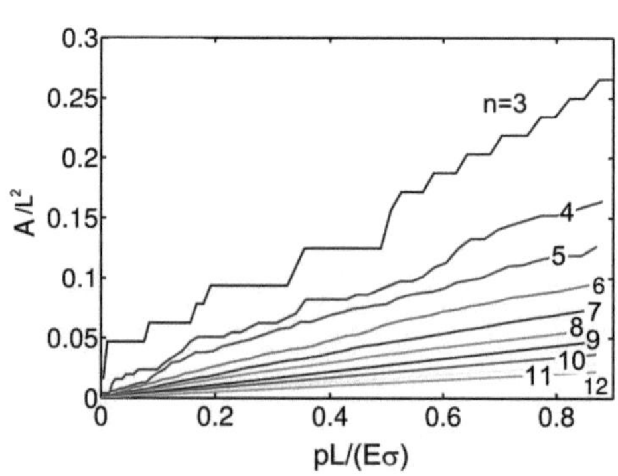

Fig. 5.14 Dependency of the dimensionless real area of contact versus dimensionless pressure for numerically generated RMD rough surfaces with $D = 2.3$ and different resolutions defined by the generation parameter n. Adapted from Paggi and He (2015)

constant radius. In the low contact regime (asymptotic approximation for low loads), the model suggested a real contact area scaling linearly with load, depending only on the composite elastic modulus E and the second moment of the spectral density function, m_2, which physically corresponds to the variance of the profile slopes. This suggested the use of the following parameter to measure the slope of the real contact area–load relationship:

$$\kappa = E\sqrt{2m_2}\frac{A}{F}, \tag{5.6}$$

where A is the real contact area and F is the total normal load. The extrapolation of the validity of this parameter, and therefore of the approximation of the asymptotic model by Bush et al. (1975) (BGT-A model) to any load level was inappropriate, as pointed out by Paggi and Ciavarella (2010), since the complete model (BGT-G model) should have been considered. Moreover, elastic interactions between asperities were not included in the model by Bush et al. (1975), and therefore, Paggi and Ciavarella (2010) proposed to improve it by accounting for this effect leading to a refined model (BGT-I model), following the approach already proposed in Ciavarella et al. (2008b) for the contact theory by Greenwood and Williamson (1966) (GW-N model), to account for elastic interactions (GW-I model).

A numerical campaign of normal contact simulations carried out on WM, RMD and SSM surfaces with generation parameter $n = 8$, different fractal dimensions D from 2.3 till 2.95, and for the same dimensionless mean plane separation $d/\sigma = 1.0$, was carried out in Paggi and Ciavarella (2010). Ten surfaces for each value of D were randomly generated and the boundary element method was employed to obtain reference numerical results not based on the model assumptions of stochastic micromechanical contact theories (asperity-based models) or of the Persson's theory, to assess the goodness of their predictions. Results revealed a strong dependency of the parameter κ upon the bandwidth parameter of the surfaces, $\alpha = m_0 m_4/m_2^2$, where m_0, m_2 and m_4 denote the variance of asperity heights, slopes and curvatures, see the bi-logarithmic plots in Figs. 5.15, 5.16 and 5.17 for WM, RMD and SSM surfaces, respectively. The use of the bandwidth parameter instead of the fractal dimension D (or the Hurst exponent H) and of the sampling interval δ was proposed in Paggi and Ciavarella (2010) by noting that α takes into account both dependencies, since, for $\delta \ll L$, the following relation holds (Sayles and Thomas 1977; Zavarise et al. 2004b) ($2 < D < 3$):

$$\alpha \cong \frac{(D-2)^2}{(D-1)(3-D)} \left(\frac{\delta}{L}\right)^{2(D-3)}. \tag{5.7}$$

Moreover, the dependency on α was also emerging from the above micromechanical contact theories, as found in Paggi and Ciavarella (2010):

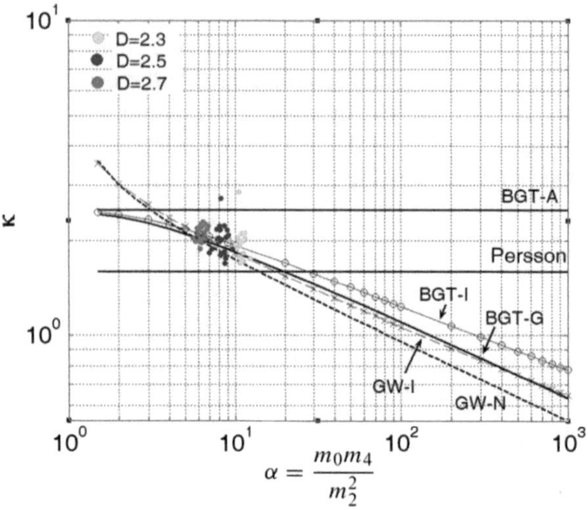

Fig. 5.15 The parameter κ for WM rough surfaces in contact with a half-plane predicted by the boundary element method (dots), and comparison with predictions by analytical theories

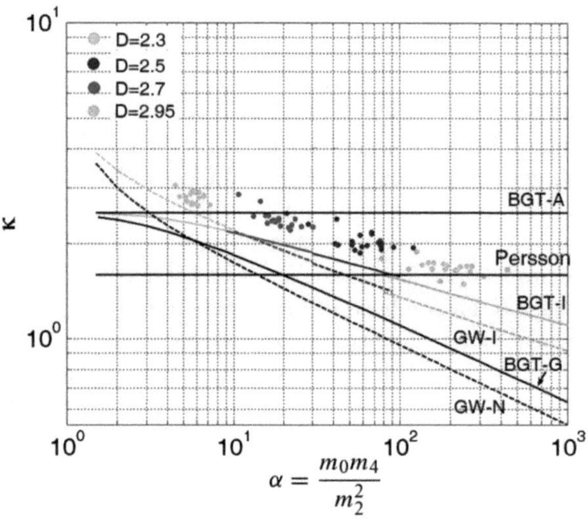

Fig. 5.16 The parameter κ for RMD rough surfaces in contact with a half-plane predicted by the boundary element method (dots), and comparison with predictions by analytical theories

$$\kappa = C_{\mathrm{GW}}(\alpha - 0.8968)^{-1/4}, \text{ for GW-N and GW-I models} \tag{5.8}$$

$$\kappa = C_{\mathrm{BGT}}\frac{3\sqrt{2}\pi}{4}\alpha^{-1/4}, \text{ for BGT-N and BGT-I models,} \tag{5.9}$$

where C_{GW} and C_{BGT} are nonlinear functions of the dimensionless mean plane separation d/σ.

Such a dependency was predicted neither by Persson's theory of contact, nor by the asymptotic BGT-A model. On the other hand, predictions by asperity contact

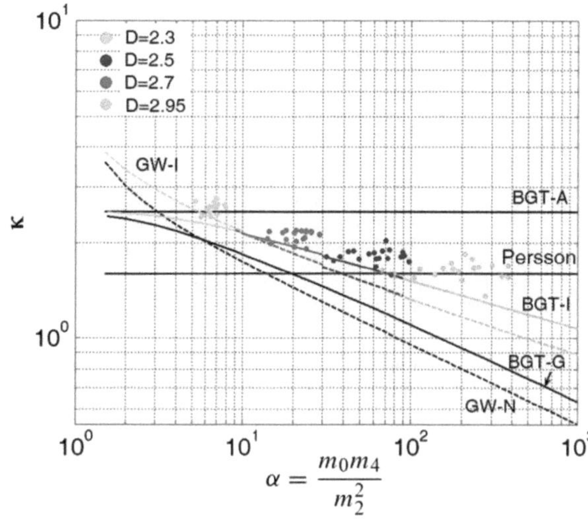

Fig. 5.17 The parameter κ for SSM rough surfaces in contact with a half-plane predicted by the boundary element method (dots), and comparison with predictions by analytical theories

theories accounting for elastic interaction effects (GW-I and BGT-I models) were quite close to the reference numerical ones shown with dots in Figs. 5.15, 5.16 and 5.17.

The Effect of the Finite Size of the Continuum

The predictions for the frictionless normal contact problem obtained using the boundary element method and summarized in the previous sections refer to half-plane contact problems. The effect of reducing the depth t of the continuum is indeed interesting in many applications and it was investigated in Paggi and Reinoso (2018) using the finite element method with the MPJR interface finite element to model and discretize a Weierstrass–Mandelbrot rough profile (see Chap. 4 for more details on the method).

In particular, a WM rough profile was generated with $D = 1.25$, $\gamma = 5$, $A_0/\lambda = 0.0025$, and $n = 2$ (Paggi and Reinoso 2018). Contact mechanics predictions for an imposed far-field displacement $w/A_0 = 3$, and for $t/\lambda = 1$, 0.5, and 0.1 were compared. The contour plots of the dimensionless stress field component σ_{yy}/E are shown in Fig. 5.18 for such three cases, highlighting an increase in the stress level for the same amount of imposed displacement w as a result of the increased stiffness of the finite-size domain obtained by reducing t. This is also evident from the comparison in terms of contact tractions along the interface shown in Fig. 5.19, which pinpoints that thin substrates are very likely to experience full contact as compared to the half-plane geometry, even in the presence of roughness that is usually promoting strong lacunarity of the contact domain.

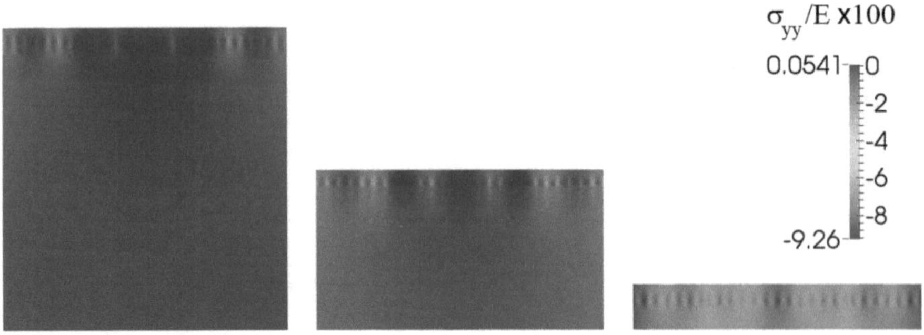

Fig. 5.18 Dimensionless vertical contact stress σ_{yy} in the continuum depending on the domain size, t, for a far-field imposed displacement $w/A_0 = 3$ (WM rough profile with $D = 1.25$, $\gamma = 5$, $n = 2$, see Fig. 5.3 for the visualization of half profile). Adapted from Paggi and Reinoso (2018)

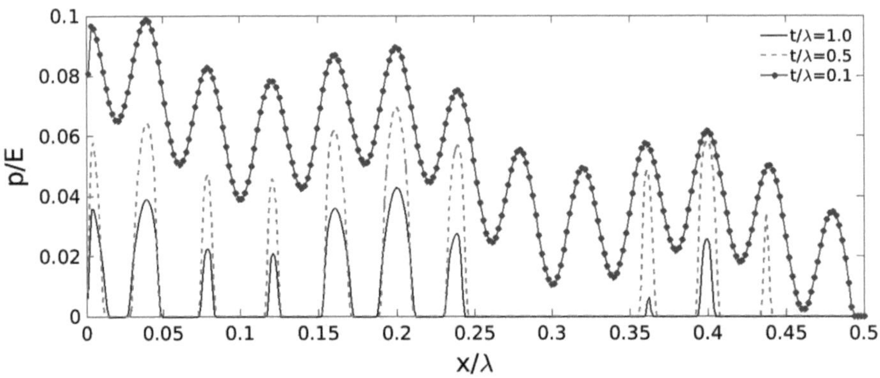

Fig. 5.19 Dimensionless contact pressure along the interface depending on the domain size, t, corresponding to the contour plots in Fig. 5.18. E and λ denote, respectively, the composite Young's modulus and the longest wavelength of the profile. Adapted from Paggi and Reinoso (2018)

Scaling of the Free Volume and Implications for Leakage

The free volume trapped between rough surfaces in contact and its evolution is of great interest for fluid lubrication and percolation. International standards (Stout et al. 1994) suggest the use of the valley fluid retention index, Svi, as a quantitative indicator for the amount of free volume. This index is deduced from the Abbott–Firestone bearing area curve (Abbott and Firestone 1933), which gives the amount of potential areas in contact estimated from the probability density function of the surface heights, neglecting elastic deformation. Specifically, the valley fluid retention index is computed as the volume V comprised between the undeformed surface and a rigid plane intersecting the surface and leaving only 20% of heights below it, divided by the root mean square of the surface heights, σ, and the nominal specimen size, L^2.

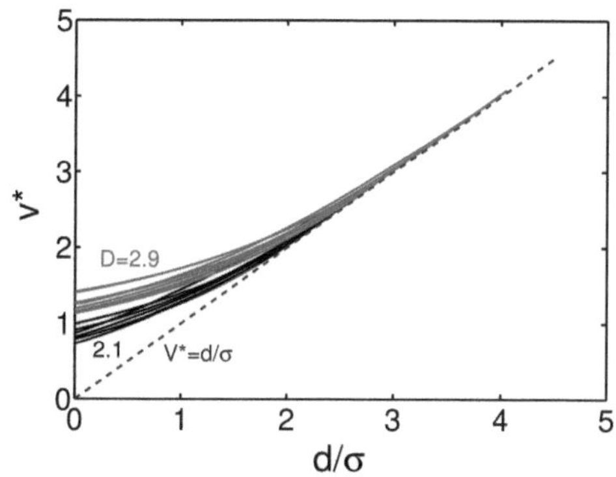

Fig. 5.20 Dependence of V^* on the mean plane separation d/σ. Note the deviation from linearity for small separations. Curves correspond to 10 randomly generated RMD rough surfaces with $n = 8$ and $D = 2.1$ (black) or $D = 2.9$ (grey). Adapted from Paggi and He (2015)

Paggi and He (2015) investigated how the actual dimensionless free volume accounting for elastic deformation, $V^* = V/(\sigma L^2)$, evolves during contact, and how it scales with the dimensionless contact pressure, $p^* = p/E$, where E is the composite Young modulus, or with the dimensionless separation between the mean planes of the rough surfaces, d/σ. Predictions from rigorous contact mechanics simulations based on the boundary element method were obtained for two sets of 10 randomly generated RMD rough surfaces with $n = 8$ and two limit values of the fractal dimension D ($D = 2.1$, very smooth, and $D = 2.9$, very rough). Results shown in Fig. 5.20 pinpoint that the simple relation $V^* \cong d/\sigma$ (depicted with a dashed line) holds only for very large separations ($d/\sigma \gtrsim 3$). For $d/\sigma \lesssim 3$, V^* and d/σ cannot be confused any longer with each other, and the free volume starts depending on the fractal dimension D.

By investigating the relation between V^* and the other contact quantities, namely the dimensionless pressure p^* and the dimensionless real contact area $A^* = A/L^2$, by varying D, the numerically predicted trends are shown in Fig. 5.21. A decay of the free volume by increasing the dimensionless contact pressure or the dimensionless real contact area is clearly observed. The relation between V^* and p^* is significantly affected by D. On the other hand, as a notable result, the relation between V^* and A^* is approximately independent of the fractal dimension, since all the curves lie in a narrow band.

To examine the role played by the surface resolution, a single surface with $D = 2.3$ and different values for the generation parameter n ranging from 3 to 8 has been also characterized in Paggi and He (2015). The trends shown in Fig. 5.22 pinpoint a convergence of the relation V^* versus p^* by increasing n. On the other hand, the relation V^* versus A^* is strongly resolution-dependent.

Numerical simulations allow also having a closer look at local quantities, such as the spatial distribution of the real contact area and the amount of the free volume $v_{i,j}$ at each grid point of the boundary element discretization defined by a pair of indices (i, j). As a general trend, due to roughness, the asperities, which are the maxima of the 3D surface, come into contact at isolated points and then progressively merge together by forming wider contact regions with vanishing free volume. Other regions

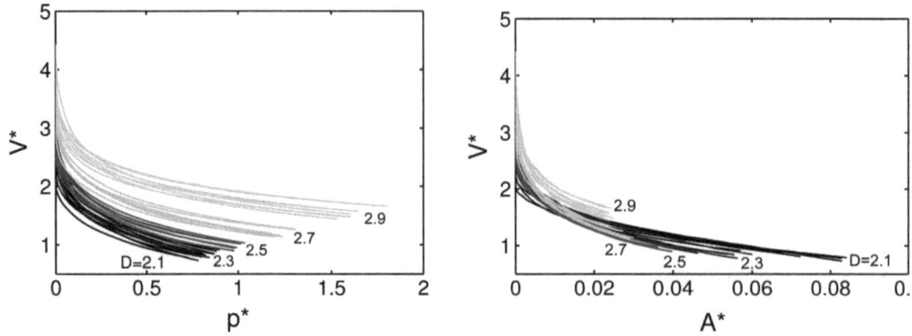

Fig. 5.21 Dimensionless volume V^* versus dimensionless nominal contact pressure p^*, or versus real contact area fraction A^*, for various fractal dimensions D (10 randomly generated RMD rough surfaces with $n = 8$ for each D). Adapted from Paggi and He (2015)

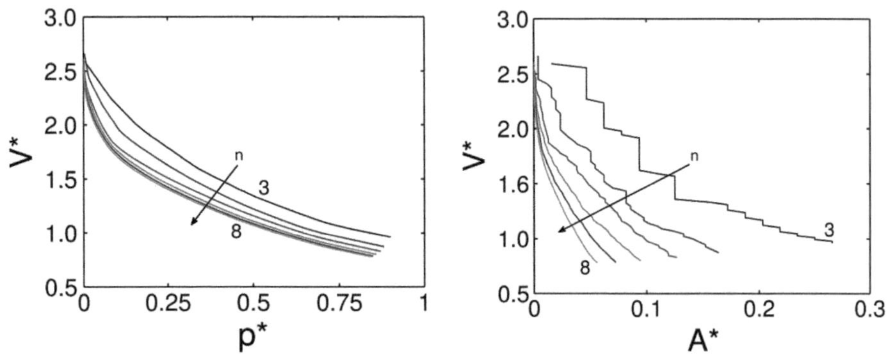

Fig. 5.22 Dimensionless volume V^* versus dimensionless nominal contact pressure p^*, or versus real contact area fraction A^*, for $D = 2.3$ and different values of the surface resolution parameter n. Adapted from Paggi and He (2015)

of the surface present free volumes $v_{i,j}$ whose size depends on the amplitude of the valleys.

A deeper insight into the morphological properties of the spatial distribution of the free volumes was made in Paggi and He (2015) by examining the contour levels corresponding to different volume thresholds, v_{th}, as shown in Fig. 5.23 for an RMD surface in contact with a half-plane with $D = 2.3$, $L = 100$, $n = 7$ and $A^* \sim 0.1$. In these contours, the black area denotes free volume domains $\mathbb{D}$ with $v_{i,j} \leq v_{th}$. Therefore, the dark islands for the limit case of $v_{th} = 0$ would simply correspond to the real contact area domain. By selecting v_{th} larger than the maximum value of the volume of the deepest valley, $v_{th} = \max(v_{i,j})$, then the picture becomes entirely black since all the grid points have $v_{i,j} \leq v_{th}$. This second limit situation corresponds to the Euclidean domain of the nominal contact area.

It has to be remarked that the contour plots in Fig. 5.23, corresponding to the same contact pressure and contact area, dynamically change during contact. At first contact,

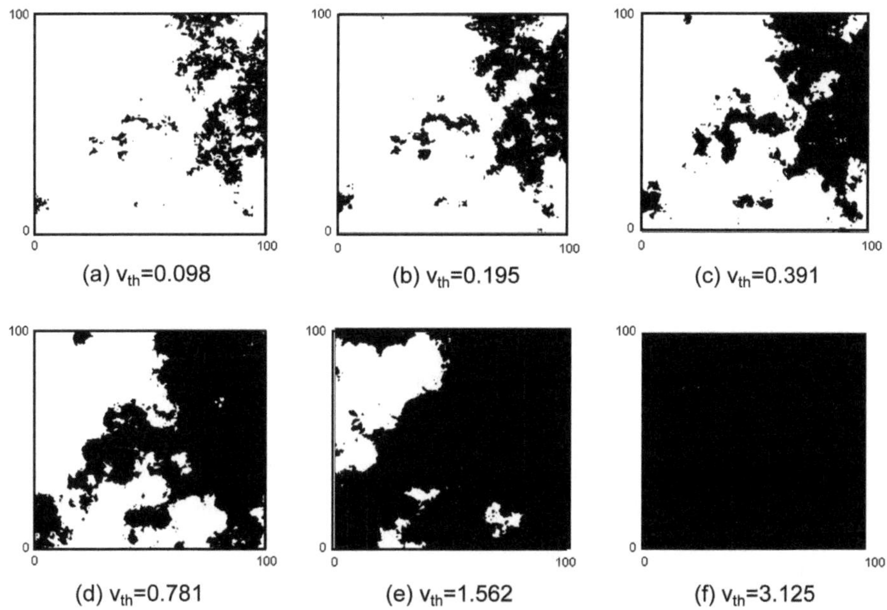

Fig. 5.23 Free volume domains (black areas) $\mathbb{D}(v_{i,j} \leq v_{th})$ corresponding to a surface with $D = 2.3$, $L = 100$, $n = 7$ and $A^* \sim 0.1$, for different free volume thresholds v_{th} (from Paggi and He (2015))

the real contact area A^* is vanishing and the free volume V^* is therefore the largest as possible. By increasing the contact pressure, the contact domain increases until the real contact area achieves $A^* = 1$ in the limit scenario of full contact. Conversely, the free volume domain progressively goes to zero in the same limit.

For each contour plot in Fig. 5.23, the topological properties of the free volume domains $\mathbb{D}(v_{i,j} \leq v_{th})$ can be investigated according to the box counting method. For $A^* = 0$, which corresponds to the undeformed rough surface, the free volume domains are expected to be self-affine as a consequence of the self-affinity of the parent surface (Bigerelle and Iost 2004). For a value $0 < A^* \leq 1$, on the other hand, the topological properties of the free volume domains have to be correlated with those of the deformed surface, whose heights have been modified by elastic deformation.

To do so, in the box counting method, for each box of lateral size r, the number N of boxes containing at least one black grid point is counted. This operation has to be repeated by varying r from 1 up to 2^n lateral size divisions, with a geometric progression of 2. The cumulative number $N(r)$ is plotted versus r in a bi-logarithmic diagram and the local fractal dimension $\mathcal{D}$ of the volume domain can be finally obtained by differentiating $\log(N)$ w.r.t. $\log(r)$.

By performing this analysis for the domains $\mathbb{D}$ in Fig. 5.23, the diagram in Fig. 5.24 on the left is obtained. The curves have a trend close to a straight line in this bi-logarithmic plot, which suggests a power-law scaling typical of fractals.

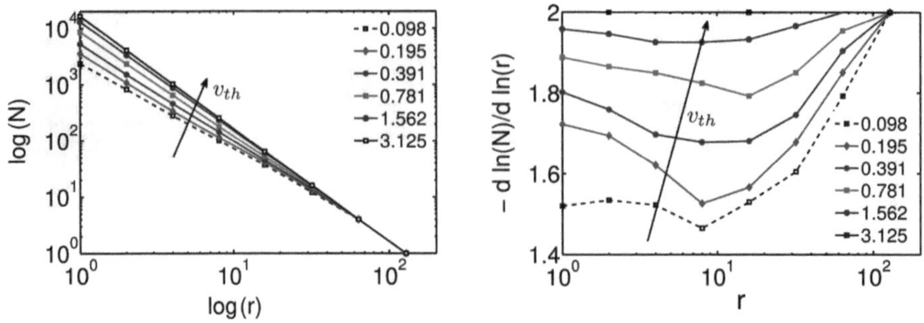

Fig. 5.24 Fractal characterization of the free volume domains $\mathbb{D}(v_{i,j} \leq v_{th})$. The local fractal dimension $\mathcal{D}$ is a function of r and v_{th}. Adapted from Paggi and He (2015)

The local fractal dimension is shown in Fig. 5.24 on the right, and it is dependent on r. In the limit case corresponding to $v_{th} = 0$, the fractal dimension of the corresponding free volume domain is equal to that of the real contact area, which is less than 2 due to the lacunarity of the contact domain. According to the results in Borri-Brunetto et al. (1999), the fractal dimension of the contact area is an increasing function of the applied pressure, but it is a decreasing function of D. In the other limit scenario of $v_{th} = \max_{i,j}(v_{i,j})$, the fractal dimension is equal to 2, i.e. it is equal to that of a Euclidean smooth surface. These limit values represent the *bounds* to the fractal dimension of the free volume contours $\mathbb{D}(v_{i,j} \leq v_{th})$ by varying v_{th}.

Scaling of the Normal Contact Stiffness

The specific contact conductance of rough surfaces, or conductance per unit nominal area, C, can be correlated to the derivative of the force–indentation curve, i.e. the incremental normal contact stiffness, according to the electrical–mechanical analogy established by Barber (2003). In formulae:

$$C = -\frac{2}{\rho E}\frac{\mathrm{d}p}{\mathrm{d}d}, \tag{5.10}$$

where p is the nominal applied pressure and d is the mean plane separation between rough surfaces in contact. The parameters E and ρ are the composite elastic modulus and the composite resistivity of the contact pair defined by continua 1 and 2, respectively. While E has been already introduced in Chap. 4, ρ is computed from the resistivities of the continua as $\rho = \rho_1 + \rho_2$.

The dependence of the contact conductance on the applied pressure for rough interfaces can therefore be investigated by focusing on the incremental normal contact stiffness. Here, it should be remarked that steady-state thermal and electric conduction through a given set of contact spots are mathematically analogous problems.

Therefore, Eq. (5.10) holds also for thermal conduction, provided that ρ is replaced by $1/k$, where k is the composite thermal conductivity of the continua, see Cooper et al. (1968). Radiation and convection contributions to heat exchange through the air gaps are neglected here, although they could have a role for a class of physical problems. In the presence of insulated surface films, or for very high temperature excursions across the interface, their contribution could become relevant over conduction.

As demonstrated in (Barber 2003; Greenwood and Wu 2001), the coarser features of the surface rule the contact conductance. Realistic surfaces often exhibit quasi-fractal properties at the fine scale, but generally there exists a cut-off to the power spectral density function at small wavenumbers (long wavelengths). If this is not the case (e.g. if the surface is fractal in the Weierstrass–Mandelbrot sense), there must nonetheless exists the largest length dimension representing the finite size of the nominal contact area. For a rigorous analysis of this issue, the long wavelength cut-off, defined by the length parameter Δ, should be in principle distinguished by the finite size of the nominal contact area, L. Following this reasoning, Paggi and Barber (2011) proposed the following functional dependence for the incremental stiffness per unit area:

$$\frac{\mathrm{d}p}{\mathrm{d}d} = \frac{\mathrm{d}p}{\mathrm{d}d}\left(E, p, m_0, m_2, m_4, \delta, L, \Delta, D\right), \tag{5.11}$$

where m_0, m_2 and m_4 are the spectral moments of the rough surface (Nayak 1971), D is the surface fractal dimension, and δ is the sampling interval. Notice incidentally that most theoretical studies of contact conductance tacitly assume that the effect of surface roughness can be decoupled from that of the macroscale conduction problem. In other words, the effect of surface roughness is to add a nominal pressure-dependent additional resistance between the macroscopic contacting bodies. However, this decoupling of scales is justified if and only if $\Delta \ll L$.

The number of independent parameters in Eq. (5.11) can be reduced by applying the Buckingham's Π theorem (Buckingham 1915), as first proposed in Paggi and Barber (2011). Most authors used the root mean square roughness $\sigma = \sqrt{m_0}$ to normalize the length parameters (Sridhar and Yovanovich 1994), but here we shall show that when the above scale separation is possible, a greater reduction in the number of parameters is achieved by normalizing with respect to the long wavelength cut-off Δ. Hence:

$$\frac{\Delta}{E}\frac{\mathrm{d}p}{\mathrm{d}d} = -\Phi\left(\frac{p}{E}, \alpha, \frac{\sqrt{m_0}}{\Delta}, \frac{\delta}{\Delta}, \frac{L}{\Delta}, D\right), \tag{5.12}$$

where the bandwidth parameter $\alpha = m_0 m_4/m_2^2$ has been introduced. This parameter is dependent on the fractal dimension and on the resolution. For $\delta \ll \Delta$, the following approximation holds (Sayles and Thomas 1977; Zavarise et al. 2004b) ($2 < D < 3$):

$$\alpha \cong \frac{(D-2)^2}{(D-1)(3-D)} \left(\frac{\delta}{\Delta}\right)^{2(D-3)} . \tag{5.13}$$

Therefore, α can be dropped from the dependencies in (5.12), retaining only D, Δ and δ in the subsequent analysis. Equation (5.10) can be used to define the *dimensionless specific contact conductance* $\widetilde{C}$ as

$$\widetilde{C} \equiv C\rho\Delta = -\frac{2\Delta}{E}\frac{\mathrm{d}p}{\mathrm{d}d} = 2\Phi\left(\frac{p}{E}, \alpha, \frac{\sqrt{m_0}}{\Delta}, \frac{\delta}{\Delta}, \frac{L}{\Delta}, D\right) . \tag{5.14}$$

An alternative description of the contact conductance can be obtained by noting that the imperfect contact at the interface is equivalent to the interposition of a fictitious layer of the same material whose thickness h is given by $\rho h = 1/C$. Normalizing this thickness with respect to Δ, one has

$$\frac{h}{\Delta} = \frac{1}{\widetilde{C}} . \tag{5.15}$$

Now consider the effect of holding D, δ, L, and Δ constant and just rescaling $\sqrt{m_0}$ to $\lambda\sqrt{m_0}$, where λ is a scalar multiplier. The contact problem is nonlinear, but the underlying elastic field is linear. Suppose that, at a given value of nominal pressure p, the actual pressure distribution is $p\,(x, y)$ and the actual contact area is $(x, y) \in \mathcal{A}$. It follows that the contact pressure distribution $\lambda p\,(x, y)$ acting over the same area $\mathcal{A}$ will produce displacements scaled in the same ratio λ with respect to the original ones. These will be exactly what is needed to establish the contact area $\mathcal{A}$ when the roughness is characterized by $\lambda\sqrt{m_0}$, D, δ, L, Δ. Furthermore, the contact conductance depends only $\mathcal{A}$, and hence, one has

$$\Phi\left(\frac{\lambda p}{E}, \frac{\lambda\sqrt{m_0}}{\Delta}, \frac{\delta}{\Delta}, \frac{L}{\Delta}, D\right) = \Phi\left(\frac{p}{E}, \frac{\sqrt{m_0}}{\Delta}, \frac{\delta}{\Delta}, \frac{L}{\Delta}, D\right) \tag{5.16}$$

for all scalar multipliers λ. Now suppose a new dimensionless parameter $p\Delta/(E\sqrt{m_0})$ is defined and used to characterize the nominal pressure p. Accordingly, Eq. (5.14) will be modified to

$$\widetilde{C} = \Phi_1\left(\frac{p\Delta}{E\sqrt{m_0}}, \frac{\sqrt{m_0}}{\Delta}, \frac{\delta}{\Delta}, \frac{L}{\Delta}, D\right), \tag{5.17}$$

and due to the result in (5.16), one has

$$\Phi_1\left(\frac{p\Delta}{E\sqrt{m_0}}, \frac{\lambda\sqrt{m_0}}{\Delta}, \frac{\delta}{\Delta}, \frac{L}{\Delta}, D\right) = \Phi_1\left(\frac{p\Delta}{E\sqrt{m_0}}, \frac{\sqrt{m_0}}{\Delta}, \frac{\delta}{\Delta}, \frac{L}{\Delta}, D\right) \tag{5.18}$$

for all λ, showing that the new function Φ_1 is independent of the dimensionless parameter $\sqrt{m_0}/\Delta$. Therefore, $\sqrt{m_0}/\Delta$ can be dropped from the functional dependence, obtaining

$$\widetilde{C} = \Phi_1 \left(\frac{p\Delta}{E\sqrt{m_0}}, \frac{\delta}{\Delta}, \frac{L}{\Delta}, D \right) , \tag{5.19}$$

where the number of dimensionless parameters has been reduced from five to four.

At this point, it is interesting to note that semi-empirical power-law correlations between the contact conductance and the applied pressure can be recovered by dimensional analysis in case of incomplete similarity in the dimensionless number $p\Delta/(E\sqrt{m_0})$. This condition usually applies in physical systems that are in an intermediate state between two limit conditions (Barenblatt and Botvina 1980). For very low pressures, a deviation from the power-law behaviour has been experimentally observed in Milanez et al. (2003a). This is attributed to the truncation of the asperity height distribution function of tested samples, which leads to an enhancement of thermal contact conductance at light contact pressures as compared to that predicted by a power-law equation. For typical values of the dimensionless truncation $d_{\max}/\sqrt{m_0}$ ranging from 3.5 to 4.5, a deviation from the power-law scaling was experimentally observed for $p/E \lesssim 1 \times 10^{-5}$. Regarding the upper threshold, this may correspond to very high pressures, when the asperities merge with each other and create large contact clusters. This limit situation is however not reached in most of the cases, since one usually explores mean plane separations $d/\sqrt{m_0} > 0$.

Assuming incomplete similarity in $p\Delta/(E\sqrt{m_0})$, Paggi and Barber (2011) postulated a power-law dependence on this dimensionless parameter, giving:

$$\widetilde{C} = \left(\frac{p\Delta}{E\sqrt{m_0}} \right)^{\beta} \Phi_2 \left(\frac{\delta}{\Delta}, \frac{L}{\Delta}, D \right), \tag{5.20}$$

where β is the incomplete similarity exponent, which may depend on the other dimensionless numbers in parenthesis and can only be obtained from real or numerical experiments. The function Φ_2 is a new dimensionless function of the remaining dimensionless parameters.

It is interesting to note that Eq. (5.20) was found to be consistent with most of the semi-empirical correlations published in the literature for the thermal contact conductance of rough surfaces (Paggi and Barber 2011; Paggi 2014). Such correlations suggested power-law dependencies between the contact conductance and the applied pressure, with coefficients of proportionality depending on various statistical parameters. Mikic (1974) used an asperity model similar to that of Greenwood and Williamson (1966) to fit experimental results, obtaining:

$$\begin{aligned} \widetilde{C} &= 2.18 \left(\frac{\sqrt{m_0}}{\Delta} \right)^{-0.06} m_2^{0.03} \left(\frac{p\Delta}{E\sqrt{m_0}} \right)^{0.94} \\ &\cong 2.43 \left(\frac{3-D}{D-2} \right)^{0.03} \left(\frac{\delta}{\Delta} \right)^{0.06(2-D)} \left(\frac{p\Delta}{E\sqrt{m_0}} \right)^{0.94} . \end{aligned} \tag{5.21}$$

Comparing Eq. (5.21) with Eq. (5.20), we note that they are of the same form with $\beta = 0.94$. Notice in particular the independence of the dimensionless ratio $\sqrt{m_0}/\Delta$, as expected from dimensional analysis arguments. The model by Mikic was subsequently reconsidered by the Yovanovich's group in a series of papers, proposing some variants. For instance, Blahey et al. (1980) established the following correlation:

$$\begin{aligned}\widetilde{C} &= 3.86\left(\frac{\sqrt{m_0}}{\Delta}\right)^{-0.07}\left(m_2\sqrt{\alpha}\right)^{0.035}\left(\frac{p\Delta}{E\sqrt{m_0}}\right)^{0.93}\\ &\cong 4.39\frac{(D-1)^{-0.035}}{\sqrt{(D-1)(3-D)}}\left(\frac{\delta}{\Delta}\right)^{-0.035(D-1)}\left(\frac{p\Delta}{E\sqrt{m_0}}\right)^{0.93}\end{aligned} \tag{5.22}$$

which is also exactly of the form (5.20). Subsequently, Sridhar and Yovanovich (1994) re-examined the theory by Greenwood and Williamson (1966) and proposed a correlation valid for $p/E > 1 \times 10^5$ and for $5 < \alpha < 100$, where a slight dependence on the bandwidth parameter α was suggested:

$$\widetilde{C} = k_{SY}\left(\frac{p}{E}\right)^{0.922\alpha^{1/205.54}}, \tag{5.23}$$

where k_{SY} is a coefficient dependent on α in its turn (see Sridhar and Yovanovich (1994) for more details). Such a dependence of β on α is also plausible according to the previous dimensional analysis arguments, since the exponent β may depend on the dimensionless numbers in Eq. (5.12).

Removing the assumption of a constant (average) radius of curvature for the asperities, Bush and Gibson (1979) derived a correlation similar to the previous ones, but with a significantly lower exponent of p/E:

$$\begin{aligned}\widetilde{C} &= 1.38\left(\frac{\sqrt{m_0}}{\Delta}\right)^{-0.11}(m_2\sqrt{\alpha})^{1/2}m_2^{-0.44}\left(\frac{p\Delta}{E\sqrt{m_0}}\right)^{0.89}\\ &\cong 1.69\frac{(D-2)^{0.49}}{(D-1)^{0.25}(3-D)^{0.19}}\left(\frac{\delta}{\Delta}\right)^{0.78(D-2)-1}\left(\frac{p\Delta}{E\sqrt{m_0}}\right)^{0.89},\end{aligned} \tag{5.24}$$

which again is of the form of Eq. (5.20), this time with $\beta = 0.89$. Notice that this model shows a significantly stronger dependence on the parameter δ/Δ.

Equation (5.20) can be recast as an ordinary differential equation with separable variables, in order to determine the relationship between pressure and separation. Doing that, one obtains

$$\frac{\mathrm{d}\tilde{p}}{\tilde{p}^{\beta}} = -\frac{\Phi_2}{2}\,\mathrm{d}\tilde{d}, \tag{5.25}$$

where the following dimensionless variables are introduced:

$$\tilde{p} = \frac{p\Delta}{E\sqrt{m_0}}; \quad \tilde{d} = \frac{d}{\sqrt{m_0}}.$$

If $\beta \neq 1$, this equation has the solution

$$\frac{\tilde{p}^{1-\beta}}{1-\beta} = \frac{\Phi_2}{2}(\tilde{d}_0 - \tilde{d}), \tag{5.26}$$

where $\tilde{d}_0$ is a constant of integration. This is a power-law relationship between the applied pressure and the normal compliance associated with the asperities, $w_a = \tilde{d}_0 - \tilde{d}$. Notice however that although the value $\tilde{d} = \tilde{d}_0$ corresponds to $\tilde{p} = 0$, we cannot identify the constant $\tilde{d}_0$ with the highest point in the surface, since the power-law behaviour depends upon incomplete similarity which breaks down at very low pressures. Indeed, the highest point of the surface $\tilde{d}_{\max}$ depends on extreme-value statistics of the surface and is likely to exhibit significant variance, making it unreliable as a parameter in a contact model.

For the special case $\beta = 1$, the ordinary differential equation (5.25) has the exponential solution:

$$\tilde{p} = K \exp\left(-\frac{\Phi_2 \tilde{d}}{2}\right), \tag{5.27}$$

where the dimensionless multiplying constant K is related to the constant of integration and hence is indeterminate. A similar expression was obtained by Greenwood and Williamson (1966), by considering an exponential distribution of asperity heights in their model derivation:

$$\tilde{p} = K_{\mathrm{GW}} \exp(-\tilde{d}), \tag{5.28}$$

where the multiplying constant K_{GW} depends on the number of asperities per unit area and the summit radii. This agrees with the form (5.27) with $\Phi_2 = 2$.

Persson's contact theory (Persson 2001b) also suggested an exponential relation between p and d:

$$p = \gamma E \exp(-d/d_0), \tag{5.29}$$

where the coefficient γ depends on the fractal dimension and on the length scales δ and Δ, and it was estimated from the best fit on numerical data. The constant d_0 was found to be of the order of magnitude of $\sqrt{m_0}$, implying Φ_2 of order 2 in Eq. (5.27).

Finally, in case of exponential relations between $\tilde{p}$ and $\tilde{d}$, it is remarkable to note that a straightforward application of Eq. (5.20) leads to a linear dependence between $\widetilde{C}$ and p/E.

To provide an independent insight into the effect of the sample lateral size and of the low-frequency cut-off to the power spectral density function, Paggi and Barber (2011) carried out contact mechanics simulations using a version of the boundary element method for multi-asperity contacts, accounting for elastic interactions between asperities. In view of uncoupling the effect of the sample lateral size from that caused by the low-frequency cut-off, synthetic rough surfaces were numerically generated by collecting together RMD surfaces (RMD patches). Supposing that each RMD patch might have its intrinsic low-frequency cut-off to the power spectral density

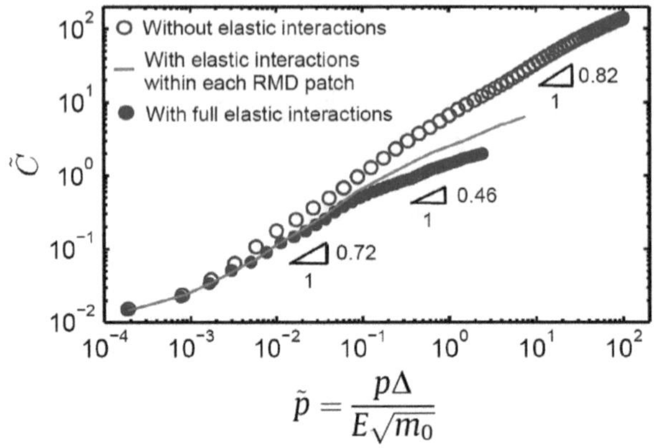

Fig. 5.25 The scaling between the dimensionless contact conductance and the dimensionless contact pressure, for rough surfaces with $D = 2.3$ obtained by collecting 8×8 RMD patches together, to separate the length scales Δ and L, i.e. $\Delta \ll L$. The role of elastic interactions is also highlighted by comparing results from different approximations

function of the same order of magnitude as the patch size, a collection of patches leads to much bigger surfaces to test, achieving $L \gg \Delta$ and therefore separation of scales. Numerical predictions in Fig. 5.25 show for such a surface what happens if elastic interactions are neglected (empty circles), if they are included between asperities within each RMD patch (solid line), or if full elastic interactions across the whole surface are considered (filled circles). Predictions obtained by neglecting elastic interactions led to a power-law dependency with an exponent $\beta = 0.82$ very close to what predicted by analytical contact theories based on statistical distributions of asperities and without elastic interactions. Full elastic interactions led to a power-law trend with two distinct regimes, one for low pressures with $\beta = 0.72$, and another for high pressures with $\beta = 0.46$. The transition from the two regimes was explained by the overcoming of a pressure threshold which leads to merging of asperities together, in analogy with the physical phenomenon of percolation.

The Interplay Between Roughness and Long-Range Adhesion

Adhesion between solids in the presence of roughness is a research topic that attracted a significant attention by the scientific community, especially for its importance in nanoscale applications, see Peressadko et al. (2005), Ciavarella (2016), Guduru (2007), Carbone et al. (2009), Pastewka and Robbins (2014, 2016), Papangelo and Ciavarella (2018) for a selection of recent references. The frictionless normal contact problem between a rigid WM rough plane strain indenter profile with parameters $A_0/\lambda = 0.0025$, $D = 1.25$, $\gamma = 5$, and $n = 1$ and a deformable half-plane in the presence of long-range adhesion was investigated in Paggi and Reinoso (2018), to assess the interplay between roughness and long-range adhesion. Adhesive tractions were simulated in the finite element framework based on the MPJR interface finite element discretization by a relation dictated by an adhesion model inspired by the interatomic Lennard-Jones potential (for $g_{\mathrm{n}} > 0$):

$$p_A = 24\varepsilon \left[\frac{\kappa^6}{(g_{\mathrm{n}} + g_{\mathrm{n},0})^7} - 2\frac{\kappa^{12}}{(g_{\mathrm{n}} + g_{\mathrm{n},0})^{13}} \right] \tag{5.30}$$

with $\varepsilon = 1 \times 10^{-3}$ N/m, $\kappa/\lambda = 2.57 \times 10^{-4}$ and $g_{\mathrm{n},0}/\lambda = 2.885 \times 10^{-4}$. The equilibrium distance $g_{\mathrm{n},0}$ is such that the condition $g_{\mathrm{n}} = 0$ leads to vanishing adhesive tractions and it correctly captures the transition from adhesion to contact.

Two ramps for the imposed far-field displacements were considered: (*i*) an approaching stage, characterized by an increasing w to put the indenter in contact with the half-plane up to $w/A_0 = 3$, and (*ii*) a separation stage, where w was reduced until the complete separation of the profile. As compared to smooth geometries, the modelling complexity regards the noncompactness of the contact domain, which imposes severe limitations in the use of semi-analytical methods. For instance, the method proposed in Guduru (2007) strictly applies to profiles whose shapes are leading to monotonically increasing gaps to avoid partial contact, which is on the other hand occurring in the present case.

Let focus the attention onto the evolution of the pressure distribution at the interface, p/E, and of the corresponding normal gap g_n/A_0. Numerical predictions for selected values of w are shown in Figs. 5.26 and 5.27 for the approaching stage, and in Figs. 5.28 and 5.29 for the separation stage. During the approach, contact takes place near the first peak at $x/\lambda = 0$, with contact pressures (positive valued in the plot) followed by adhesive tractions (negative valued) significant within a small distance after the end of the contact strip (solid curve A in Fig. 5.26). A further increase in the closing displacement w leads to an increase in the contact and adhesive domains in a self-similar manner, leading to curves intermediate from the curve A and the

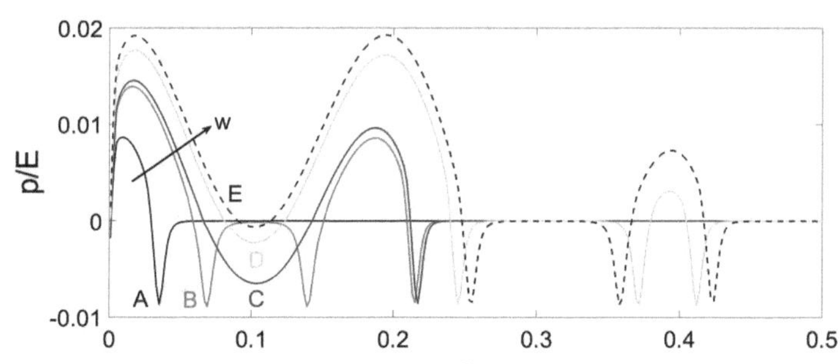

Fig. 5.26 Dimensionless contact pressure along the interface for different far-field imposed displacements w during the approaching stage. Adapted from Paggi and Reinoso (2018)

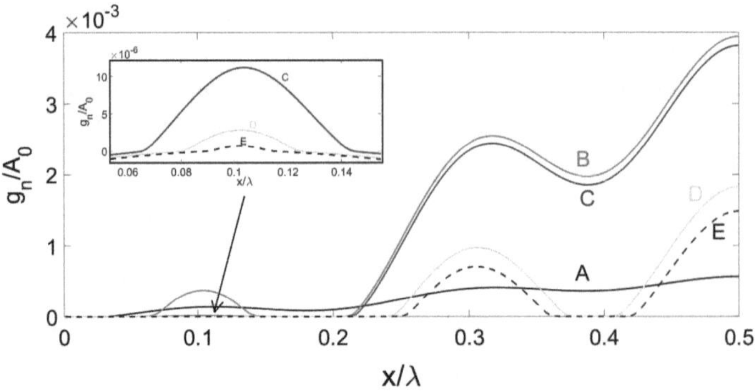

Fig. 5.27 Dimensionless gap along the interface for different far-field imposed displacements w during the approaching stage. Adapted from Paggi and Reinoso (2018)

curve B. Within that evolution, another peak comes into contact near $x/\lambda \cong 0.17$ (see the curve B which highlights the presence of another contact strip along the rough profile). A further increase in w leads to the curve C which now looks quite different from the previous traction profiles in the region between the two peaks near $x/\lambda \cong 0.1$. The normal gap is significantly diminished in that region, as it can be noticed from the magnification in Fig. 5.27, and the adhesive zones of the two peaks in between the two contact strips start interacting. Such an interaction leads to adhesive tractions in the ascending branch of the adhesive law based on the Lennard-Jones potential in between $g_n = 0$ and g_n corresponding to the maximum adhesive traction. As a result, the corresponding adhesive traction distribution along the interface near $x/\lambda \cong 0.1$ presents now a bell-shaped form, with a fully pressurized gap. The size of such pressurized gap is then diminished by further increasing w (curve D) due to the growth of the strips in contact. We also observe that another peak comes in contact at $x/\lambda \cong 0.38$. This proceeds up to curve E, which corresponds to the maximum imposed far-field displacement $w = 3A_0$.

The unloading stage, whose numerical predictions are shown in Figs. 5.28 and 5.29, starts from the curve E and proceeds, by progressively reducing w, to the sequence of curves F, G, H and I, just before the complete separation of the entire profile. During the separation stage, the contact strips shrink and the adhesive ones increase.

Once the fundamental mechanisms taking place during adhesive contact in the presence of roughness have been elucidated, the role played by the length scales of roughness is finally discussed by comparing the pressure distributions for the same far-field imposed displacement $w/A_0 = 2.4$, but for WM rough profiles with different resolution parameters n ranging from 0 up to 3, see Fig. 5.30.

Numerical predictions show that the addition of length scales of roughness leads to an increase in the maximum tractions in the noncompact contact regions, a trend noticed also for the adhesiveless scenario. On the other hand, the adhesive contact pressure is always bounded by the fact that the adhesive constitutive relation has maximum adhesive traction that cannot be overcome.

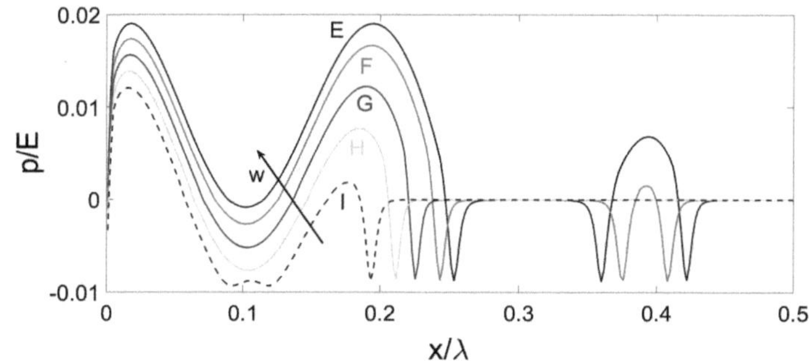

Fig. 5.28 Dimensionless contact pressure along the interface for different far-field imposed displacements w during the separation stage. Adapted from Paggi and Reinoso (2018)

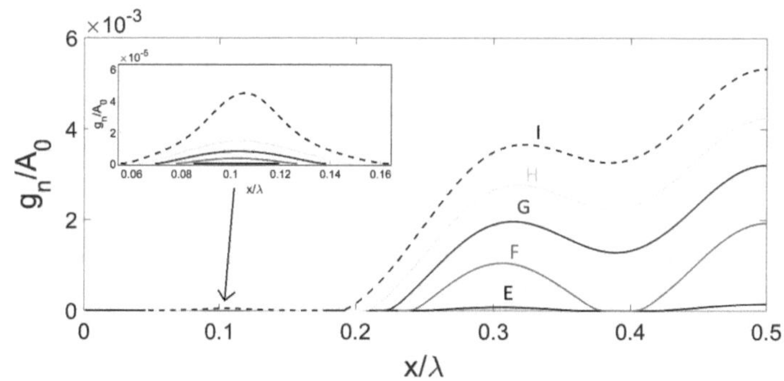

Fig. 5.29 Dimensionless gap along the interface for different far-field imposed displacements w during the separation stage. Adapted from Paggi and Reinoso (2018)

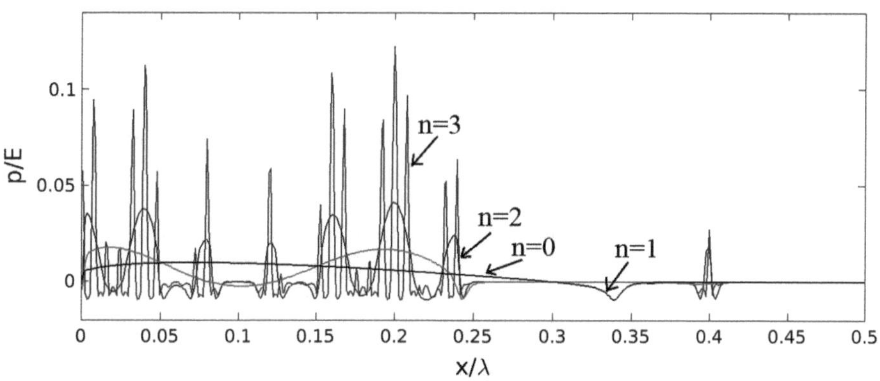

Fig. 5.30 Dimensionless contact pressure along the interface for different values of the resolution parameter n, for the same far-field imposed displacement $w/A_0 = 2.4$

The key effect of roughness can be well appreciated by comparing tractions for $n = 0$ and the other predictions for larger n. For $n = 0$, a single compact contact domain occurs, with a corresponding zone of the interface ahead of the edge of the contact strip with adhesive tractions rapidly decaying in space. For $n = 1$, two contact strips occur and the contact domain becomes noncompact. As a result, a fully pressurized region between the two peaks in contact takes place. The same trend, even more pronounced, can be observed by a further increase in n, which leads to very rarefied noncompact contact strips along the profile, and many fully pressurized adhesive gaps in between the peaks in contact.

The Effect of Roughness on the Transition from Full Stick to Full Slip in Frictional Contacts

When a rigid rough surface is pressed against a deformable linear elastic half-plane without adhesion, contact takes place across the so-called real contact area, as discussed in the previous sections. Let's hold the normal contact force and impose a monotonically increasing displacement to the rough surface in a given tangential direction. If one postulates the existence of a constant static friction coefficient μ at the local level, then the shearing displacement will lead to a progressive slip of the points in contact. Overall, the state of contact will change from *full stick* in absence of shearing displacement, when the rough surface adheres to the half-plane through the real contact area, to *full slip*, when gross sliding takes place with a macroscopic relative displacement between the bodies. The transition from these two regimes is referred to as *partial slip*. As discussed in Chap. 4, the most general scenario with a Poisson ratio $\nu \neq 1/2$ leads to coupling between the normal and the tangential contact problems, and therefore shearing tractions activated by friction would modify the real contact area. However, in many cases of engineering relevance, this coupling is negligible, and it is often neglected for simplicity.

The solution of the contact problem during the partial slip regime requires finding the shearing contact tractions and surface deflections for any imposed far-field displacement. This problem was solved in closed form for smooth geometries of the indenter leading to compact contact domains, and in particular for indenters of parabolic shape by Cattaneo (1938) and Mindlin (1949), who established independently a mathematical analogy between the tangential contact problems and the normal one.

The transition from full stick to full slip for spheres in contact is schematically depicted in Fig. 5.31. At the beginning, for $u_x = 0$, the real contact area has a radius $a_0 = \sqrt{Rd_0}$ given by the imposed normal interference d_0, which corresponds to the normal contact force $F_{z,0} = \dfrac{4}{3} E R^{1/2} d_0^{3/2}$ according to Hertz theory, where R is the radius of curvature and E is the composite Young's modulus. The increase in the shearing displacement leads to a progressive slip at the points of the contact domain for which the ratio between the shearing traction and the normal pressure reaches the value of the local friction coefficient μ. Therefore, the slip domain is represented by an annulus, developing from the outer border of the circular contact area towards

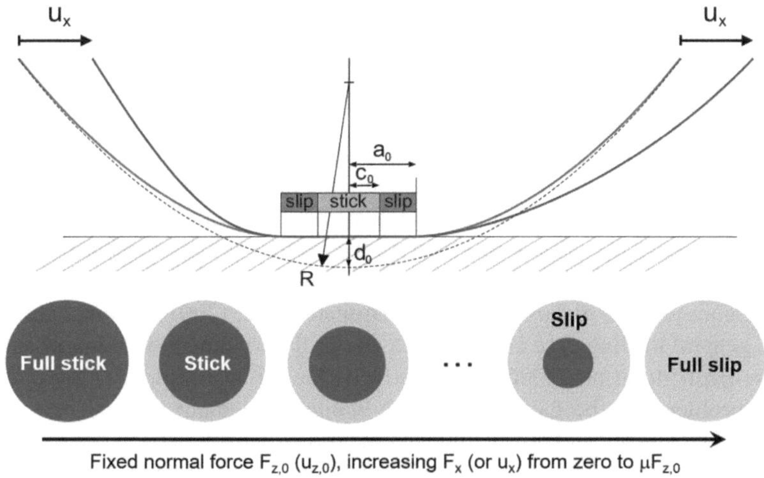

Fig. 5.31 Partition and evolution of the contact domain into stick and slip portions, by holding the normal force and increasing the shearing displacement

its centre. The contact domain in stick condition, for which the ratio between the shearing traction and the normal pressure is less than μ, progressively shrinks from the circle of radius a_0 to a point. The radius c_0 of the circle defining the stick contact domain can be computed from the following equation (Johnson 1985, Chap. 7):

$$u_x = \mu d_0 \frac{(2-\nu)E}{4G}\left(1 - \frac{c_0^2}{a_0^2}\right). \tag{5.31}$$

The displacement corresponding to the onset of full-sliding is therefore given by the condition $c_0 = a_0$ and it is equal to $u_x = \mu d_0(2-\nu)E/(4G)$.

The investigation of the evolution of the stick and slip parts of the real contact area is indeed relevant for many applications. For example, the differential tangential stiffness is only dependent on the current configuration of the stick region of contact. If bodies are subjected to tangential oscillation with a small amplitude (smaller than the maximum slip amplitude for complete sliding), then wear will occur only in the slip-region (fretting). As shown by Ciavarella and Hills (1999) for arbitrary two-dimensional contacts and by Popov (2014) for three-dimensional axis-symmetrical contacts, during fretting the initial stick region always stays in the stick state (both at a constant normal force and a constant indentation), while in the initial sliding region the surfaces finally loose contact completely. This leads to changes in stiffness as well as in the electrical and thermal conductivities, which in the final state will only depend on the initial configuration of the stick region. The area of the stick region is also of interest for instance for impurity-film resistance of contacts. The decomposition of the contact area into stick and slip parts plays an important role in other practical problems such as frictional energy dissipation (Dini and Hills 2009; Barber et al. 2011), geotechnical engineering (Desai et al. 1985), damping and stiffness of joints (Kirsanova 1967), and control engineering (Harnoy et al. 1994).

The investigation of partial slip contact problems with arbitrary rough contacts can be made by exploiting the generalization of the Cattaneo–Mindlin analogy between the tangential contact problem and the normal one for noncompact contact domains, proposed independently by Ciavarella (1998a, b) and Jaeger (1998). They have proven that the solution of the tangential contact problem can be gained from the linear superposition of two normal solutions, one corresponding to the limiting frictional traction distribution and a negative correction due to a distribution equal to the coefficient of friction multiplied by the normal contact pressure distribution corresponding to some lower value of the normal load. Numerical implementation of this technique was discussed in Borri-Brunetto et al. (2001) and its extension to a more general loading case consisting in an oscillatory tangential force is available in Borri-Brunetto et al. (2006). Alternatively, the direct solution of the tangential contact problem can be gained using the boundary element method, see Chap. 4, section "The Contact Problem with Friction" of this book.

From an operative point of view, the application of the Cattaneo–Mindlin analogy requires two steps. First, the application of a displacement u_z in the normal direction, with a value from 0 up to $u_{z,0}$, and the solution of the corresponding normal contact problem, determining F_z for the sequence of displacements. We call $F_{z,0}$ the normal contact force corresponding to $u_{z,0}$. This step can be solved analytically for smooth convex contact domains, or numerically using the boundary element method or the finite element method for rough surfaces. Next, the normal force is held constant and a progressively increasing shearing displacement u_x is applied. The solution of the tangential problem can be deduced from the solution of two normal contact problems: one associated to the problem for $u_{z,0}$, and another corresponding to smaller

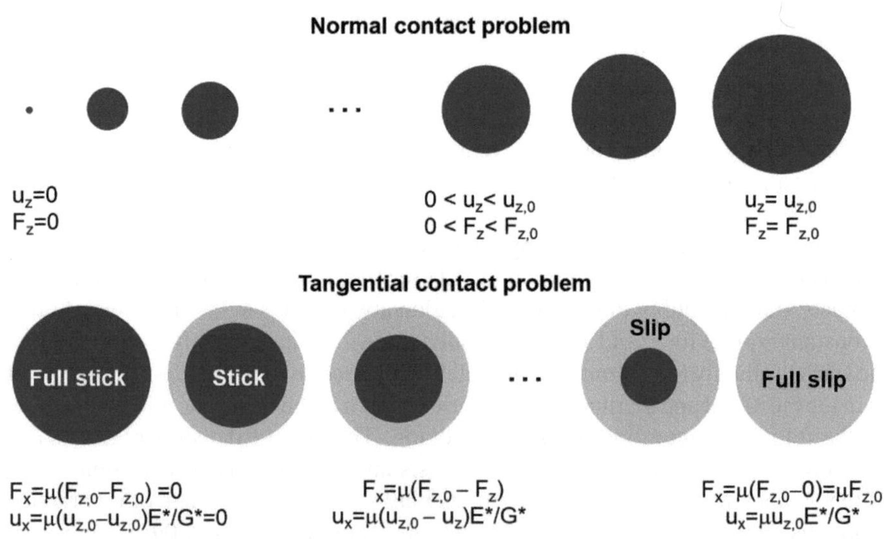

Fig. 5.32 Evolution of the real contact area and of the stick and slip portions of the contact domain based on the application of the Cattaneo–Mindlin analogy

displacement u_z, using the same formulae shown in Fig. 5.32 for the contact problem involving a sphere and a half-plane.

For smooth surfaces like spheres or cylinders, it has been shown in Carpinteri et al. (2009) that the tangential force component acting in the stick region and the stick contact area are both nonlinear functions of the total tangential force.

For rough surfaces, semi-analytical predictions based on the Greenwood and Williamson contact theory (Greenwood and Williamson 1966) have been made in Paggi et al. (2014), integrating the individual asperity contributions (treated as Hertzian spheres in contact with a half-plane with a constant radius of curvature and an elevation obeying the exponential probability distribution function $\Phi(z) = C\exp(-z/\lambda)$) to the stick and slip components of the contact domain. After some algebra, the following simple result was achieved:

$$\frac{F_x}{F_{x,\max}} = \left[1 - \exp\left(-\frac{d_{\min}}{\lambda}\right)\right], \tag{5.32}$$

$$\frac{F_x}{F_{x,\max}} = 1 - \frac{A_{\text{stick}}}{A_0} = \frac{A_{\text{slip}}}{A_0}, \tag{5.33}$$

where $F_{x,\max} = \mu F_z$. The former equation leads to a smooth regularization of the Coulomb friction law of the overall interface, as an emerging property from the local partial slip evolution of the asperities, see Fig. 5.33. The latter is a simple linear dependency between the portion of the real contact area in stick condition

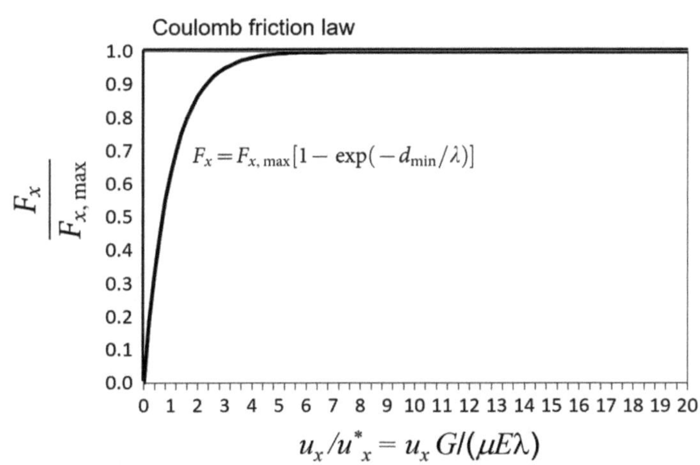

Fig. 5.33 Shearing force–displacement relation analytically obtained from the progressive slip of asperities for the Greenwood and Williamson contact model with an exponential probability distribution function of asperity heights. The Coulomb friction law is also superimposed to the diagram and it corresponds to a sharp transition from full stick to full slip, as for a flat plane bearing a uniform normal pressure. The emergent response due to roughness and partial slip leads to a smooth transition between the two regimes

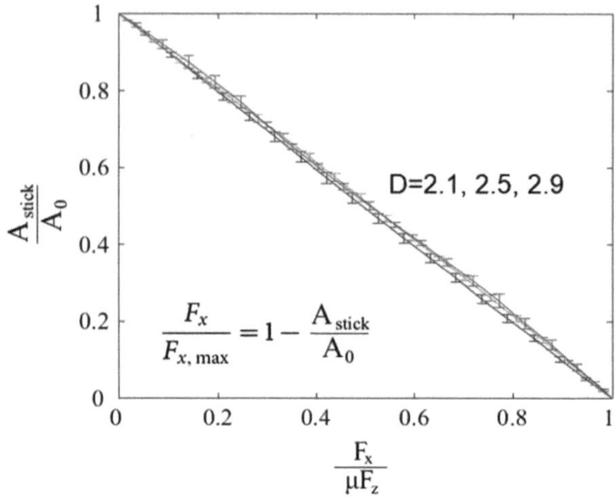

Fig. 5.34 Dependence of the real contact area in stick conditions and the shearing force for RMD rough surfaces. All the results present the same linear curve as for the Greenwood and Williamson contact theory with an exponential probability distribution function of asperity heights, suggesting a universal trend

and the applied shearing force. Surprisingly, such a linear trend was confirmed in Paggi et al. (2014) by numerical simulations on RMD fractal surfaces with $n = 8$ and $D = 2.1, 2.5$ and 2.9. This suggests a universal trend independent of the actual probability density functions of the asperity elevations and curvatures, or on fractality, see Fig. 5.34.

Conclusions

Contact mechanics between rough surfaces is a prominent example of a complex system, where understanding the role of the multi-scale features of roughness on the overall interface mechanical properties is a key research question. As shown in the examples collected in this chapter, even in the simplest case of homogeneous and isotropic linear elastic continua, the local interactions between elastic contact spots lead to a collective behaviour with non-trivial emergent features.

Such emergent properties regard the relation between the real contact area and the load, the scaling of the thermal/electric contact conductance with the normal pressure, the evolution of the free volume network trapped between rough surfaces in contact, as well as the evolution of the stick and slip portions of the contact domain under the application of a shearing load. For all of such properties, the application of numerical methods, such as the boundary element method or the finite element method as detailed in Chap. 4, allows understanding of the role of roughness and identifying possible universal trends. In this regard, algorithms for the numerical generation of synthetic rough surfaces, such as the WM, RMD and SSM, can be used to provide the height field to be used as input for computational contact mechanics simulations, so far mostly focused on multi-scale roughness with self-affine scaling typical of fractals.

Recent trends opening further research directions regard the study of other complex surface topologies, as, for instance, textured surfaces. It is in fact known that some natural surfaces present specific textures for the promotion of hydrophobicity. Similarly, special textures can be used to enhance wave absorbance or lubrication. Many other forms are yet unexplored and the connection between the surface topology and the physico-mechanical response is yet unrevealed. Moreover, a key issue in the realization of bio-inspired surfaces is the understanding of how many length scales of roughness should be included in the mimicking artificial surfaces, in order to guarantee the achievement of the desired response. Machine learning algorithms to select and identify optimal topologies to achieve a target response could be a method to purse, following the pioneering approach in Cinat (2018), Cinat et al. (2019).

Finally, to foster sharing of real surface data in the scientific community, an international collaborative project called Wiki Surface (http://musam.imtlucca.it/wikisurf.html) has recently been launched to create the first open access database of rough surfaces from nature and technology. The database is expected to significantly contribute to the research in surface physics, surface chemistry and tribology, extending the investigation on the role of roughness to a much wider range of complex topologies and surface textures.

Acknowledgements This chapter is derived in part from an article published in the International Journal of Heat and Mass Transfer (Elsevier), available online 2 July 2011, doi: 10.1016/j.ijheatmasstransfer.2011.06.011; an article published in Wear (Elsevier), available online 5 May 2015, doi: 10.1016/j.wear.2015.04.021; and an article published in Mechanics of Advanced Materials and Structures (Taylor & Francis), available online 5 Nov. 2018, doi: 10.1080/15376494.2018.1525454. The author would like to acknowledge the discussion and fruitful collaboration over the years with Prof. J. R. Barber (University of Michigan, US), Prof. M. Borri-Brunetto (Politecnico di Torino, Italy), Prof. M. Ciavarella (Politecnico di Bari, Italy), Dr. J. A. Greenwood (University of Cambridge, UK), Prof. Q.-C. He (University of Paris-Est, France), Prof. V. Popov, Dr. R. Pohrt (Technical University of Berlin, Germany), Dr. J. Reinoso (University of Seville, Spain) and Prof. G. Zavarise (Politecnico di Torino, Italy).

References

Abbott, E. J., & Firestone, F. A. (1933). Specifying surface quality: A method based on accurate measurement and comparison. *Mechanical Engineering*, *55*, 569–572.

Almqvist, A., & Dasht, J. (2006). The homogenization process of the Reynolds equation describing compressible liquid flow. *Tribology International*, *39*, 994–1002.

Almqvist, A., Fabricius, J., Larsson, R., & Wall, P. (2014). A new approach for studying cavitation in lubrication. *Proceedings of the Royal Society London, Series A, 136*, 011706.

Bandis, S., Lumsden, A. C., & Barton, N. R. (1981). Experimental studies of scale effects on the shear behaviour of rock joints. *International Journal of Rock Mechanics and Mining Sciences & Geomechanics Abstracts*, *18*, 1–21.

Barber, J. R. (2003). Bounds on the electrical resistance between contacting elastic rough bodies. *Proceedings of the Royal Society of London, Series A, 459*, 53–66.

Barber, J. R. (2018). *Contact mechanics*. Springer International Publishing.

Barber, J. R., Davies, M., & Hills, D. A. (2011). Frictional elastic contact with periodic loading. *International Journal of Solids and Structures*, *48*, 2041–2047.

Barenblatt, G. I., & Botvina, L. R. (1980). Incomplete self-similarity of fatigue in the linear range of fatigue crack growth. *Fatigue and Fracture of Engineering Materials and Structures*, *3*, 193–202.
Berkowitz, B. (2002). Characterizing flow and transport in fractured geological media: A review. *Advances in Water Resources*, *25*, 861–884.
Bhushan, B., & Majumdar, A. (1992). Elastic-plastic contact model for bifractal surfaces. *Wear*, *153*, 53–64.
Bigerelle, M., & Iost, A. (2004). Statistical artefacts in the determination of the fractal dimension by the slit island method. *Engineering Fracture Mechanics*, *71*, 1081–1105.
Blahey, A., Tevaarwerk, J. L., & Yovanovich, M. M. (1980). Contact conductance correlations of elastically deforming flat rough surfaces. *AIAA Paper No. 80-1470 Presented at The AIAA 5th Thermo-Physics Conference, Snowmass, Colorado*.
Borodich, F. M. (1997). Some fractal models of fracture. *Journal of the Mechanics and Physics of Solids*, *45*, 239–259.
Borodich, F. M., & Mosolov, A. B. (1992). Fractal roughness in contact problems. *Journal of Applied Mathematics and Mechanics*, *56*, 681–690.
Borri, C., & Paggi, M. (2015). Topological characterization of antireflective and hydrophobic rough surfaces: Are random process theory and fractal modeling applicable? *Journal of Physics D: Applied Physics*, *48*, 045301.
Borri, C., & Paggi, M. (2016). Topology simulation and contact mechanics of bifractal rough surfaces. *Proceedings of the Institution of Mechanical Engineers, Part J: Journal of Engineering Tribology*, *230*, 1345–1358.
Borri-Brunetto, M., Carpinteri, A., & Chiaia, B. (1999). Scaling phenomena due to fractal contact in concrete and rock fractures. *International Journal of Fracture*, *95*, 221–238.
Borri-Brunetto, M., Chiaia, B., & Ciavarella, M. (2001). Incipient sliding of rough surfaces in contact: A multiscale numerical analysis. *Computer Methods in Applied Mechanics and Engineering*, *190*, 6053–6073.
Borri-Brunetto, M., Carpinteri, A., Invernizzi, S., & Paggi, M. (2006). Micro-slip of rough surfaces under cyclic tangential loading. In P. Wriggers & U. Nackenhorst (Eds.), *Analysis and simulation of contact problems*. Lecture notes in applied and computational mechanics (Vol. 27, pp. 191–200). Berlin, Heidelberg: Springer.
Bouchaud, E. (1997). Scaling properties of cracks. *Journal of Physics Condensed Matter*, *9*, 4319–4344.
Bowden, F. P., & Tabor, D. (1964). *The friction and lubrication of solids, Part II*. Oxford, UK: Clarendon Press.
Buckingham, E. (1915). Model experiments and the form of empirical equations. *ASME Transactions*, *37*, 263–296.
Bush, A. W., & Gibson, R. D. (1979). A theoretical investigation of thermal contact conductance. *Applied Energy*, *5*, 11–22.
Bush, A. W., Gibson, R. D., & Thomas, T. R. (1975). The elastic contact of a rough surface. *Wear*, *35*, 87–111.
Bush, A. W., Gibson, R. D., & Keogh, G. P. (1976). The limit of elastic deformation in the contact of rough surfaces. *Mechanical Resources Communications*, *3*, 169–174.
Campaña, C., Persson, B. N. J., & Mueser, M. H. (2001). Transverse and normal interfacial stiffness of solids with randomly rough surfaces. *Journal of Physics: Condensed Matter*, *23*, 085001.
Carbone, G., & Bottiglione, F. (2008). Asperity contact theories: Do they predict linearity between contact area and load? *Journal of the Mechanics and Physics of Solids*, *56*, 2555–2572.
Carbone, G., & Mangialardi, L. (2004). Adhesion and friction of an elastic half-space in contact with a lightly wavy rigid surface. *Journal of the Mechanics and Physics of Solids*, *52*, 1267–1287.
Carbone, G., & Putignano, C. (2013). A novel methodology to predict sliding and rolling friction of viscoelastic materials: Theory and experiments. *Journal of the Mechanics and Physics of Solids*, *61*, 1822–1834.

Carbone, G., Scaraggi, M., & Tartaglino, U. (2009). Adhesive contact of rough surfaces: Comparison between numerical calculations and analytical theories. *The European Physical Journal E, Soft Matter*, *30*, 65–74.

Carpinteri, A. (1994). Fractal nature of material microstructure and size effects on apparent mechanical properties. *Mechanics of Materials*, *18*, 89–101.

Carpinteri, A., & Chiaia, B. (1995). Multifractal nature of concrete fracture surfaces and size effects on nominal fracture energy. *RILEM Materials & Structures*, *28*, 435–443.

Carpinteri, A., & Paggi, M. (2005). Size-scale effects on the friction coefficient. *International Journal of Solids and Structures*, *42*, 2901–2910.

Carpinteri, A., & Paggi, M. (2008). Size-scale effects on strength, friction and fracture energy of faults: A unified interpretation according to fractal geometry. *Rock Mechanics and Rock Engineering*, *41*, 735–746.

Carpinteri, A., & Paggi, M. (2009). A fractal interpretation of size-scale effects on strength, friction and fracture energy of faults. *Chaos, Solitons & Fractals*, *39*, 540–546.

Carpinteri, A., Paggi, M., & Zavarise, G. (2009). Cusp-catastrophe interpretation of the stick-slip behaviour of rough surfaces. *Computational Modelling in Engineering Science*, *53*, 1–23.

Cartwright, D. E., & Longuet-Higgins, M. S. (1956). The distribution of the maxima of a random function. *Philosophycal Transaction of the Royal Society of London, Series A, 237*, 212–232.

Cattaneo, C. (1938). Sul contatto di due corpi elastici: Distribuzione locale degli sforzi. *Rendiconti dell'Accademia Nazionale dei Lincei, 6*, 342–348, 434–436, 474–478.

Ciavarella, M. (1998a). The generalized Cattaneo partial slip plane contact problem. I-Theory, II-Examples. *International Journal of Solids and Structures, 35*, 2349–2378.

Ciavarella, M. (1998b). Tangential loading of general three-dimensional contacts. *ASME Journal of Applied Mechanics*, *64*, 998–1003.

Ciavarella, M. (2016). On roughness-induced adhesion enhancement. *The Journal of Strain Analysis for Engineering Design*, *51*, 473–481.

Ciavarella, M., & Demelio, G. (2000). Elastic multiscale contact of rough surfaces: Archard's model revisited and comparisons with modern fractal models. *ASME Journal of Applied Mechanics*, *68*, 496–498.

Ciavarella, M., & Hills, D. A. (1999). Brief note: Some observations on the oscillating tangential forces and wear in general plane contacts. *European Journal of Mechanics - A/Solids*, *18*, 491–497.

Ciavarella, M., Demelio, G., Barber, J. R., & Jang, Y. H. (2000). Linear elastic contact of the Weierstrass profile. *Proceedings of the Royal Society of London, Series A, 456*, 387–405.

Ciavarella, M., Murolo, G., & Demelio, G. (2004a). The electrical/thermal conductance of rough surfaces: The Weierstrass-Archard multiscale model. *International Journal of Solids and Structures*, *41*, 4107–4120.

Ciavarella, M., Murolo, G., Demelio, G., & Barber, J. R. (2004b). Elastic contact stiffness and contact resistance for the Weierstrass profile. *Journal of the Mechanics and Physics of Solids*, *52*, 1247–1265.

Ciavarella, M., Delfine, V., & Demelio, G. (2006). A "re-vitalized" Greenwood and Williamson model of elastic contact between fractal surfaces. *Journal of the Mechanics and Physics of Solids*, *54*, 2569–2591.

Ciavarella, M., Dibello, S., & Demelio, G. (2008a). Conductance of rough random profiles. *International Journal of Solids and Structures*, *45*, 879–893.

Ciavarella, M., Greenwood, J. A., & Paggi, M. (2008b). Inclusion of "interaction" in the Greenwood and Williamson contact theory. *Wear*, *265*, 729–734.

Cinat, P. (2018). *Surface roughness genomics in contact mechanics: A new method enabling roughness design towards surface prototyping*. Ph.D. Thesis, IMT School for Advanced Studies Lucca, Lucca, Italy.

Cinat, P., Paggi, M., & Gnecco, G. (2019). Identification of roughness with optimal contact response with respect to real contact area and normal stiffness. *Mathematical Problems in Engineering*, 7051512.

Cooper, M. G., Mikic, B. B., & Yovanovich, M. M. (1968). Thermal contact conductance. *International Journal of Heat and Mass Transfer*, *12*, 279–300.
Desai, C. S., Drumm, E. C., & Zaman, M. M. (1985). Cyclic interface and joint shear device including pore pressure effects. *ASCE Journal of Geotechnical Engineering*, *111*, 793–815.
Dini, D., & Hills, D. A. (2009). Frictional energy dissipation in a rough Hertzian contact. *ASME Journal of Tribology*, *131*, 021401.
Feder, J. (1988). *Fractals*. New York: Plenum Press.
Gagliardi, M., Lenarda, P., & Paggi, M. (2017). A reaction-diffusion formulation to simulate EVA polymer degradation in environmental and accelerated ageing conditions. *Solar Energy Materials and Solar Cells*, *164*, 93–106.
Goryacheva, I. G. (1998). *Contact mechanics in tribology* (Vol. 61). Netherlands, Dordrecht: Springer.
Green, C. K. (2007). *Development of a leakage model for solid oxide fuel cells compressive seals*. Ph.D. Thesis, Georgia Institute of Technology, Atlanta, GA, USA.
Greenwood, J. A. (1984). A unified theory of surface roughness. *Proceedings of the Royal Society of London, Series A*, *393*, 133–157.
Greenwood, J. A. (2006). A simplified elliptic model of rough surface contact. *Wear*, *261*, 191–200.
Greenwood, J.A., & Williamson, J. B. P. (1966). Contact of nominally flat surfaces. *Proceedings of the Royal Society of London, Series A*, *295*, 300–319.
Greenwood, J. A., & Wu, J. J. (2001). Surface roughness and contact: An apology. *Meccanica*, *36*, 617–630.
Guduru, P. R. (2007). Detachment of a rigid solid from an elastic wavy surface: Theory. *Journal of the Mechanics and Physics of Solids*, *55*, 445–472.
Guduru, P. R., & Bull, C. (2007). Detachment of a rigid solid from an elastic wavy surface: Experiments. *Journal of the Mechanics and Physics of Solids*, *55*, 473–488.
Han, B. (2012). Measurements of true leak rates of MEMS packages. *Sensors*, *12*, 3082–3104.
Harnoy, A., Friedland, B., & Rachoor, H. (1994). Modeling and simulation of elastic and friction forces in lubricated bearings for precise motion control. *Wear*, *172*, 155–165.
Holm, R. (1958). *Electric contact. Theory and applications*. Berlin, Germany: Springer.
Jaeger, J. (1998). A new principle in contact mechanics. *ASME Journal of Tribology*, *120*, 677–684.
Johnson, K. L. (1985). *Contact mechanics*. Cambridge, UK: Cambridge University Press.
Jones, R., Chen, F., Pitt, S., Paggi, M., & Carpinteri, A. (2016). From NASGRO to fractals: Representing crack growth in metals. *International Journal of Fatigue*, *82*, 540–549.
Kirsanova, V. N. (1967). The shear compliance of flat joints. *Machine and Tooling*, *38*, 30–34.
Leachman, W. J., Li, H., Flynn, T. J., Stephens, L. S., & Trinkle, C. A. (2014). Statistical analysis of wear of biplanar deterministically-arrayed surfaces for load bearing applications. *Wear*, *311*, 137–148.
Lenarda, P., & Paggi, M. (2016). A geometrical multi-scale numerical method for coupled hygro-thermo-mechanical problems in photovoltaic laminates. *Computational Mechanics*, *57*, 947–963.
Li, Q., Argatov, I., & Popov, V. (2018). Onset of detachment in adhesive contact of an elastic half-space and flat-ended punches with non-circular shape: Analytic estimates and comparison with numeric analysis. *Journal of Physics D: Applied Physics*, *51*, 145601.
Longuet-Higgins, M. S. (1957a). The statistical analysis of a random moving surface. *Philosophycal Transaction of the Royal Society of London, Series A*, *249*, 321–387.
Longuet-Higgins, M. S. (1957b). Statistical properties of an isotropic random surface. *Philosophycal Transaction of the Royal Society of London, Series A*, *250*, 157–174.
Luan, B., & Robbins, M. O. (2005). The breakdown of continuum models for mechanical contacts. *Nature*, *435*, 929–932.
Majumdar, A. (1989). *Fractal surfaces and their applications to surface phenomena*. Ph.D. Thesis, University of California at Berkeley, Berkeley, California, USA.
Majumdar, A., & Bhushan, B. (1990). Role of fractal geometry in roughness characterization and contact mechanics of surfaces. *ASME Journal of Tribology*, *112*, 205–216.

Majumdar, A., & Bhushan, B. (1991). Fractal model of elastic-plastic contact between rough surfaces. *ASME Journal of Tribology*, *113*, 1–11.
Mandelbrot, B. B., Passoja, D. E., & Paullay, A. J. (1984). Fractal character of fracture surfaces of metals. *Nature*, *308*, 721–722.
Mikic, B. B. (1974). Thermal contact conductance: Theoretical considerations. *International Journal of Heat and Mass Transfer*, *205*, 416–417.
Milanez, F. H., Yovanovich, M. M., & Culham, J. R. (2003a). Effect of surface asperity truncation on thermal contact conductance. *IEEE Transactions on Components and Packaging Technologies*, *26*, 48–54.
Milanez, F. H., Yovanovich, M. M., & Culham, J. R. (2003b). Effect of surface asperity truncation on thermal contact conductance. *IEEE Transactions on Components and Packaging Technologies*, *26*, 48–54.
Mindlin, R. D. (1949). Compliance of elastic bodies in contact. *ASME Journal of Applied Mechanics*, *16*, 259–268.
Nayak, P. R. (1971). Random process model of rough surfaces. *ASME Journal of Lubrication Technology*, *93*, 398–407.
Nayak, P. R. (1973). Random process model of rough surfaces in plastic contact. *Wear*, *26*, 305–333.
Nosonovsky, M. (2007). Multiscale roughness and stability of superhydrophobic biomimetic interfaces. *Langmuir*, *23*, 3157–3161.
Nosonovsky, M., & Bhushan, B. (2008). *Multiscale dissipative mechanisms and hierarchical surfaces: Friction, superhydrophobicity, and biomimetics*. Springer.
Onions, R. A., & Archard, J. F. (1973). The contact of surfaces having a random structure. *Journal of Physics D*, *289*, 416.
Paggi, M. (2014). *Thermal contact conductance of rough surfaces* (pp. 4948–4957). Netherlands, Dordrecht: Springer. ISBN 978-94-007-2739-7.
Paggi, M., & Barber, J. R. (2011). Contact conductance of rough surfaces composed of modified RMD patches. *International Journal of Heat and Mass Transfer*, *54*, 4664–4672.
Paggi, M., & Ciavarella, M. (2010). The coefficient of proportionality κ between real contact area and load, with new asperity models. *Wear*, *268*, 1020–1029.
Paggi, M., & He, Q.-C. (2015). Evolution of the free volume between rough surfaces in contact. *Wear*, *336–337*, 86–95.
Paggi, M., & Hills, D. A. (2016a). Special issue on EUROMECH 575. *Proceedings of the Institution of Mechanical Engineers, Part C: Journal of Mechanical Engineering Science*, *230*(9), 1373–1373.
Paggi, M., & Hills, D. A. (2016b). Editorial of the special issue on the EUROMECH colloquium 575. *The Journal of Strain Analysis for Engineering Design*, *51*(4), 239–239.
Paggi, M., & Reinoso, J. (2018). A variational approach with embedded roughness for adhesive contact problems. *Mechanics of Advanced Materials and Structures*. https://doi.org/10.1080/15376494.2018.1525454.
Paggi, M., Pohrt, R., & Popov, V. L. (2014). Partial-slip frictional response of rough surfaces. *Scientific Reports*, *4*, 5178.
Papangelo, A., & Ciavarella, M. (2018). Adhesion of surfaces with wavy roughness and a shallow depression. *Mechanics of Materials*, *118*, 11–16.
Pastewka, L., & Robbins, M. O. (2014). Contact between rough surfaces and a criterion for macroscopic adhesion. *Proceedings of the National Academy of Sciences*, *111*, 3298–3303.
Pastewka, L., & Robbins, M. O. (2016). Contact area of rough spheres: Large scale simulations and simple scaling laws. *Applied Physics Letters*, *108*, 221601.
Peitgen, H. O., & Saupe, D. (1988). *The science of fractal images*. New York: Springer-Verlag.
Peressadko, A. G., Hosoda, N., & Persson, B. N. J. (2005). Influence of surface roughness on adhesion between elastic bodies. *Physical Review Letters*, *95*, 124301.
Persson, B. N. J. (2000). *Sliding friction, physical principles and applications*. Springer.
Persson, B. N. J. (2001a). Elastoplastic contact between randomly rough surfaces. *Physical Review Letters*, *87*, 116101.

Persson, B. N. J. (2001b). Theory of rubber friction and contact mechanics. *Journal of Chemical Physics*, *115*, 3840–3861.
Persson, B. N. J. (2002a). Adhesion between elastic bodies with randomly rough surfaces. *Physical Review Letters*, *89*, 245502.
Persson, B. N. J. (2002b). Adhesion between elastic bodies with randomly rough surfaces. *European Physical Journal E*, *8*, 385.
Persson, B. N. J. (2006). Contact mechanics for randomly rough surfaces. *Surface Science Reports*, *261*, 201–227.
Persson, B. N. J., Bucher, F., & Chiaia, B. (2002). Elastic contact between randomly rough surfaces: Comparison of theory with numerical results. *Physical Review B*, *65*, 184106.
Persson, B. N. J., Albohr, O., Tartaglino, U., Volokitin, A. I., & Tosatti, E. (2005). On the nature of surface roughness with application to contact mechanics, sealing, rubber friction and adhesion. *Journal of Physics: Condensed Matter*, *17*, R1.
Popov, V. L. (2010). *Contact mechanics and friction*. Berlin, Heidelberg: Springer.
Popov, V. L. (2014). Analytic solution for the limiting shape of profiles due to fretting wear. *Scientific Reports*, *4*, 3749.
Popov, V. L., & Hess, M. (2015). *Method of dimensionality reduction in contact mechanics and friction*. Berlin, Heidelberg: Springer.
Popov, V. L., Pohrt, R., & Li, Q. (2017). Strength of adhesive contacts: Influence of contact geometry and material gradients. *Friction*, *5*, 308–325.
Rabinowicz, E. (1965). *Friction and wear of materials*. New York: Wiley.
Raja, J., Muralikrishnan, B., & Fu, S. (2002). Recent advances in separation of roughness, waviness and form. *Precision Engineering*, *26*, 222–235.
Rey, V., Anciaux, G., & Molinari, J.-F. (2017). Normal adhesive contact on rough surfaces: efficient algorithm for FFT-based BEM resolution. *Computational Mechanics*, *60*, 69–81.
Russ, J. C. (1994). *Fractal surfaces*. New York: Plenum Press.
Sayles, R. S., & Thomas, T. R. (1977). The spatial representation of surface roughness by means of the structure function: A practical alternative to correlation. *Wear*, *42*, 263–276.
Scaraggi, M. (2012). Lubrication of textured surfaces: A general theory for flow and shear stress factors. *Physical Review E*, *86*, 026314.
Scaraggi, M., & Persson, B. N. J. (2012). Time-dependent fluid squeeze-out between soft elastic solids with randomly rough surfaces. *Tribology Letters*, *47*, 409–416.
Sherge, M., & Gorb, S. (2001). *Biological micro- and nano-tribology & nature's solutions*. Berlin, Germany: Springer.
Sridhar, M. R., & Yovanovich, M. M. (1994). Review of elastic and plastic contact conductance models: Comparison with experiments. *Journal of Thermophysics and Heat Transfer*, *8*, 633–640.
Sridhar, M. R., & Yovanovich, M. M. (1996a). Elastoplastic contact conductance model for isotropic, conforming rough surfaces and comparison with experiments. *Journal of Heat Transfer*, *118*, 3–16.
Sridhar, M. R. & Yovanovich, M. M. (1996b). Contact conductance correlations based on Greenwood and Williamson surface model. In *ASME National Heat Transfer Conference, Houston, Texas* (pp. 1–11).
Stout, K. J., Sullivan, P. J., Dong, W. P., Mainsah, E., Luo, N., Mathia, T., & Zahouani, H. (1994). *The development of methods for the characterization of roughness on three dimensions*. Publication no. EUR 15178 EN of the Commission of the European Communities, Luxembourg.
Tarabay, A. (2014). Advanced computation models for the evolution of fracture networks in shale during hydraulic fracturing. In *Proceedings of the 1st International Symposium on Energy Challenges and Mechanics, Aberdeen, Scotland, UK*.
Vakis, A.I., Yastrebov, V.A., Scheibert, J., Nicola, L., Dini, D., Minfray, C., et al. (2018). Modeling and simulation in tribology across scales: An overview. *Tribology International*, *125*, 169–199.
Waters, J. F., Lee, S., & Guduru, P. R. (2009). Mechanics of axisymmetric wavy surface adhesion: JKR-DMT transition solution. *International Journal of Solids and Structures*, *46*, 1033–1042.

Whitehouse, D. J. & Archard, D. J. (1970). The properties of random surfaces of significance in their contact. *Proceedings of the Royal Society of London, Series A*, *316*, 97–121.
Yastrebov, V. A., Anciaux, G., & Molinari, J.-F. (2015). From infinitesimal to full contact between rough surfaces: Evolution of the contact area. *International Journal of Solids and Structures*, *52*, 83–102.
Yu, N., & Polycarpou, A. A. (2004). Adhesive contact based on the Lennard-Jones potential: A correction to the value of the equilibrium distance as used in the potential. *Journal of Colloid and Interface Science*, *278*, 428–435.
Zavarise, G., Borri-Brunetto, M., & Paggi, M. (2004a). On the reliability of microscopical contact models. *Wear*, *257*, 229–245.
Zavarise, G., Borri-Brunetto, M., & Paggi, M. (2004b). On the resolution dependence of micromechanical contact models. *Wear*, *262*, 42–54.

Chapter 6
Modelling Flows in Lubrication

Andreas Almqvist and Francesc Pérez-Ràfols

Abstract This chapter introduces the reader to lubrication theory and describes the governing equations, models and methods that can be used to simulate various types of lubricated systems. It starts with an introduction to the tribological contact and to the different lubrication regimes. The basis for the classical lubrication theory is then given and thereafter follows a presentation of how to obtain the Reynolds equation by means of scaling and asymptotic analysis of the Navier–Stokes equations. After having obtained the Reynolds equation, a quite elaborate presentation of cavitation algorithms is given. It includes discretisation and presents the analytical solution for a pocket bearing as a benchmark model problem. Then, the concept of homogenisation of surface roughness is introduced. This starts from the simplest iso-viscous and incompressible case, expands to include compressibility with a constant bulk modulus constitutive relation and then also addresses the case of ideal gases. Thereafter, the relation between homogenised coefficients and the Patir and Cheng flow factors is described and finally it is shown how to incorporate the effect of mixed lubrication into the model.

Introduction

Machines consist of machine elements and their safe and efficient operation relies on carefully designed interfaces between these elements. Many of these are lubricated and the material herein is meant to provide an understanding of established models and numerical solution procedures that can be used to study the behaviour of lubricated interfaces such as bearings and seals. Hopefully, it also inspires and encourages the reader to contribute to further development thereof.

A. Almqvist (✉) · F. Pérez-Ràfols
Luleå University of Technology, Luleå, Sweden
e-mail: andreas.almqvist@ltu.se

M. Paggi and D. Hills (eds.), *Modeling and Simulation of Tribological Problems in Technology*, CISM International Centre for Mechanical Sciences 593,
https://doi.org/10.1007/978-3-030-20377-1_6

The Tribological Contact

At start-ups, at stops as well as during operation, most machine elements experience varying contact conditions. Take, for example, the (axial) tilted pad thrust bearing illustrated in Fig. 6.1. This type of bearing belongs to the class of *hydrodynamic fluid film bearings*, which are designed to generate a fluid film pressure build-up that separates the rotating and stationary surfaces so that contact-less rotation is achieved while carrying the load on the shaft. In fact, it is the relative motion of the surfaces, as the lubricant is pulled into the converging geometry between the collar and the pad, that creates the necessary fluid film pressure. Typically, the collar in a tilting pad thrust bearing is made of steel while the pads have a soft (compliant) facing made of Babbitt (metal alloy) or Teflon® (polytetrafluoroethylene (PTFE)). This means that the smallest direct contact the collar makes with the shaft while rotating, will cause severe wear on the facing surface and it is of crucial importance to have a system that separates the surfaces during initiation of start-up and stop. A common solution is to implement a system that pressurises the supplied lubricant, generating hydrostatic lift.

Another example is the piston with its reciprocating motion inside the cylinder of a heavy-duty diesel engine, such as the one depicted in Fig. 6.2. In this case, the lubricated ring interfaces never see stationary conditions as they are decelerating from maximum speed at midstroke to fully stopped at the dead centres, reversing and then accelerating to reach maximum speed when back at midstroke again. Together with the speed, the lift generated by the resulting fluid pressure distribution will also vary from a maximum value at the midstroke to almost nothing at the dead centres.

As revealed by these two examples, the conditions seen by the interfaces may be substantially different. Therefore, depending on the application and the operating conditions it is common to characterise the tribological contact by the lubrication regime it operates in. The lubricant regimes are often divided into: Boundary Lubrication (BL), Mixed Lubrication (ML) and Full Film Lubrication (FL). Note that some systems may operate in various regimes. One such example is the aforemen-

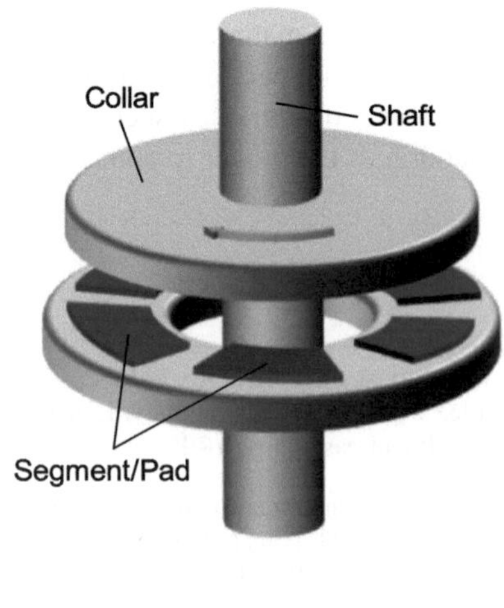

Fig. 6.1 Schematics of a tilting pad thrust bearing including the shaft connecting, e.g. the turbine to the generator in a hydro-power machine

Fig. 6.2 Piston with rings inside cylinder liner. Illustration courtesy Söderfjäll (2017)

tioned piston rings of a heavy- duty diesel engine, in which the load that the interface between the compression ring and liner surface see comes from ring tension and possibly the gas pressure behind the compression ring. During operation, the contact between ring and the linear varies and it is understood that it can be in the full-film regime at some parts in the mixed at others and sometimes it may even enter the boundary lubrication regime.

The Boundary Lubrication Regime

In the boundary lubrication (BL) regime, the lubricant's hydrodynamic action is negligible and the load is carried directly by surface asperities or by surface-active additives (a so-called tribofilm). Here, the surface topography is preferably chosen to optimise the frictional behaviour without increasing the rate of wear. To do this, one must also understand how the chemical processes are affected by the actual contact conditions, in terms of, e.g. heat generation, pressure peaks or real area of contact, and vice versa. This book does not describe any comprehensive BL-model incorporating all these features. It can be pointed out, however, that there exists nowadays a well-known elasto-plastic contact mechanics model Tian and Bhushan (1996), Almqvist et al. (2007b), Sahlin et al. (2010a) with the corresponding numerical approach grounded on a variational formulation, expressed in terms of total complementary potential energy, with acceleration relying on the fast Fourier transform (FFT) Liu et al. (2000, 2007), Wang et al. (2003). This approach has proved to ensure a stable and effective simulation of (rough) contact mechanics and it can help to increase the understanding of how the surface roughness influences the elastic deflection, the plastic deformation (and plasticity index), the pressure build- up and the real area of contact. An in-depth understanding of this connection is required to refine the design of interfaces operating under these circumstances.

As the hydrodynamical action of the lubricant increases, the contact mechanical response becomes less influential in terms of pressure and real contact area, and a transition from the BL- to the ML-regime may therefore occur.

The Mixed Lubrication Regime

What characterizes the ML-regime is that the load is mostly carried by the lubricant's hydrodynamical action, which may be influenced by the elastic deflection of the surfaces, the tribofilm, directly by surface asperities, or a combination thereof. Indeed, in this regime, a certain degree of contact between the interfaces is still expected.

This means that the objectives of the surface topography are in this case to support the hydrodynamic action of the lubricant, aid the elastic deflection in rendering a smoother surface, enable bonding of the surface active additives and optimise friction in the contact spots without increasing wear.

Modelling mixed lubrication has turned out to be a true challenge and the models available are built upon assumptions simplifying the physics involved in the transition from the BL- and the FL-regime. As indicated above, a contact mechanics model may be used to indicate a possible transition between the BL- and the ML-regimes. Similarly, modelling performed regarding full-film lubrication has lead to numerical approaches that may be used to increase the understanding of the transition from the FL- to the ML-regimes. One well-known example of an ML-model, is the Luleå mixed lubrication model by Sahlin et al. (2010a), in which partitioning between lubricant- carried load and load carried by direct contact, is determined by the separation. More precisely, when the separation between the interfaces becomes smaller than a chosen measure of the surface roughness height, the lubricant load is alleviated with the amount that the corresponding unlubricated interface would carry at that separation. The main constituents of this type models are a contact mechanics model and a thin film flow model based on a homogenised two-scale formulation, which incorporates the influence of the surface roughness in an averaged sense.

The Full-Film Lubrication Regime

When the hydrodynamic action of the lubricant fully separates the surfaces and the load is no longer carried by the contact between the surfaces, the interface enters the full-film lubrication (FL) regime. In the FL regime, friction may still be reduced by carefully chosen topographies. Even though there is no direct contact, the lubricant pressure may lead to stress concentrations high enough to cause fatigue, likely leading to excessive wear in the form of spalling in highly loaded situations.

This regime is commonly subdivided into hydrodynamic lubrication (HL) and elastohydrodynamic lubrication (EHL), since the performance is greatly affected by the presence of elastic deflections, i.e. fluid–structure interaction, at the lubricated interface.

Hydrodynamic Lubrication

Slider bearings are typical examples of applications that, under certain conditions, operate in the hydrodynamic lubrication (HL) regime where the elastic deformations of the bearing surfaces are sufficiently small to be neglected. For example, the tilting pad thrust bearing, as depicted in Fig. 6.1, exhibits a conformal interface between the pad and the collar and is designed to for operation in the hydrodynamic lubrication regime. Note that the angle of inclination of the pads, which is generally only a fraction of a degree, has been greatly exaggerated in the figure. One problem that arise when modelling conformal interfaces like this one, comes from the large differences in scales. More precisely, the global scale describing the geometry, pad–collar interface, is several orders of magnitude larger than the local scale describing the surface topography/roughness. This situation can be approached by means of homogenisation. This is also a subject further discussed in section 'Homogenisation of the Reynolds Equation'.

Elastohydrodynamic Lubrication

Elastohydrodynamic lubrication (EHL) is the type of hydrodynamic lubrication where the fluid–structure interaction (FSI), caused by elastic deformations of the contacting surfaces, plays a major role. This situation may occur when lubricating interacting non-conformal bodies. This leads to highly localised (concentrated) contacts, and it is the lubricant's piezo-viscous response combined with the elastic flattening of surface roughness features that facilitate the separation of the interacting surfaces. An example where EHL is typically found is at the interface between the roller and the raceway in a typical roller bearing, as shown in Fig. 6.3, which are most commonly designed to operate in the full-film elastohydrodynamic lubrication regime.

Fig. 6.3 Schematics of a typical rolling element bearing

The apparent contact zone for a rolling bearing is, in general, elliptic in shape. Depending on the design parameters previously mentioned and the actual running conditions, the shape of the ellipse will change. In any case, the contact region is small and the concentrated load implies a severe surface—as well as sub-surface stress condition that may lead to both elastic—but also plastic deformation. For a bearing in operation, high stresses eventually cause fatigue, which in turn can lead to shortened service life due to, for example, spalling. When the contact is starved of lubricant, or when running conditions do not allow for a hydrodynamic action that fully separates the surfaces, the risk for plastic deformation increases.

If the width of the contact ellipse exceeds the minimum width of the raceway and the roller, the contact will be then truncated and this leads lead to increased stresses in the material. In the case of a contact ellipse which is more than 4 times wider than its length in the rolling direction, the pressure at the centerline in the rolling direction can be approximated to the pressure corresponding to a line contact, Evans et al. (2001), Shirzadegan et al. (2016). This motivates describing the problem with a two-dimensional instead of a three-dimensional domain. Moreover, it has been shown that the one-dimensional Reynolds equation can give highly accurate estimates of deformations and stresses inside the interface. Still, as with most tribological problems, this is a very demanding problem that requires advanced mathematical descriptions as well as highly efficient numerical solution procedures, see e.g. Venner and Lubrecht (2000), Holmes (2002), Holmes et al. (2003a, b), Hooke and Li (2006), Persson (2010), Scaraggi and Carbone (2012), Ahmed et al. (2012), Zhu et al. (2015), Habchi (2018, 2019). Homogenisation of roughness, Fast Fourier Transformation (FFT), and multilevel techniques are examples of such. This usually renders quite complex methods that often require end users with rather specialised background.

Introduction to Lubrication Theory

In the celebrated paper Reynolds (1886), Reynolds presented an analysis of hydrodynamic flows in thin gaps. Examples of such are the gap between the rolling element and the raceway in a bearing, between the contacting surfaces in a seal, between the eye and the contact lens, in our joints and when a water film is generated between a car's tyre and the road surface. Figure 6.4 illustrates the typical schematics of a flow domain in lubrication, which is representative of all aforementioned cases. What these situations have in common is that Reynolds' thin film approximation applies, implying that, $h_0 \ll l_0$ and $h_0 \ll b_0$, where h_0 represents the gap between the surfaces h_u and h_l, and l_0 and b_0 represent the dimensions of the surfaces. The main advantage of using this approach is that the resulting equation is a partial differential equation for the fluid pressure, p, which is of one dimension less than the physical problem, viz.,

$$\frac{\partial(\rho h)}{\partial t} = \nabla \cdot \left(\frac{\rho h^3}{12\mu} \nabla p - \frac{1}{2} \begin{bmatrix} u_u + u_l \\ v_u + v_l \end{bmatrix} \rho h, \right) \tag{6.1}$$

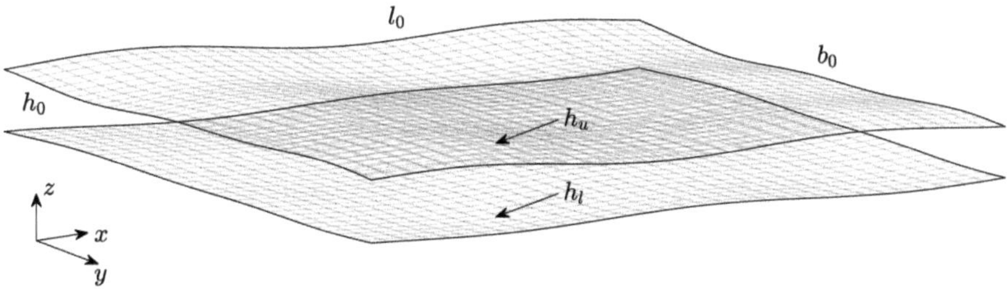

Fig. 6.4 Schematic illustration of a thin gap between the two impermeable surfaces h_u and h_l

where $h = h_u - h_l$ models the gap between the surfaces, ρ and μ are the density and viscosity of the fluid, the velocity in the x-direction of the upper and the lower surfaces are denoted by u_u and u_l, respectively, and v_u and v_l denote the surfaces' velocities in the y-direction. This equation, known as Reynolds equation, was derived originally by Reynolds for incompressible and iso-viscous fluids by simplifying the Navier–Stokes equation under the thin film assumption and neglecting inertia.

Common examples of this use of Reynolds equation can be found in the field of hydrodynamic lubrication of thrust and journal bearings, elastohydrodynamic lubrication and leakage in seals, where (6.1) is often used together with semi-empirical expressions for density– and viscosity–pressure relationships.

In the following, a simplistic scaling and straightforward analysis will be applied to the Navier–Stokes equations for flows in thin gaps. It will be demonstrated that it is possible to obtain the set of reduced equations for the classical lubrication approximation governing incompressible and iso-viscous flows. The reduced set of equations will thereafter be integrated to obtain an explicit formulation of the velocity field. Finally, the continuity equation will be integrated rendering the well-known Reynolds equation. We start by introducing the continuity equation describing conservation of mass and the Navier–Stokes momentum equations, for compressible and viscous flow in 3D, i.e.

$$0 = \frac{\partial \rho}{\partial t} + \frac{\partial (\rho u)}{\partial x} + \frac{\partial (\rho v)}{\partial y} + \frac{\partial (\rho w)}{\partial z}, \tag{6.2a}$$

$$\rho \frac{Du}{Dt} = \rho g_x - \frac{\partial p}{\partial x} - \frac{2}{3}\frac{\partial}{\partial x}(\mu \xi) + 2\frac{\partial}{\partial x}\left(\mu \frac{\partial u}{\partial x}\right) + \frac{\partial}{\partial y}\left(\mu\left(\frac{\partial u}{\partial y} + \frac{\partial v}{\partial x}\right)\right) + \frac{\partial}{\partial z}\left(\mu\left(\frac{\partial u}{\partial z} + \frac{\partial w}{\partial x}\right)\right), \tag{6.2b}$$

$$\rho \frac{Dv}{Dt} = \rho g_y - \frac{\partial p}{\partial y} - \frac{2}{3}\frac{\partial}{\partial y}(\mu \xi) + \frac{\partial}{\partial x}\left(\mu\left(\frac{\partial u}{\partial y} + \frac{\partial v}{\partial x}\right)\right) + 2\frac{\partial}{\partial y}\left(\mu \frac{\partial v}{\partial y}\right) + \frac{\partial}{\partial z}\left(\mu\left(\frac{\partial v}{\partial z} + \frac{\partial w}{\partial y}\right)\right), \tag{6.2c}$$

$$\rho\frac{Dw}{Dt} = \rho g_z - \frac{\partial p}{\partial z} - \frac{2}{3}\frac{\partial}{\partial z}(\mu\xi) + \frac{\partial}{\partial x}\left(\mu\left(\frac{\partial u}{\partial z} + \frac{\partial w}{\partial x}\right)\right) + \frac{\partial}{\partial y}\left(\mu\left(\frac{\partial v}{\partial z} + \frac{\partial w}{\partial y}\right)\right) + 2\frac{\partial}{\partial z}\left(\mu\frac{\partial u}{\partial z}\right), \tag{6.2d}$$

where D/Dt is the material derivative

$$\frac{D}{Dt} = \frac{\partial}{\partial t} + u\frac{\partial}{\partial x} + v\frac{\partial}{\partial y} + w\frac{\partial}{\partial z}$$

and ξ is the dilation

$$\xi = \frac{\partial u}{\partial x} + \frac{\partial v}{\partial y} + \frac{\partial w}{\partial z}. \tag{6.3}$$

This form of the Navier–Stokes equations is based on the Stokes' assumption that the bulk viscosity is zero, which is an assumption that is justified for ideal gases and for monatomic liquids, see e.g. Buresti (2015). It is here highlighted that, although a fluid (in general) exhibits both compressibility and nonlinear viscosity–pressure and temperature dependence, the analysis presented herein is restricted to isothermal conditions.

In the following the fact that fluid film is very thin relative to the other dimensions will be used, in order to derive simplified forms of (6.2). More precisely, this will be done by scaling and dimensional analysis.

Scaling and Asymptotic Analysis of the Navier–Stokes Equations

The derivation of Reynolds equation can be started by transforming (6.2) into dimensionless from via scaling. To this end, a set of scaling parameters is introduced that applies to flow situations were the typical length and speed in the x- and y-directions, respectively, are of the same order. In order to write the independent and dependent variables in dimensionless form, the following scaling is introduced:

$$\begin{gathered}\bar{x} = x/l_0, \quad \bar{y} = y/l_0, \quad \bar{z} = z/h_0, \quad \bar{t} = t/t_0, \\ \bar{u} = u/u_0, \quad \bar{v} = v/u_0, \quad \bar{w} = w/w_0, \\ \bar{p} = p/p_0, \quad \bar{\rho} = \rho/\rho_0, \quad \bar{\mu} = \mu/\mu_0.\end{gathered} \tag{6.4}$$

Among the scaling parameters (subscripted 0), l_0 and h_0 are problem specific and so is u_0 for the situation where the flow is shear driven and caused by relative motion of the surfaces. The scaling parameters t_0, w_0 and p_0 are *a priori* unknown. The fluid is, at this point, regarded compressible and piezo-viscous meaning that $\rho = \rho(p)$ and

$\mu = \mu(p)$, respectively, and it should be noted that ρ_0 and μ_0 should not be directly used in the dimensional analysis as the scaling of the density and viscosity will depend on terms of p_0 and specific inputs to the constitutive relations (6.5) and (6.6)

$$\rho = \rho_a f_\rho(p), \tag{6.5}$$

and

$$\mu = \mu_a f_\mu(p), \tag{6.6}$$

where f_ρ and f_μ are strictly increasing functions with $f_\rho(p_a) = f_\mu(p_a) = 1$ and ρ_a and μ_a are the density and the viscosity at the ambient pressure p_a.

By means of (6.4), the non-dimensional form of Eq. (6.2a) becomes

$$\frac{1}{t_0}\frac{\partial\bar{\rho}}{\partial\bar{t}} + \frac{u_0}{l_0}\frac{\partial(\bar{\rho}\bar{u})}{\partial\bar{x}} + \frac{u_0}{l_0}\frac{\partial(\bar{\rho}\bar{v})}{\partial\bar{y}} + \frac{w_0}{h_0}\frac{\partial(\bar{\rho}\bar{w})}{\partial\bar{z}} = 0. \tag{6.7}$$

The current objective is to analyse the asymptotic behaviour of (6.2) as h_0/l_0 goes to zero. The notation

$$\varepsilon = h_0/l_0 \tag{6.8}$$

is, therefore, convenient. In this notation, (6.7) becomes

$$\varepsilon\frac{l_0}{t_0 u_0}\frac{\partial\bar{\rho}_\varepsilon}{\partial\bar{t}} + \varepsilon\frac{\partial(\bar{\rho}_\varepsilon\bar{u}_\varepsilon)}{\partial\bar{x}} + \varepsilon\frac{\partial(\bar{\rho}_\varepsilon\bar{v}_\varepsilon)}{\partial\bar{y}} + \frac{w_0}{u_0}\frac{\partial(\bar{\rho}_\varepsilon\bar{w}_\varepsilon)}{\partial\bar{z}} = 0, \tag{6.9}$$

where the subscript ε indicates dependent variables' parameterisation in ε. For steady or incompressible flow, it is realised that (6.9) has little if any meaning if not at least the three last terms are of the same order. In case not, $\bar{\rho}_\varepsilon\bar{w}_\varepsilon$ would not depend on z, which is unreasonable for most types of flow situations, e.g. for an incompressible fluid confined in a (narrow) converging gap. This is the motivation for the scaling $w_0 = \varepsilon u_0$. In the case of compressible and unsteady flow there are two cases, i.e. (i) the time scale for the variations in the density is larger or of the same order as l_0/u_0 or (ii) the density oscillates extremely fast in comparison to l_0/u_0. In the first case, we can choose $t_0 = l_0/u_0$ and $w_0 = \varepsilon u_0$ and all terms become of the same order. In the latter case, $t_0 = h_0/u_0$ and the first term will dominate the flow. This situation will not be considered here and thus the following scaling can be adopted:

$$w_0 = \varepsilon u_0, \quad t_0 = l_0/u_0. \tag{6.10}$$

In terms of this scaling, the continuity equation (6.9) becomes

$$\frac{\partial\bar{\rho}_\varepsilon}{\partial\bar{t}} + \frac{\partial(\bar{\rho}_\varepsilon\bar{u}_\varepsilon)}{\partial\bar{x}} + \frac{\partial(\bar{\rho}_\varepsilon\bar{v}_\varepsilon)}{\partial\bar{y}} + \frac{\partial(\bar{\rho}_\varepsilon\bar{w}_\varepsilon)}{\partial\bar{z}} = 0. \tag{6.11}$$

Now, (6.2b)–(6.2d) can be scaled using (6.4) and (6.10). These read

$$\begin{aligned}
\varepsilon^2\frac{\rho_0 u_0 l_0}{\mu_0}\bar{\rho}_\varepsilon\left(\frac{\partial\bar{u}_\varepsilon}{\partial\bar{t}}+\bar{u}_\varepsilon\frac{\partial\bar{u}_\varepsilon}{\partial\bar{x}}+\bar{v}_\varepsilon\frac{\partial\bar{u}_\varepsilon}{\partial\bar{y}}+\bar{w}_\varepsilon\frac{\partial\bar{u}_\varepsilon}{\partial\bar{z}}\right)&=\varepsilon^2\frac{\rho_0 l_0^2}{\mu_0 u_0}\bar{\rho}_\varepsilon g_x\\
-\varepsilon^2\frac{p_0 l_0}{\mu_0 u_0}\frac{\partial\bar{p}_\varepsilon}{\partial\bar{x}}+-\varepsilon^2\frac{2}{3}\frac{\partial}{\partial\bar{x}}&\left(\bar{\mu}_\varepsilon\left(\frac{\partial\bar{u}_\varepsilon}{\partial\bar{x}}+\frac{\partial\bar{v}_\varepsilon}{\partial\bar{y}}+\frac{\partial\bar{w}_\varepsilon}{\partial\bar{z}}\right)\right)\\
+2\varepsilon^2\frac{\partial}{\partial\bar{x}}\left(\bar{\mu}_\varepsilon\frac{\partial\bar{u}_\varepsilon}{\partial\bar{x}}\right)+\varepsilon^2\frac{\partial}{\partial\bar{y}}&\left(\bar{\mu}_\varepsilon\left(\frac{\partial\bar{u}_\varepsilon}{\partial\bar{y}}+\frac{\partial\bar{v}_\varepsilon}{\partial\bar{x}}\right)\right)\\
+\frac{\partial}{\partial\bar{z}}&\left(\bar{\mu}_\varepsilon\left(\frac{\partial\bar{u}_\varepsilon}{\partial\bar{z}}+\varepsilon^2\frac{\partial\bar{w}_\varepsilon}{\partial\bar{x}}\right)\right),
\end{aligned}\tag{6.12}$$

$$\begin{aligned}
\varepsilon^2\frac{\rho_0 u_0 l_0}{\mu_0}\bar{\rho}_\varepsilon\left(\frac{\partial\bar{v}_\varepsilon}{\partial\bar{t}}+\bar{u}_\varepsilon\frac{\partial\bar{v}_\varepsilon}{\partial\bar{x}}+\bar{v}_\varepsilon\frac{\partial\bar{v}_\varepsilon}{\partial\bar{y}}+\bar{w}_\varepsilon\frac{\partial\bar{v}_\varepsilon}{\partial\bar{z}}\right)&=\varepsilon^2\frac{\rho_0 l_0^2}{\mu_0 u_0}\bar{\rho}_\varepsilon g_y\\
-\varepsilon^2\frac{p_0 l_0}{\mu_0 u_0}\frac{\partial\bar{p}_\varepsilon}{\partial\bar{y}}-\frac{2}{3}\varepsilon^2\frac{\partial}{\partial\bar{x}}&\left(\bar{\mu}_\varepsilon\left(\frac{\partial\bar{u}_\varepsilon}{\partial\bar{x}}+\frac{\partial\bar{v}_\varepsilon}{\partial\bar{y}}+\frac{\partial\bar{w}_\varepsilon}{\partial\bar{z}}\right)\right)\\
+\varepsilon^2\frac{\partial}{\partial\bar{x}}\left(\bar{\mu}_\varepsilon\left(\frac{\partial\bar{u}_\varepsilon}{\partial\bar{y}}+\frac{\partial\bar{v}_\varepsilon}{\partial\bar{x}}\right)\right)&+2\varepsilon^2\frac{\partial}{\partial\bar{y}}\left(\bar{\mu}_\varepsilon\frac{\partial\bar{v}_\varepsilon}{\partial\bar{y}}\right)\\
+\frac{\partial}{\partial\bar{z}}&\left(\bar{\mu}_\varepsilon\left(\frac{\partial\bar{v}_\varepsilon}{\partial\bar{z}}+\varepsilon^2\frac{\partial\bar{w}_\varepsilon}{\partial\bar{y}}\right)\right),
\end{aligned}\tag{6.13}$$

and

$$\begin{aligned}
\varepsilon^2\frac{\rho_0 u_0 l_0}{\mu_0}\bar{\rho}_\varepsilon\left(\frac{\partial\bar{w}_\varepsilon}{\partial\bar{t}}+\bar{u}_\varepsilon\frac{\partial\bar{w}_\varepsilon}{\partial\bar{x}}+\bar{v}_\varepsilon\frac{\partial\bar{w}_\varepsilon}{\partial\bar{y}}+\bar{w}_\varepsilon\frac{\partial\bar{w}_\varepsilon}{\partial\bar{z}}\right)&=\varepsilon\frac{\rho_0 l_0}{\mu_0 u_0}\bar{\rho}_\varepsilon g_z\\
-\frac{p_0 l_0}{\rho_0 u_0}\frac{\partial\bar{p}_\varepsilon}{\partial\bar{z}}-\frac{2}{3}\frac{\partial}{\partial\bar{z}}&\left(\bar{\mu}_\varepsilon\left(\frac{\partial\bar{u}_\varepsilon}{\partial\bar{x}}+\frac{\partial\bar{v}_\varepsilon}{\partial\bar{y}}+\frac{\partial\bar{w}_\varepsilon}{\partial\bar{z}}\right)\right)\\
+\frac{\partial}{\partial\bar{x}}\left(\bar{\mu}_\varepsilon\left(\frac{\partial\bar{u}_\varepsilon}{\partial\bar{z}}+\varepsilon^2\frac{\partial\bar{w}_\varepsilon}{\partial\bar{x}}\right)\right)&+\frac{\partial}{\partial\bar{y}}\left(\bar{\mu}_\varepsilon\left(\frac{\partial\bar{v}_\varepsilon}{\partial\bar{z}}+\varepsilon^2\frac{\partial\bar{w}_\varepsilon}{\partial\bar{y}}\right)\right)\\
+2\frac{\partial}{\partial\bar{z}}&\left(\bar{\mu}_\varepsilon\frac{\partial\bar{w}_\varepsilon}{\partial\bar{z}}\right),
\end{aligned}\tag{6.14}$$

after multiplication with

$$\frac{l_0^2}{\mu_0 u_0}.$$

It should be noted that the modified Reynolds number defined as

$$\mathcal{R}_\varepsilon := \varepsilon^2 \frac{\rho_0 u_0 l_0}{\mu_0} \tag{6.15}$$

appears in the left-hand sides of (6.12)–(6.14).

The importance of incorporating the precise relationship $\mu(p)$ before carrying out the asymptotic analysis, rather than after, was highlighted in Rajagopal and Szeri (2003). There, the asymptotic analysis was carried out by adopting the most commonly used viscosity–pressure constitutive relation in EHL modelling, known as Barus' law, i.e. $\mu = \exp(\alpha p)$, where α is the so-called pressure–viscosity constant. This was also addressed in other works, Bayada et al. (2013), Gustafsson et al. (2015), all of which highlighted the importance of incorporating the precise constitutive relationships before carrying out the asymptotic analysis, rather than after. Similarly, the derivation of a model for a gas- lubricated system, needs to incorporate the well-known linear constitutive relation between density and pressure for an ideal gas, i.e. $\rho = kp$, where k is a constant, before the asymptotic analysis is carried out. The reason for this need comes from the fact that the typical values of density and viscosity to be used for the scaling, ρ_0 and μ_0, depend on the pressure in a manner determined by the precise constitutive relation. Without knowing it, the relative magnitude of the different terms in (6.12)–(6.14) cannot be assessed.

In the following, a recapitulation is given of the derivation of the reduced system for an incompressible and iso-viscous fluid, i.e. a fluid with constant density and viscosity, leading to the well-known Reynolds equation. With incompressible is hereby meant that the density is constant and without loss of generality $\rho = \rho_a$ is specified. Similarly, by iso-viscous, it is meant that the viscosity is constant and $\mu = \mu_a$ is defined. Clearly, in this case the density and viscosity are no longer dependent variables and $\rho_0 = \rho_a$ and $\mu_0 = \mu_a$ can be chosen. This means that $\bar{\rho}_\varepsilon = 1$ and $\bar{\mu}_\varepsilon = 1$ and that the continuity equation (6.11) reduces to

$$\frac{\partial \bar{u}_\varepsilon}{\partial \bar{x}} + \frac{\partial \bar{v}_\varepsilon}{\partial \bar{y}} + \frac{\partial \bar{w}_\varepsilon}{\partial \bar{z}} = 0. \tag{6.16}$$

Further, by using (6.16) it follows that (6.12)–(6.14) become

$$\begin{aligned} \varepsilon^2 \frac{\rho_a u_0 l_0}{\mu_a} \left(\frac{\partial \bar{u}_\varepsilon}{\partial \bar{t}} + \bar{u}_\varepsilon \frac{\partial \bar{u}_\varepsilon}{\partial \bar{x}} + \bar{v}_\varepsilon \frac{\partial \bar{u}_\varepsilon}{\partial \bar{y}} + \bar{w}_\varepsilon \frac{\partial \bar{u}_\varepsilon}{\partial \bar{z}} \right) &= \varepsilon^2 \frac{\rho_a l_0^2}{\mu_a u_0} g_x \\ &- \varepsilon^2 \frac{p_0 l_0}{\mu_a u_0} \frac{\partial \bar{p}_\varepsilon}{\partial \bar{x}} + 2\varepsilon^2 \frac{\partial^2 \bar{u}_\varepsilon}{\partial \bar{x}^2} \\ + \varepsilon^2 \frac{\partial}{\partial \bar{y}} \left(\frac{\partial \bar{u}_\varepsilon}{\partial \bar{y}} + \frac{\partial \bar{v}_\varepsilon}{\partial \bar{x}} \right) &+ \frac{\partial}{\partial \bar{z}} \left(\frac{\partial \bar{u}_\varepsilon}{\partial \bar{z}} + \varepsilon^2 \frac{\partial \bar{w}_\varepsilon}{\partial \bar{x}} \right), \end{aligned} \tag{6.17}$$

$$\varepsilon^2 \frac{\rho_a u_0 l_0}{\mu_a}\left(\frac{\partial \bar{v}_\varepsilon}{\partial \bar{t}} + \bar{u}_\varepsilon \frac{\partial \bar{v}_\varepsilon}{\partial \bar{x}} + \bar{v}_\varepsilon \frac{\partial \bar{v}_\varepsilon}{\partial \bar{y}} + \bar{w}_\varepsilon \frac{\partial \bar{v}_\varepsilon}{\partial \bar{z}}\right) = \varepsilon^2 \frac{\rho_a l_0^2}{\mu_a u_0} g_y$$
$$-\varepsilon^2 \frac{p_0 l_0}{\mu_a u_0}\frac{\partial \bar{p}_\varepsilon}{\partial \bar{y}} + \varepsilon^2 \frac{\partial}{\partial \bar{x}}\left(\frac{\partial \bar{u}_\varepsilon}{\partial \bar{y}} + \frac{\partial \bar{v}_\varepsilon}{\partial \bar{x}}\right) \tag{6.18}$$
$$+2\varepsilon^2 \frac{\partial^2 \bar{v}_\varepsilon}{\partial \bar{y}^2} + \frac{\partial}{\partial \bar{z}}\left(\frac{\partial \bar{v}_\varepsilon}{\partial \bar{z}} + \varepsilon^2 \frac{\partial \bar{w}_\varepsilon}{\partial \bar{y}}\right),$$

and

$$\varepsilon^2 \frac{\rho_a u_0 l_0}{\mu_a}\left(\frac{\partial \bar{w}_\varepsilon}{\partial \bar{t}} + \bar{u}_\varepsilon \frac{\partial \bar{w}_\varepsilon}{\partial \bar{x}} + \bar{v}_\varepsilon \frac{\partial \bar{w}_\varepsilon}{\partial \bar{y}} + \bar{w}_\varepsilon \frac{\partial \bar{w}_\varepsilon}{\partial \bar{z}}\right) = \varepsilon \frac{\rho_a l_0^2}{\mu_a u_0} g_z$$
$$-\frac{p_0 l_0}{\mu_a u_0}\frac{\partial \bar{p}_\varepsilon}{\partial \bar{z}} + \frac{\partial}{\partial \bar{x}}\left(\frac{\partial \bar{u}_\varepsilon}{\partial \bar{z}} + \varepsilon^2 \frac{\partial \bar{w}_\varepsilon}{\partial \bar{x}}\right) \tag{6.19}$$
$$+\frac{\partial}{\partial \bar{y}}\left(\frac{\partial \bar{v}_\varepsilon}{\partial \bar{z}} + \varepsilon^2 \frac{\partial \bar{w}_\varepsilon}{\partial \bar{y}}\right) + 2\frac{\partial^2 \bar{w}_\varepsilon}{\partial \bar{z}^2}.$$

Without considering the physics of the flow that a reduced model would reflect, one could just neglect the terms of order ε and higher in (6.17)–(6.19). This leads to the system

$$\frac{\partial \bar{u}_\varepsilon}{\partial \bar{x}} + \frac{\partial \bar{v}_\varepsilon}{\partial \bar{y}} + \frac{\partial \bar{w}_\varepsilon}{\partial \bar{z}} = 0, \tag{6.20a}$$
$$0 = \frac{\partial^2 \bar{u}_\varepsilon}{\partial \bar{z}^2}, \tag{6.20b}$$
$$0 = \frac{\partial^2 \bar{v}_\varepsilon}{\partial \bar{z}^2}, \tag{6.20c}$$
$$\frac{\partial^2 \bar{u}_\varepsilon}{\partial \bar{x}\partial \bar{z}} + \frac{\partial^2 \bar{v}_\varepsilon}{\partial \bar{y}\partial \bar{z}} + 2\frac{\partial^2 \bar{w}_\varepsilon}{\partial \bar{z}^2} = \frac{p_0 l_0}{\mu_a u_0}\frac{\partial \bar{p}_\varepsilon}{\partial \bar{z}}, \tag{6.20d}$$

which after using (6.16) reads

$$\frac{\partial \bar{u}_\varepsilon}{\partial \bar{x}} + \frac{\partial \bar{v}_\varepsilon}{\partial \bar{y}} + \frac{\partial \bar{w}_\varepsilon}{\partial \bar{z}} = 0, \tag{6.21a}$$
$$0 = \frac{\partial^2 \bar{u}_\varepsilon}{\partial \bar{z}^2}, \tag{6.21b}$$
$$0 = \frac{\partial^2 \bar{v}_\varepsilon}{\partial \bar{z}^2}, \tag{6.21c}$$
$$\frac{\partial^2 \bar{w}_\varepsilon}{\partial \bar{z}^2} = \frac{p_0 l_0}{\mu_a u_0}\frac{\partial \bar{p}_\varepsilon}{\partial \bar{z}}. \tag{6.21d}$$

From (6.21b) and (6.21c), it is clear that this model implies that $\bar{u}_\varepsilon$ and $\bar{v}_\varepsilon$ vary linearly with z, as in the case with moving parallel plates, in other words, plain Couette type of flow. In this case, the pressure is constant and equal to the atmospheric pressure, p_a, specified at the boundaries and thus $p_0 = p_a$ is a perfectly reasonable choice. In other cases, however, a pressure built up occurs and the reference pressure to choose p_0 is unknown. In order to model more general flow situations the pressure needs to be scaled relative to the velocity so that the pressure is preserved as a dependent variable as ε goes to zero. Indeed, this motivates the scaling $p_0 = \varepsilon^{-2}\mu_a u_0 / l_0$, which leads to that the system (6.17)–(6.19) is reduced to

$$\frac{\partial \bar{u}_\varepsilon}{\partial \bar{x}} + \frac{\partial \bar{v}_\varepsilon}{\partial \bar{y}} + \frac{\partial \bar{w}_\varepsilon}{\partial \bar{z}} = 0, \tag{6.22a}$$

$$\frac{\partial \bar{p}_\varepsilon}{\partial \bar{x}} = \frac{\partial^2 \bar{u}_\varepsilon}{\partial \bar{z}^2}, \tag{6.22b}$$

$$\frac{\partial \bar{p}_\varepsilon}{\partial \bar{y}} = \frac{\partial^2 \bar{v}_\varepsilon}{\partial \bar{z}^2}, \tag{6.22c}$$

$$\frac{\partial \bar{p}_\varepsilon}{\partial \bar{z}} = 0, \tag{6.22d}$$

when terms of order ε and higher are neglected. Note that this is actually the classical set of lubrication equations, which in dimensions reads

$$\frac{\partial u}{\partial x} + \frac{\partial v}{\partial y} + \frac{\partial w}{\partial z} = 0. \tag{6.23a}$$

$$\frac{\partial p}{\partial x} = \mu_a \frac{\partial^2 u}{\partial z^2}, \tag{6.23b}$$

$$\frac{\partial p}{\partial y} = \mu_a \frac{\partial^2 v}{\partial z^2}, \tag{6.23c}$$

$$\frac{\partial p}{\partial z} = 0. \tag{6.23d}$$

Remember that the underlying assumption for arriving at (6.23) is that p scales with ε^{-2}. It is also important to have in mind that even though the explicit time dependence has disappeared from the original set of equations, the dependent variables are still functions of time. For example, the domain may change with time as in the case with moving surfaces.

Derivation of the Reynolds Equation

Proceeding with the derivation of the Reynolds equation, (6.23) can be integrated to obtain an analytical description of the velocity field. First, it should be noted that (6.23d) implies that p does not depend on z, hence (6.23b) and (6.23c) may be integrated twice with respect to z. This leads to

$$u(x, y, z, t) = \frac{1}{2\mu_a}\frac{\partial p}{\partial x}z^2 + A(x, y, t)z + B(x, y, t), \tag{6.24a}$$

$$v(x, y, z, t) = \frac{1}{2\mu_a}\frac{\partial p}{\partial y}z^2 + C(x, y, t)z + D(x, y, t). \tag{6.24b}$$

The functions A, B, C and D may be found via boundary conditions on the surfaces. Indeed, consider the case when it can be assumed that the fluid stick to the surfaces (i.e. no-slip boundary conditions). This means that the velocity boundary condition at the lower surface is

$$\mathbf{u}_l = \begin{pmatrix} u(x, y, h_l(x, y, t), t) \\ v(x, y, h_l(x, y, t), t) \\ w(x, y, h_l(x, y, t), t) \end{pmatrix} = \begin{pmatrix} u_l(x, y, t) \\ v_l(x, y, t) \\ \dfrac{\partial h_l}{\partial t}(x, y, t) \end{pmatrix}, \tag{6.25}$$

and at the upper surface it is

$$\mathbf{u}_u = \begin{pmatrix} u(x, y, h_u(x, y, t), t) \\ v(x, y, h_u(x, y, t), t) \\ w(x, y, h_u(x, y, t), t) \end{pmatrix} = \begin{pmatrix} u_u(x, y, t) \\ v_u(x, y, t) \\ \dfrac{\partial h_u}{\partial t}(x, y, t) \end{pmatrix}, \tag{6.26}$$

where u_l, v_l, u_u and v_u are explicitly specified. The boundary condition at the lower surface (6.25) together with (6.24) gives

$$u(x, y, h_l, t) = \frac{1}{2\mu_a}\frac{\partial p}{\partial x}h_l^2 + Ah_l + B = u_l, \tag{6.27a}$$

$$v(x, y, h_l, t) = \frac{1}{2\mu_a}\frac{\partial p}{\partial y}h_l^2 + Ch_l + D = v_l \tag{6.27b}$$

and for the upper surface we obtain

$$u(x, y, h_u, t) = \frac{1}{2\mu_a}\frac{\partial p}{\partial x}h_u^2 + Ah_u + B = u_u, \tag{6.28a}$$

$$v(x, y, h_u, t) = \frac{1}{2\mu_a}\frac{\partial p}{\partial y}h_u^2 + Ch_u + D = v_u. \tag{6.28b}$$

It is straightforward to find A, B, C and D by solving the system (6.27)–(6.28). Inserting the result into (6.24) gives

$$\mathbf{u} = \begin{pmatrix} \dfrac{(z - h_l)(z - h_u)}{2\mu_a}\dfrac{\partial p}{\partial x} + (u_u - u_l)\dfrac{z - h_l}{h} + u_l \\[2ex] \dfrac{(z - h_l)(z - h_u)}{2\mu_a}\dfrac{\partial p}{\partial y} + (v_u - v_l)\dfrac{z - h_l}{h} + v_l \end{pmatrix}, \tag{6.29}$$

The final step in the derivation is to integrate the continuity equation (6.23a) from $z = h_l$ to $z = h_u$, i.e.

$$\int_{h_l}^{h_u} \left(\frac{\partial u}{\partial x} + \frac{\partial v}{\partial y} + \frac{\partial w}{\partial z} \right) dz = 0. \tag{6.30}$$

It follows from straightforward calculations that

$$\int_{h_l}^{h_u} \frac{\partial u}{\partial x} dz = \frac{\partial}{\partial x} \left(-\frac{h^3}{12\mu_a} \frac{\partial p}{\partial x} + \frac{(u_u + u_l)}{2} h \right)$$

$$\int_{h_l}^{h_u} \frac{\partial v}{\partial y} dz = \frac{\partial}{\partial y} \left(-\frac{h^3}{12\mu_a} \frac{\partial p}{\partial y} + \frac{(v_u + v_l)}{2} h \right)$$

$$\int_{h_l}^{h_u} \frac{\partial w}{\partial z} dz = w(x, y, h_u, t) - w(x, y, h_u, t) = \frac{\partial h}{\partial t},$$

where $h = h_u - h_l$. The resulting equation is indeed the well-known Reynolds equation for an incompressible and iso-viscous fluid;

$$\frac{\partial h}{\partial t} = \frac{\partial}{\partial x} \left(\frac{h^3}{12\mu_a} \frac{\partial p}{\partial x} - \frac{(u_u + u_l)}{2} h \right) + \frac{\partial}{\partial y} \left(\frac{h^3}{12\mu_a} \frac{\partial p}{\partial y} - \frac{(v_u + v_l)}{2} h \right). \tag{6.31}$$

In the literature, the Reynolds equation is commonly derived in another way. Indeed, it starts from the Navier–Stokes momentum equations without considering the density– and viscosity–pressure relationship prior carrying out the asymptotic analysis. More precisely, it is typically assumed that $p_0 \propto \varepsilon^{-2}$ and without considering the density–pressure relationship, the inertial- and body-force terms become of the order ε^2 and are then neglected. Taking the example of an ideal gas, for which $\rho = kp$, and with $p_0 \propto \varepsilon^{-2}$ the modified Reynolds number (6.15) becomes $ku_0 l_0/\mu_0$, which is, obviously, not dependent on ε. A similar consideration can be made for viscosity–pressure relationships. When overlooking this, the analysis leads to the same expression for the velocity field as in (6.29), with $\mu = \mu(p)$ replacing μ_a. However, as already pointed out by Rajagopal and Szeri (2003), this is not correct, since the viscosity–pressure relationship (and the density–pressure relationship) must be considered *a priori* carrying out the asymptotic analysis. Disregarding this, the analysis typically continues based on the velocity field (6.29), with, $\mu = \mu(p)$ replacing μ_a, and by considering mass conservation. That is, the continuity equation (6.2a) is integrated with respect to z (across the fluid film, from h_l to h_u) viz.

$$0 = \int_{h_l}^{h_u} \left(\frac{\partial \rho}{\partial t} + \frac{\partial (\rho u)}{\partial x} + \frac{\partial (\rho v)}{\partial y} + \frac{\partial (\rho w)}{\partial z} \right) dz, \tag{6.32}$$

where it is assumed that density is independent of z, i.e. $\rho = \rho(x, y)$. This leads to a Reynolds equation including density– and viscosity–pressure relationships that describe fluid compressibility and piezo-viscous behaviour. In this way, the Reynolds' type of equation (6.1), implying that both the density and the viscosity do not need to be constants and thus may depend on the pressure, viz.

$$\frac{\partial(\rho h)}{\partial t} = \frac{\partial}{\partial x}\left(\frac{\rho h^3}{12\mu}\frac{\partial p}{\partial x} - \frac{(u_u + u_l)}{2}\rho h\right) + \frac{\partial}{\partial y}\left(\frac{\rho h^3}{12\mu}\frac{\partial p}{\partial y} - \frac{(v_u + v_l)}{2}\rho h\right).$$

This type of derivation, that overlooks the aforementioned difficulties, may lead to the impression that Reynolds equation is universally valid for general fluids. It might be noted, however, that already in his original derivation Reynolds (1886), Reynolds considered the viscosity to be 'nearly constant'. He also assumed that 'the forces arising from weight and inertia are altogether small compared with the stresses arising from viscosity'. Under these assumptions, all the terms involving the density in the momentum equations may be omitted and safely arrived at (6.31).

It is clear from this discussion, however, that the use of (6.1) together with density– and viscosity–pressure relationships other than unity, may lead to erroneous results. In the case of mildly compressible or piezo-viscous fluids, this error may be sufficiently small for practical purposes. In general, however, the inconsistency in the derivation must be noted so that one is aware of the possible occurrence of these errors and can thus assess it influence before using Reynolds equation as in (6.1).

Modelling Mass-Conserving Hydrodynamic Cavitation

Hydrodynamic cavitation is found in various lubrication situations. For example, at the divergent section between the shaft and the bushing in a plain journal bearing, where the fluid film is subjected to a tensile stress situation. The fluid cannot withstand these stresses and thus the fluid film ruptures. In this situation, Reynolds equation will fail to capture this phenomenon. Instead, it will wrongly predict negative (tensile) pressures. An early approach to avoid this issue was simply to ignore the negative pressures. This, however, does not preserve mass continuity and thus leads to inaccurate results. The first attempts to model mass-conserving hydrodynamic cavitation were presented by three authors Jacobson, Floberg, and Olsson, see Jakobsson and Floberg (1957), Floberg (1960, 1961), Olsson (1965). They described the so-called rupture and reformation boundary conditions and showed how they could be incorporated in the Reynolds equation leading to a mathematical model of hydrodynamic cavitation.

Elrod and Adams (1975) developed a cavitation algorithm using a single equation throughout the lubrication region without the need for explicit equations to locate the cavitation boundaries, and used a switch function to terminate the pressure gradient in the region of cavitation. In Elrod (1981), Elrod presented a variant of the cavitation algorithm proposed in Elrod and Adams (1975), in which he used a different constitutive relationship between pressure and density. Vijayaraghavan and Keith introduced a more rigorous derivation and presented their contribution to cavitation modelling in Vijayaraghavan and Keith Jr (1989, 1990). As in Elrod and Adams (1975), their algorithm was derived based on a constant bulk modulus type of compressibility. For real lubricants, the bulk modulus varies with pressure. The importance of using more realistic models for the compressibility was investigated further in the paper by Sahlin et al. (2007). However, treating the bulk modulus as a constant could produce good results in a narrow pressure range. Examples of other work addressing the difficulties associated with cavitation modelling are Woods and Brewe (1989), Boukrouche and Bayada (1993), Ausas et al. (2009).

JFO Theory

Jakobsson and Floberg (1957) developed a mass-preserving cavitation theory. They assumed a constant pressure in the cavitation region, i.e. the pressure gradient is zero. They also derived a set of conditions to locate the cavitation boundaries. Later, Floberg and Olsson (Floberg 1960, 1961; Olsson 1965), extended the theory of cavitation and implemented it in the numerical solution procedures for numerous bearing types. The JFO theory is based on the complementary assumption that the fluid is either fully saturated, i.e. $\theta = 1$ and the fluid pressure is larger than the cavitation pressure $p > p_c$, or cavitated, i.e. $\theta < 1$ and the pressure equals the cavitation pressure $p = p_c$. They formulated this mathematically as a boundary condition for the location x_c of the cavitation inception or rupture, i.e.

$$p(x_c) = p_c \quad \text{and} \quad \left.\frac{\partial p}{\partial \mathbf{n}}\right|_{x=x_c} \tag{6.33}$$

and as a condition for preservation of mass flow at the point x_r of where the fluid film reforms

$$\left.\frac{h^3}{12\mu_a}\frac{\partial p}{\partial \mathbf{n}}\right|_{x=x_r} = \left.\frac{V_n}{2}(1-\theta)\right|_{x=x_r}. \tag{6.34}$$

Though the JFO-rupture and reformation conditions can be applied to various lubrication problems, such as the ones for journal bearings, piston ring–cylinder liner conjunctions and rolling element bearings, it is difficult to handle situations where rupture and cavitation occurs many times inside the interface, as would be the case with, e.g. a textured bearing surface. This lead successors to develop 'universal' cavitation algorithms, as will be presented below.

Elrod's and Adams' Universal Cavitation Algorithm

It is not so easy to describe the universal cavitation algorithm developed by Elrod and Adams. One reason is that they did not use a consistent density–pressure relationship. Indeed, their work starts in Elrod and Adams (1975), with a paper in which they employ a constant bulk modulus type of constitutive relationship between density and pressure, i.e.

$$p = p_c + \beta g(\theta) \ln \theta, \tag{6.35}$$

where θ is the saturation or the dimensionless density given by

$$\theta(x) = \frac{\rho(p(x))}{\rho_c},$$

and where $\rho_c = \rho(p_c)$ and $g(\theta)$ is a so-called switch function

$$g(\theta) = \begin{cases} 1, & p > p_c, \\ 0, & p = p_c. \end{cases} \tag{6.36}$$

The expression (6.35) comes from

$$\rho = \rho_a e^{(p-p_a)/\beta} \tag{6.37}$$

with $p_a = p_c$. By means of precisely this constitutive relationship between p and θ, Elrod and Adams presented a 'universal differential equation' originating from (6.1), which reads

$$\frac{\partial(\theta h)}{\partial t} = \nabla_x \cdot \left(\frac{\beta h^3}{12\mu_a} g(\theta) \nabla_x \theta - \frac{\mathbf{u}_s}{2} \theta h \right). \tag{6.38}$$

Elrod then continues the development in Elrod (1981) but there another constitutive relationship between density and pressure is adopted, i.e. (6.39). Elrod describes his model like this: 'Within the cavitated zone, the liquid everywhere possesses the density, ρ_c, but the actual mass content is $\rho_c \theta h$ per unit area. Here $1 - \theta$, then, is the same as the void fraction. Within the complete film, due to variation in pressure, the fluid density also varies. By reason of slight compression, the film mass content exceeds the content that would exist if the pressure were p_c. In other words, $\theta = \rho/\rho_c$ and the corresponding film pressure is:

$$p = p_c + \beta g(\theta)(\theta - 1) \tag{6.39}$$

where $\beta = \mathcal{O}\left(10^9\right)$ in SI units for a typical lubricating oil. It is realised, of course, that θ will be very nearly unity in the full-film zone, but when Eq. (6.39) constitutes an analytical convenience, it will be used'. In a more condensed form, this means that the fluid behaves more or less as incompressible in the full-film zones and that it expands as a homogeneous blend in the cavitated zones.

Based on (6.39), the corresponding Reynolds equation (6.1) would become

$$\frac{\partial(\theta h)}{\partial t} = \nabla_x \cdot \left(\frac{\beta \theta h^3}{12\mu_a} \nabla_x \left(g(\theta)(\theta - 1) \right) - \frac{\mathbf{u}_s}{2} \theta h \right). \tag{6.40}$$

The finite difference scheme presented in Elrod (1981) implies that it would originate from

$$\frac{\partial(\theta h)}{\partial t} = \nabla_x \cdot \left(\frac{\beta h^3}{12\mu_a} \nabla_x \left(g(\theta)(\theta - 1) \right) - \frac{\mathbf{u}_s}{2} \theta h \right), \tag{6.41}$$

which differs from (6.40) as it reflects a flow situation where the fluid behaves like an incompressible liquid in the full-film zones and as a homogeneous compressible gas–liquid mixture in the cavitated zones. This is, also, to some extent consistent with Elrod's description of his model, reprinted above. However, they found, and so did successors, that obtaining a converged numerical solution to this equation can be quite challenging if not even impossible sometimes.

In the following sections, the inconsistency of the introduction of the constitutive expression in (6.39) and the controversy that Reynolds equation, (6.1), actually reduces to (6.40) when adopting the constitutive relation given by (6.35) will be elaborated upon.

Vijayaraghavan's and Keith Jr's Cavitation Model

As done in Elrod and Adams (1975), Vijayaraghavan and Keith Jr (1989, 1990) also used the constant bulk modulus type of compressibility as a starting point in their derivation of a cavitation model. They, however, presented a more rigorous derivation, which finally lead to a cavitation algorithm similar to the one presented in Elrod (1981). More precisely, they use the density–pressure relation defined in (6.37) and reach again

$$\frac{\partial(\theta h)}{\partial t} = \nabla_x \cdot \left(\frac{\beta h^3}{12\mu_a} g(\theta) \nabla_x \theta - \frac{\mathbf{u}_s}{2} \theta h \right).$$

Studying this equation, they conclude

$$g(\theta)\nabla_x \cdot (\theta) = g(\theta)\nabla_x \cdot (\theta - 1) = \nabla_x \cdot (g(\theta)(\theta - 1)) - (\theta - 1)\nabla_x \cdot (g(\theta)). \tag{6.42}$$

However, in the full-film zones $\theta = 1$ and in the cavitated ones $g(\theta) = 0$ thus the last term in (6.42) vanishes everywhere except at the point of rupture and we have

$$g(\theta)\nabla_x \cdot (\theta) = \nabla_x \cdot (g(\theta)(\theta - 1)) \tag{6.43}$$

so that (6.38) remarkably reduces to (6.41), which is the continuous interpretation of the cavitation algorithm that Elrod proposed in Elrod and Adams (1975). Moreover, they introduce the concept of type differencing from transonic flow computations to obtain a finite difference stencil for the shear flow term that effectively considers that the governing equation goes from elliptic to hyperbolic when flow goes from fully flooded to cavitated. That is,

$$\frac{\partial(\theta h)}{\partial x} = \frac{\partial}{\partial x}\left(\theta h - (1 - g(\theta))\frac{\partial(\theta h)}{\partial x}\frac{\Delta x}{2}\right), \tag{6.44}$$

where Δx is the element size in the x-direction. When discretised this leads to the finite difference scheme

$$\left.\frac{\partial(\theta h)}{\partial x}\right|_{x_i} = \frac{(\theta h)|_{x_i} - (\theta h)|_{x_{i-1}}}{\Delta x} + \mathcal{O}(\Delta x), \tag{6.45}$$

for the shear flow term within the cavitated zone, while it becomes

$$\left.\frac{\partial(\theta h)}{\partial x}\right|_{x_i} = \frac{(\theta h)|_{x_{i+1}} - (\theta h)|_{x_{i-1}}}{2\Delta x} + \mathcal{O}(\Delta x)^2, \tag{6.46}$$

in the full-film zone. In comparison, the scheme presented in Elrod (1981) yields

$$\left.\frac{\partial(\theta h)}{\partial x}\right|_{x_i} = \frac{h|_{x_i} - h|_{x_{i-1}}}{\Delta x} + \mathcal{O}(\Delta x), \tag{6.47}$$

which clearly represents shear flow for an incompressible fluid. The scheme (6.45) is of first order, while (6.46) is of second order. In Vijayaraghavan and Keith Jr (1990), the remedy for this was presented by extending the type differencing scheme to

$$\frac{\partial E}{\partial x} = \frac{\partial}{\partial x}\left(E - (1 - g(\theta))\left(\frac{\partial^2 E}{\partial x^2}\frac{(\Delta x)^2}{2} - \frac{\partial^3 E}{\partial x^3}\frac{(\Delta x)^3}{8}\right)\right), \tag{6.48}$$

where $E = \theta h$. This leads to the second-order finite difference scheme

$$\begin{aligned}\left.\frac{\partial E}{\partial x}\right|_{x_i} &= \frac{1}{2\Delta x}\left(g_{i+1/2}E_{i+1} - (2 - g_{i+1/2} - g_{i-1/2})E_i - (2 - g_{i-1/2})E_{i-1}\right) + \\ &= \frac{1}{2\Delta x}\left((1 - g_{i+1/2})E_i - (2 - g_{i+1/2} - g_{i-1/2})E_{i-1} + (1 - g_{i-1/2})E_{i-2}\right) + \\ &\qquad \mathcal{O}(\Delta x)^2.\end{aligned} \tag{6.49}$$

The pressure-driven flow can then be discretised with a standard second-order accurate finite difference scheme. This method leads to a much more robust solution than previous approaches.

Arbitrary Compressibility Switch Function Based Cavitation Algorithm

In the paper Sahlin et al. (2007), they adopted the expression (6.5), viz.

$$\rho = \rho_a f(p),$$

as the constitutive relationship for the fluid compressibility. We recall, from above, that f is a strictly increasing function with $f(p_a) = 1$ and that ρ_a is the density at the ambient pressure p_a. This means that

$$\nabla \rho = \rho_a f'(p) \nabla p, \tag{6.50}$$

which can be inverted to express the pressure gradient as

$$\nabla p = \frac{1}{f'(p)} \nabla \theta, \tag{6.51}$$

where $\theta(x) = \rho(p(x))/\rho_a$. This means that (6.1) including the switch function $g(\theta)$ (6.36) can be posed as

$$\frac{\partial (\theta h)}{\partial t} = \nabla_x \cdot \left(\frac{h^3}{12 f'(p) \mu_a} g(\theta) \nabla_x \theta - \frac{\mathbf{u}_s}{2} \theta h \right). \tag{6.52}$$

In Sahlin et al. (2007), they analyse the resulting predictions based on both the constant bulk modulus type of compressibility (6.37) and another kind of density–pressure relationship, the well-known relation,

$$\rho = \rho_a \frac{C_1 + C_2 (p - p_a)}{C_1 + p - p_a}, \tag{6.53}$$

deduced by Dowson and Higginson and presented in Dowson and Higginson (1966), together with the constants $C_1 = 0.59$ GPa and $C_2 = 1.34$ that they found to be best fit to mineral oil density–pressure data they had access to. Sahlin et al. Sahlin et al. (2007) fitted the constants to another set of mineral oil density-pressure data for pressures up to 1 GPa, and found a close fit for $C_1 = 2.22$ GPa and $C_2 = 1.66$.

The Linear Complementarity Problem Formulation

In 2005, Bayada presented in Bayada et al. (2005b) a continuous complementarity formulation of Elrod's and Adams' cavitation algorithm Elrod and Adams (1975). Later, Giacopini et al. (2010) presented the same formulation together with the discretised linear complementarity problem formulation of the same cavitation algorithm on the basis that the fluid in the fully flooded regions behaves as incompressible. Independent of each other, two groups of authors, Bertocchi et al. (2013) and Almqvist et al. (2014), Almqvist and Wall (2016) developed this further. The work in Almqvist et al. (2014) is built upon the Reynolds equation, expressed in its most fundamental form:

$$\nabla \cdot q = 0, \tag{6.54}$$

where q is the mass flow. It proceeds by stating that in the full- film zones, the density can be expressed as $\rho = \rho_c e^{(p-p_c)/\beta}$. In the cavitation zones the density, or the saturation, is an unknown, here denoted by δ, hence

$$\rho(p) = \rho_c \begin{cases} e^{(p-p_c)/\beta} & , \; p > p_c \\ \delta & , \; p = p_c \end{cases}. \tag{6.55}$$

The unknown saturation function δ satisfies $0 \leq \delta \leq 1$. Since $\nabla p = 0$ in the region where $p = p_c$, the mass flow is

$$q = \rho_c \begin{cases} \dfrac{e^{(p-p_c)/\beta} h}{2} U - \dfrac{e^{(p-p_c)/\beta} h^3}{12\mu} \nabla p & , \; p > p_c \\ \dfrac{\delta h}{2} U & , \; p = p_c \end{cases}, \tag{6.56}$$

which is a nonlinear expression in p. By introducing the following change of variables

$$u = e^{(p-p_c)/\beta} - 1, \quad u \geq 0, \tag{6.57}$$

the mass flow can be expressed as

$$q = \rho_c \begin{cases} \dfrac{uh}{2} U + \dfrac{h}{2} U - \dfrac{\beta h^3}{12\mu} \nabla u & , \; u > 0 \\ \dfrac{\delta h}{2} U & , \; u = 0 \end{cases}. \tag{6.58}$$

A key point in the derivation of the cavitation model presented in Almqvist et al. (2014) is that (6.58) is rewritten by introducing a new unknown variable η, which is complementary to u in the whole domain, i.e. $u\eta = 0$.The variable η is defined as

$$\eta = 1 - \delta = \begin{cases} 0 & , \; u > 0 \\ 1 - \delta & , \; u = 0 \end{cases}. \tag{6.59}$$

This means that, if $u > 0$ then $\eta = 0$ and if $u = 0$ then $0 \leq \eta \leq 1$. The expression for the mass flow (6.58) can now be rewritten as

$$q = \rho_c \begin{cases} \dfrac{hu}{2}U + \dfrac{h}{2}U - \dfrac{\beta h^3}{12\mu}\nabla u \ , & u > 0 \\ \dfrac{h}{2}U - \dfrac{\eta h}{2}U & , \ u = 0 \end{cases} ,$$

or alternatively

$$q = \rho_c \left(\frac{hu}{2}U + \frac{h}{2}U - \frac{\beta h^3}{12\mu}\nabla u - \frac{\eta h}{2}U \right), \quad u \geq 0. \tag{6.60}$$

With this expression for the mass flow, the continuity equation becomes

$$\nabla \cdot q = \rho_c \nabla \cdot \left(\frac{hu}{2}U + \frac{h}{2}U - \frac{\beta h^3}{12\mu}\nabla u - \frac{\eta h}{2}U \right) = 0,$$

and summing up, the mass preserving cavitation model is expressed as

$$\begin{aligned} &\nabla \cdot \left(\frac{\beta h^3}{12\mu}\nabla u - \frac{hu}{2}U \right) = \nabla \cdot \left(\frac{h}{2}U \right) - \nabla \cdot \left(\frac{\eta h}{2}U \right), \\ &u \geq 0, \quad 0 \leq \eta \leq 1, \quad u\eta = 0. \end{aligned} \tag{6.61}$$

The beauty of this formulation is that it permits a subsequent numerical LCP analysis by means of readily available methods, such as Lemke's algorithm, see e.g. Cottle et al. (2009). It should also be noted that MATLAB code for the numerical solution of this cavitation algorithm has been made available at MATLAB file central, Almqvist et al. (2013). This model was later generalised in to include elastic deformation as well as the situation where the distance between the surfaces varies with time in Almqvist and Wall (2016).

By varying the bulk modulus, the compressibility of the lubricant is varied. A low value of the bulk modulus corresponds to a highly compressible lubricant, while a high value corresponds to a nearly incompressible lubricant. In fact, in the limit $\beta \to \infty$, (6.61) becomes

$$\begin{aligned} &\nabla \cdot \left(\frac{h^3}{12\mu}\nabla p \right) = \nabla \cdot \left(\frac{h}{2}U \right) - \nabla \cdot \left(\frac{\eta h}{2}U \right), \\ &p \geq 0, \quad 0 \leq \eta \leq 1, \quad p\eta = 0, \end{aligned} \tag{6.62}$$

which is the same cavitation algorithm as in Giacopini et al. (2010). It should be noted that, the system (6.62) can also be obtained by starting from the assumption $\rho = \rho_c$ and thereafter following the procedure presented above.

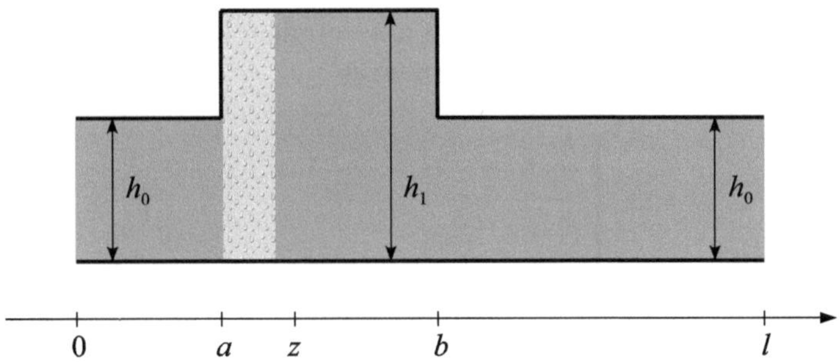

Fig. 6.5 Schematic illustration of the modelled pocket bearing

Analytical Solution

The analytical solution to the cavitation algorithm (6.61) can be obtained for some simple cases. One of these is the two-dimensional pocket slider bearing, defined as

$$h(x) = \begin{cases} h_0 \; 0 \leq x \leq a \\ h_1 \; a < x < b \\ h_0 \; b \leq x \leq l \end{cases}, \quad \text{with} \quad h_1 > h_0, \tag{6.63}$$

with graphical representation as presented in Fig. 6.5. The analytical solution to this pocket bearing, for an the incompressible case (6.62) can be found in Olver et al. (2006), see also Fowell et al. (2007). For 2D pocket bearing geometry, the continuity equation becomes one-dimensional and reads

$$\frac{dq}{dx} = \rho_c \frac{d}{dx}\left(\frac{hu}{2}U + \frac{h}{2}U - \frac{\beta h^3}{12\mu}\frac{du}{dx} - \frac{\eta h}{2}U\right) = 0.$$

Together with the boundary conditions $p(0) = p_{in}$ and $p(l) = p_{out}$, for the inlet and outlet, respectively, the cavitation model can be formulated as

$$\begin{aligned} &\frac{d}{dx}\left(\frac{\beta h^3}{12\mu}\frac{du}{dx} - \frac{U}{2}hu\right) = \frac{U}{2}\frac{dh}{dx} - \frac{U}{2}\frac{d}{dx}(\eta h), \\ &u \geq 0, \quad 0 \leq \eta \leq 1, \quad u\eta = 0. \end{aligned} \tag{6.64}$$

It is assumed that the fluid cavitates inside the pocket between the point of rupture at $x = a$ and the point of reformation at $x = z$, where $a \leq z \leq b$. This means that the bearing can be subdivided into three liquid-phase zones; $0 \leq x \leq a$, $z \leq x \leq b$, and $b \leq x \leq l$ and one gas-phase zone $a \leq x \leq z$, where $u = 0$ and $0 \leq \eta \leq 1$. In each of the liquid-phase zones, $\eta = 0$ and u are given by

$$\frac{d}{dx}\left(\frac{\beta h^3}{12\mu}\frac{du}{dx} - \frac{U}{2}hu\right) = \frac{U}{2}\frac{dh}{dx}.$$

from which an explicit solution for u can be obtained. Summing up, for each of the zones, the solution explicitly reads

$$u = C_1 + C_2 \exp\left(\frac{6\mu U}{\beta h_0^2}x\right), \qquad \eta = 0, \qquad 0 \le x \le a, \tag{6.65}$$

$$u = 0, \qquad \eta = C, \qquad a \le x \le z, \tag{6.66}$$

$$u = C_3 + C_4 \exp\left(\frac{6\mu U}{\beta h_1^2}x\right), \qquad \eta = 0, \qquad z \le x \le b, \tag{6.67}$$

$$u = C_5 + C_6 \exp\left(\frac{6\mu U}{\beta h_0^2}x\right), \qquad \eta = 0, \qquad b \le x \le l. \tag{6.68}$$

The boundary conditions $u\,(0) = e^{(p_{in}-p_c)/\beta} - 1$ and $u\,(a) = 0$, can be used to determine the constants C_1 and C_2, i.e.

$$C_1 = -\frac{1 - \exp\left(\left(p_{in} - p_c\right)/\beta\right)}{1 - \exp\left(-\dfrac{6\mu U a}{\beta h_0^2}\right)}, \tag{6.69}$$

$$C_2 = -C_1 \exp\left(-\frac{6\mu U a}{\beta h_0^2}\right). \tag{6.70}$$

Knowing unambiguously the solution u in $0 \le x \le a$, it can be used to compute the mass flow,

$$q = \frac{\rho_c U h_0}{2}\left(1 - \frac{1 - \exp\left(\left(p_{in} - p_c\right)/\beta\right)}{1 - \exp\left(-\dfrac{6\mu U a}{\beta h_0^2}\right)}\right). \tag{6.71}$$

which is preserved throughout the whole domain. According to Eq. (6.60), the mass flow in $a < x < z$ is given by

$$q = \frac{\rho_c U h_0}{2}\left(1 - C\right), \tag{6.72}$$

and the constant C can be determined using the condition that the mass flow is constant. More precisely,

$$C = \frac{1 - \exp\left(\left(p_{in} - p_c\right)/\beta\right)}{1 - \exp\left(-\dfrac{6\mu U a}{\beta h_0^2}\right)} \quad (= -C_1)\,.$$

The remaining constants $C_i, i = 3, \ldots, 6$ and z, can be found by using the conditions that both u and the mass flow are continuous at $x = z$ and $x = b$ and the boundary condition $u\,(l) = e^{(p_{out}-p_c)/\beta} - 1$. Summing up

$$C_3 = \frac{h_0}{h_1}(1 - C) - 1, \tag{6.73}$$

$$C_5 = -C, \tag{6.74}$$

$$C_6 = \frac{\exp\left((p_{out} - p_c)/\beta\right) - 1 - C_5}{\exp\left(\dfrac{6\mu U l}{\beta h_0^2}\right)}, \tag{6.75}$$

$$C_4 = \frac{-C_3 + C_5 + C_6 \exp\left(\dfrac{6\mu U b}{\beta h_0^2}\right)}{\exp\left(\dfrac{6\mu U b}{\beta h_1^2}\right)}. \tag{6.76}$$

The point of reformation (if it exists) reads

$$z = \frac{\beta h_1^2}{6\mu U} \ln\left(-\frac{C_3}{C_4}\right) - a. \tag{6.77}$$

Note that C_3 in (6.73) is the correct value. In Almqvist et al. (2014), it was not correct but an errata with the expression in (6.73) was published shortly after.

Numerical Solution Procedures

In the following subsections, a discretization of (6.61) is presented such that the discretised problem is in the form of a standard linear complementary problem (LCP). Notice that once written as a LCP, the problem can be solved using various methods. Lemke's algorithm can, for example, be used. Although a two-dimensional formulation is straightforward, a one-dimensional version is presented here in order not to complicate the notation unnecessarily. Two different types of differencing schemes are discussed. In the first type, central and upwind differencing is combined, while the second type is based on central differences only.

Combined Central and Upwind Differencing

The one-dimensional form of (6.61) reads

$$\frac{d}{dx}\left(a\frac{du}{dx} + bu\right) = \frac{dF}{dx} - \frac{d}{dx}(\eta F), \tag{6.78}$$
$$u \geq 0, \quad 0 \leq \eta \leq 1, \quad u\eta = 0.$$

where

$$a = \frac{\beta h^3}{12\mu}, \; b = -\frac{U}{2}h, \; F = \frac{U}{2}h. \tag{6.79}$$

The problem here is discretised using finite differences. First, the domain $0 < x < l$ is divided into a uniform grid with $N+2$ points, thus having elements of size $\Delta x = l/(N+1)$. Note that, by interpreting u as pressure and choosing coefficients as

$$a = \frac{h^3}{12\mu},\ b = 0,\ F = \frac{U}{2}, \tag{6.80}$$

the model obtained corresponds to the one presented in Giacopini et al. (2010), where the lubricant is assumed to be incompressible in the full-film regions. The following notation is adopted $x_i := il/N$, where $i = 0, \ldots, N+2$ and

$$u_i := u\left(x_i\right).$$

The problem at hand is elliptic in the full-film domain, where $\eta = 0$. A central difference scheme is, therefore, used to approximate the derivatives in (6.78). Using the notation

$$a_{i\pm 1/2} = \frac{a_{i\pm 1} + a_i}{2},$$

and the approximation

$$\left.\frac{du}{dx}\right|_{i+1/2} \approx \frac{u_{i+1} - u_i}{\Delta x} \quad \text{and} \quad \left.\frac{du}{dx}\right|_{i-1/2} \approx \frac{u_i - u_{i-1}}{\Delta x},$$

the left-hand side of (6.78) can be approximated as

$$\frac{d}{dx}\left(a\frac{du}{dx} + bu\right) \approx \frac{a_{i+1/2}\left.\frac{du}{dx}\right|_{i+1/2} - a_{i-1/2}\left.\frac{du}{dx}\right|_{i-1/2}}{\Delta x} + \frac{(bu)_{i+1} - (bu)_{i-1}}{2\Delta x}$$

$$\approx \frac{\left(\frac{a_{i+1} + a_i}{2}\right)\left(\frac{u_{i+1} - u_i}{\Delta x}\right) - \left(\frac{a_{i-1} + a_i}{2}\right)\left(\frac{u_i - u_{i-1}}{\Delta x}\right)}{\Delta x} + \frac{b_{i+1}u_{i+1} - b_{i-1}u_{i-1}}{2\Delta x}$$

$$= \frac{1}{2\Delta x^2}\left[\left(a_i + a_{i-1}\right)u_{i-1} - \left(a_{i-1} + 2a_i + a_{i+1}\right)u_i + \left(a_i + a_{i+1}\right)u_{i+1}\right] + \\ + \frac{b_{i+1}u_{i+1} - b_{i-1}u_{i-1}}{2\Delta x}.$$

The first term in the right-hand side becomes

$$\frac{dF}{dx} \approx \frac{F_{i+1} - F_{i-1}}{2\Delta x}.$$

In the cavitated regions, where $u = 0$, the equation is hyperbolic in η and an upwind difference scheme should be employed accordingly, i.e.

$$\frac{d}{dx}(\eta F) \approx \frac{\eta_i F_i - \eta_{i-1} F_{i-1}}{\Delta x}.$$

To simplify the presentation, the following notation is introduced:

$$\begin{aligned}
e_i^w &= \frac{a_{i-1} + a_i}{2\Delta x^2} - \frac{b_{i-1}}{2\Delta x},\\
e_i^c &= -\frac{a_{i-1} + 2a_i + a_{i+1}}{2\Delta x^2},\\
e_i^e &= \frac{a_i + a_{i+1}}{2\Delta x^2} + \frac{b_{i+1}}{2\Delta x},\\
z_i &= \frac{F_{i+1} - F_{i-1}}{2\Delta x},\\
g_i^c &= -\frac{F_i}{\Delta x},\\
g_i^w &= \frac{F_{i-1}}{\Delta x}.
\end{aligned}$$

Using this notation, the following matrices and vectors are defined:

$$A = \begin{bmatrix}
e_1^c & e_1^e & 0 & 0 & 0 & \cdots & & & \\
e_2^w & e_2^c & e_2^e & 0 & 0 & \cdots & & & \\
0 & e_3^w & e_3^c & e_3^e & 0 & \cdots & & & \\
\vdots & \vdots & \ddots & \ddots & \ddots & & & & \\
 & & & \ddots & \ddots & \ddots & & & \\
 & & & & \ddots & \ddots & \ddots & & \\
 & & & & & & e_{N-2}^w & e_{N-2}^c & e_{N-2}^e \\
 & & & & & & 0 & e_{N-1}^w & e_{N-1}^e
\end{bmatrix}, \tag{6.81}$$

$$f = \begin{bmatrix}
z_1 - e_1^w u_0 - g_1^w \eta_0 \\
z_2 \\
z_3 \\
\\
\\
\\
z_{N-2} \\
z_{N-1} - e_{N-1}^e u_N - g_{N-1}^c \eta_N
\end{bmatrix}, \tag{6.82}$$

where the values of η on the boundaries are computed from the complementarity conditions $u_0\eta_0 = 0$ and $u_N\eta_N = 0$, and

$$B = \begin{bmatrix} g_1^c & 0 & 0 & 0 & 0 & \cdots & & \\ g_2^w & g_2^c & 0 & 0 & 0 & \cdots & & \\ 0 & g_3^w & g_3^c & 0 & 0 & \cdots & & \\ \vdots & \vdots & \ddots & \ddots & & & & \\ & & & \ddots & \ddots & & & \\ & & & & \ddots & \ddots & & \\ & & & & & g_N^w & g_{N-2}^c & 0 \\ & & & & & 0 & g_{N-1}^w & g_{N-1}^c \end{bmatrix}. \tag{6.83}$$

The discretised form of (6.78) can now be written as

$$Au = f + B\eta, \quad u_i, \eta_i \geq 0, \quad u_i\eta_i = 0. \tag{6.84}$$

Solving this system for u gives

$$u = q + M\eta, \quad u_i, \eta_i \geq 0, \quad u_i\eta_i = 0, \tag{6.85}$$

where $q = A^{-1}f$ and $M = A^{-1}B$.

The linear complementarity problem (6.85) can readily be solved by employing standard numerical methods. One which is frequently used is Lemke's pivoting algorithm, see Cottle et al. (2009). One advantage, is that Lemke's pivoting algorithm finds the solution in a finite number of steps. Hence, the solution obtained is numerically exact. The method chosen here to compute a numerical example is a vectorized MATLAB version of a pivoting algorithm solving linear complementarity problems given by Almqvist et al. (2013).

Note that in the Lemke algorithm it not explicitly stated that $\eta_i \leq 1$. However, by using the same ideas as in Bayada et al. (2005b), it can be proved that any solution to (6.61) with Dirichlet boundary conditions, even without the condition $\eta \leq 1$, still satisfies $0 \leq \eta \leq 1$. This implies that the numerical solution found with the Lemke algorithm automatically satisfies $\eta_i \leq 1$. This agrees with the physical interpretation that the saturation $(1 - \eta)$ must be positive and cannot be larger 1.

Elliptic Formulation and Central Differencing

In the previous subsection, central differences were used in the full-film region and upwind differences in the cavitated regions. However, it is possible to use central differences throughout the whole domain by introducing a small perturbation, which makes the problem (6.78) elliptic also in η and not only in u. Indeed,

$$\frac{d}{dx}\left(a\frac{du^{\varepsilon}}{dx}+bu^{\varepsilon}\right)=\frac{dF}{dx}-\frac{d}{dx}\left(\varepsilon\frac{d\eta^{\varepsilon}}{dx}+\eta^{\varepsilon}F\right), \qquad (6.86)$$
$$u^{\varepsilon}\geq 0, \quad 0\leq \eta^{\varepsilon}\leq 1, \quad u^{\varepsilon}\eta^{\varepsilon}=0.$$

where $\varepsilon > 0$ is a small parameter. Discretised by central differences, in the same manner as described above, this can be written as

$$Au^{\varepsilon}=f+(\varepsilon D+B)\,\eta^{\varepsilon}, \quad u_i^{\varepsilon},\eta_i^{\varepsilon}\geq 0, \quad u_i^{\varepsilon}\eta_i^{\varepsilon}=0, \qquad (6.87)$$

where the matrix D is computed in a similar manner as the other matrices. In order to get the standard form for linear complementarity problems, we can rewrite this as

$$u^{\varepsilon}=q^{\varepsilon}+M^{\varepsilon}\eta^{\varepsilon}, \quad u_i^{\varepsilon},\eta_i^{\varepsilon}\geq 0, \quad u_i^{\varepsilon}\eta_i^{\varepsilon}=0,$$

where

$$q^{\varepsilon}=A^{-1}f \;\text{ and }\; M^{\varepsilon}=A^{-1}\left(\varepsilon D+B\right).$$

For small values of ε, u^{ε} and η^{ε} in (6.86) are good approximations of u and η in (6.78). In practice, relatively small means small compared to $(\Delta x)^2$. The idea of adding an extra term is inspired from regularity theory for partial differential equations, which in this context is known as artificial viscosity, see e.g. Evans (2010).

Numerical versus Analytical Solution

In this example, the numerical simulation procedure described above is verified against the analytical solution for the pocket bearing problem earlier in this section. The selected set of input parameters are presented in Table 6.1.

Figure 6.6 depicts the analytical and the numerical pressure solutions for the pocket bearing with input parameters given Table 6.1. The solutions are obtained for $\beta = 5\times 10^8$ and $N = 512$. The relative error, defined as

$$\sum_{i=0}^{N}\left|p_i^{\text{analytical}}-p_i^{\text{numerical}}\right| \Bigg/ \sum_{i=0}^{N} p_i^{\text{analytical}},$$

was 6%, 2.5% and 1% (and the computational times were 3 ms, 22 ms and 106 ms on a standard laptop) for $N = 128$, $N = 256$ and $N = 512$, respectively. The numerical difficulties are concentrated to the reformation occurring close to the discontinuity in h.

Table 6.1 Input parameters for the pocket bearing problem

a	b	l	h_0	h_1	p_c	U	μ	$p_{in} = p_{out}$
2 mm	5 mm	20 mm	1 μm	10 μm	0 Pa	1m/s	0.01 Pa s	100 kPa

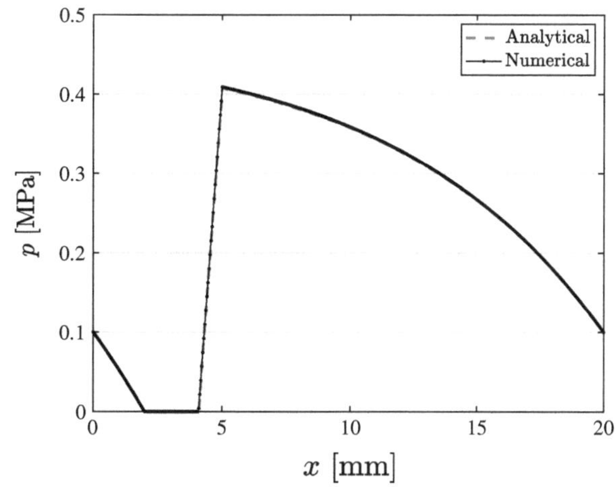

Fig. 6.6 The analytical and the numerical pressure solutions for the pocket bearing with input parameters given in Table 6.1, $\beta = 5 \times 10^8$ and $N = 512$

Homogenisation of the Reynolds Equation

Up until this point, the surfaces have been implicitly assumed to be smooth. This is, however, not generally the case. An illustration of a 3D step bearing with two rough surfaces is depicted in Fig. 6.7. This roughness poses a problem when solving the problem numerically. Indeed, a very fine discretisation would be needed to resolve the roughness with sufficient accuracy, which would result to very long computations. An alternative is to use a two-scale formulation. Roughly speaking, a two-scale formulation works by solving the problem at two distinct scales. The smooth geometry is considered at the global scale. Since the geometry is smooth, a coarse discretisation can be used in this scale thus avoiding lengthy computation times. To incorporate the effect of roughness, the equation is modified by introducing the so-called 'flow factors'. These are computed by solving a similar flow problem at the local scale. In this scale, the domains are very small and thus can be resolved in the detail required by the roughness. The translation of these loose concepts to mathematical terms is known as homogenization, which is the topic of this section.

Homogenisation is a type of averaging which has been found applicable for two-scale problems with highly varying coefficients. There are many papers reporting the successful application of Homogenisation in the field of lubrication, where the fluid flow may be governed by the Reynolds equation, see Bayada and Chambat (1988), Bayada et al. (2005a), Almqvist and Dasht (2006), Wall (2007), Almqvist et al. (2007a, 2012), Lukkassen et al. (2009), Fabricius et al. (2014), Almqvist (2011). This has lead to highly effective numerical tools where the effects caused by the surface roughness are embedded in the derived homogenised equations. Moreover, the equations are unambiguously determined and their nature allow for straightforward parallelisation. These tools enable studies of hydrodynamically lubricated problems involving rough surfaces such as that arising in the bearing configuration visualised in Fig. 6.1. For this, the theoretical model must consider different types of the non-stationary Reynolds equation in two dimensions.

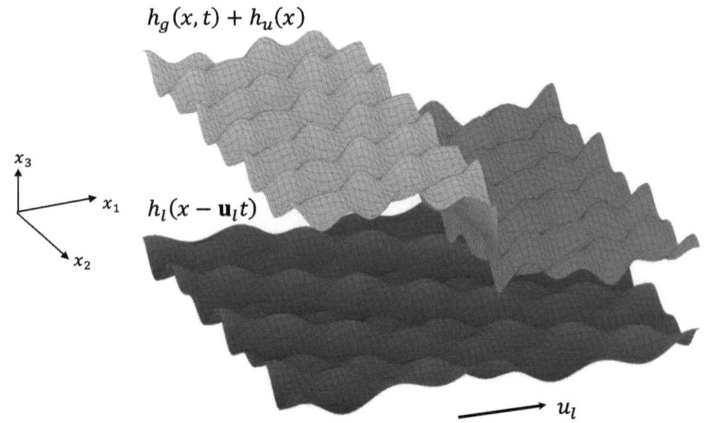

Fig. 6.7 A step-shaped bearing including idealised periodic surface roughness at a given time t. The lower surface is moving in the x_1-direction with speed u_l and the upper surface is stationary

The key ingredient when homogenising the effect of surface roughness in the Reynolds equation is the two-scale description of the gap between the surfaces. This is done via a mathematical description of the film thickness, including the surface roughness, that appears in the Reynolds equation. More precisely, it is assumed that the film thickness can be modelled by means of the following auxiliary function:

$$h(x, t, y, \tau) = h_g(x, t) + h_u(y - \mathbf{u}_u\tau) - h_l(y - \mathbf{u}_l\tau), \tag{6.88}$$

where h_g, h_u and h_l are (mathematically) smooth functions, expressed in the global-scale variables x and t and the local-scale variables y and τ. The expression $\mathbf{u}_u = [u_u, v_u]^T$ defines the velocity of the upper surface and $\mathbf{u}_l = [u_l, v_l]^T$ defines the velocity of the lower surface. Note that the velocities may depend on position, x. The global-scale film thickness is modelled by the function h_g, which describes the geometry of the problem, and the local-scale film thickness is modelled via the y- and τ-periodic functions h_u and h_l that describes the surface roughness of the upper and the lower surfaces, respectively. Indeed, the functions h_u and h_l are assumed to be periodic in their second argument and the cell of periodicity is denoted by Y for both h_u and h_l. It is also assumed that u_l, u_u, v_l and v_u are such that h is periodic in τ and the cell of periodicity, in τ, is denoted by Z. By means of this auxiliary function, a simplified physical description of the gap h_ε between the surfaces can be achieved. That is

$$h_\varepsilon(x, t) = h(x, t, x/\varepsilon, t/\varepsilon), \quad \varepsilon > 0. \tag{6.89}$$

In this expression for the film thickness, ε is a (small) parameter which moderates the wavelength of the surface roughness. As mentioned above (6.89) admits a velocity field that varies with position x (encountered in, e.g. a rotating application). Moreover, since h_l and h_u are functions of x, this also allows for different roughness descriptions on different positions. An illustration of a bearing with step-shaped geometry (global-scale) and idealised periodic surface roughness (local-scale) at a given time t, is presented in Fig. 6.7. The lower surface is moving in the x_1-direction with speed u_l, i.e. $\mathbf{u}_l = [u_l, 0]^T$, and the upper surface is stationary. Assuming iso-viscous and compressible flow, then, in terms of the small wavelength parameter ε, the Reynolds equation (6.1), may be stated as

$$\frac{\partial}{\partial t}\left(c_\varepsilon(x,t)\right) = \nabla_x \cdot \left(A_\varepsilon(x,t)\nabla_x p_\varepsilon - B_\varepsilon(x,t)\right), \tag{6.90}$$

where

$$c_\varepsilon(x,t) = \rho(p_\varepsilon)h_\varepsilon,\ A_\varepsilon(x,t) = \frac{\rho(p_\varepsilon)h_\varepsilon^3}{12\mu_a},\ B_\varepsilon(x,t) = \frac{\mathbf{u}_s}{2}\rho(p_\varepsilon)h_\varepsilon. \tag{6.91}$$

Due to roughness, the coefficients $c_\varepsilon(x,t)$, $A_\varepsilon(x,t)$ and $B_\varepsilon(x,t)$ are rapidly oscillating functions that require high spatial and time resolution for a mesh-independent numerical analysis.

The main idea in homogenization is to prove that there exists a solution p_0, solving a so-called homogenised equation that does not involve rapidly oscillating coefficient functions, such that

$$p_\varepsilon \to p_0, \quad \text{as} \quad \varepsilon \to 0.$$

This means that for small values of epsilon—which is the case for realistic surfaces, p_0 is a good approximation of p_ε.

In the subsections below, the multiple scale expansion homogenisation procedure will be applied to the Reynolds equation in order to obtain the equation that governs p_0. This will be done for the flow of incompressible fluids, as well as compressible fluids governed by constant bulk modulus and ideal gas type of density–pressure relationships, will be presented. The fluid is in all three of these cases assumed to be iso-viscous. It should be noted that a more mathematically rigorous derivation of the cases below can be found using the two-scale convergence method Lukkassen et al. (2002). The presentation here is, however, more intuitive.

Incompressible Fluid

The stationary form of the Reynolds equation for iso-viscous and compressible flow of ideal gases, in both cartesian and polar coordinates, admits the following generalisation:

$$\nabla \cdot (A(x)\nabla u(x)) - \nabla \cdot B(x) = 0, \tag{6.92}$$

where u is the dependent variable, A and B are known functions and $x = (x_1, x_2)^T$. In a non-dimensional cartesian coordinate formulation, $A = h^3$ and $B = h$ are scalar functions but in polar coordinates, for the fluid film formation in a rotating device, as discussed Almqvist (2011), they are a matrix and a vector, respectively. That is,

$$A(x) = h^3 \begin{bmatrix} x_2 & 0 \\ 0 & 1/x_2 \end{bmatrix}, \tag{6.93}$$

$$B(x) = h \begin{bmatrix} x_2 \\ 0 \end{bmatrix}, \tag{6.94}$$

where h describes the gap between the surfaces (film thickness function) and where x_1 is the angular coordinate and x_2 is the radial coordinate. For the subsequent homogenisation process, we consider the following auxiliary equation

$$\nabla \cdot (A(x, y)\nabla u(x, y)) - \nabla \cdot (B(x, y)) = 0, \tag{6.95}$$

where, again, u is the dependent variable and A and B are known functions. The variables $x = (x_1, x_2)^T$ and $y = (y_1, y_2)^T$ refer to the global- and local-scale domains, respectively. It holds that $y = x/\varepsilon$.

It is assumed that $A(x, y)$ and $B(x, y)$ are periodic in y and also that the following expansion holds:

$$u(x, y) = u_0(x, y) + \varepsilon u_1(x, y) + \varepsilon^2 u_2(x, y) + \cdots, \tag{6.96}$$

where u_i are periodic in y. Note that Eq. (6.96) reflexes that $u(x, y)$ can be expanded with respect with the perturbation ε. Indeed, roughness is treated as a perturbation of the smooth geometry. What Eq. (6.96) claims is therefore that, when $\varepsilon = 0$ (i.e. when the surface is smooth), $u = u_0$. It further claims that, when there is some roughness ($\varepsilon > 0$, small), the solution can be approximated by adding to u_0 correction terms of the form $\varepsilon^i u_i$, which become very small as the index i increases. This approach is commonly used in mathematics under the name of perturbation theory. In this terminology, (6.95) becomes

$$\begin{aligned} \left(\nabla_x + \frac{1}{\varepsilon}\nabla_y\right) \cdot \left(A(x, y)\left(\nabla_x + \frac{1}{\varepsilon}\nabla_y\right)\left(u_0 + \varepsilon u_1 + \varepsilon^2 u_2 + \cdots\right)\right) \\ - \left(\nabla_x + \frac{1}{\varepsilon}\nabla_y\right) \cdot B(x, y) = 0, \end{aligned} \tag{6.97}$$

where ∇_x and ∇_y indicate that the ∇ operator is applied to the set of variables x and y, respectively. In the following, the asymptotic behaviour of (6.97) as ε goes to zero will be considered. It can be realised that obtaining a reasonable result when expanding (6.97) requires that the coefficients for ε^{-2}, ε^{-1} and ε^0 are equivalent to zero independently of each other. Indeed, as $\varepsilon \to 0$ the coefficients of higher orders of ε, i.e. ε^k with $k \geq 1$, will not influence the result and the following set of determining equations ca be obtained:

$$\varepsilon^{-2}: \ 0 = \nabla_y \cdot \left(A \nabla_y u_0\right), \tag{6.98a}$$

$$\varepsilon^{-1}: \ 0 = \nabla_y \cdot \left(A \nabla_y u_1\right) + \nabla_y \cdot (A \nabla_x u_0) + \nabla_x \cdot \left(A \nabla_y u_0\right) - \nabla_y \cdot B, \tag{6.98b}$$

$$\varepsilon^{0}: \ 0 = \nabla_y \cdot \left(A \nabla_y u_2\right) + \nabla_y \cdot (A \nabla_x u_1) + \nabla_x \cdot \left(A \nabla_y u_1\right) + \nabla_x \cdot (A \nabla_x u_0) - \nabla_x \cdot B, \tag{6.98c}$$

where the dependency to x and y has dropped for the readers convenience.

From (6.98a), it follows that $u_0 = u_0(x)$. By means of this fact, (6.98b) simplifies to

$$\nabla_y \cdot \left(A \nabla_y u_1\right) + \nabla_y \left(A \nabla_x u_0\right) - \nabla_y \cdot B = 0. \tag{6.99}$$

Moreover, if $\nabla_y A$ denotes the gradient of each of the columns of A independently, then (6.99) can be written as

$$\nabla_y \cdot \left(A \nabla_y u_1\right) + \nabla_y A \cdot \nabla_x u_0 - \nabla_y \cdot B = 0. \tag{6.100}$$

This means that the solution u_1 to this equation must be on the form

$$u_1 = \chi_0 + \chi_1 \frac{\partial u_0}{\partial x_1} + \chi_2 \frac{\partial u_0}{\partial x_2}, \tag{6.101}$$

where χ_i are the solutions of the following (periodic) local problems:

$$0 = \ \nabla_y \cdot \left(A \nabla_y \chi_0\right) - \nabla_y \cdot B \ \text{ in } Y, \tag{6.102a}$$

$$0 = \ \nabla_y \cdot \left(A \nabla_y \chi_1\right) + \nabla_y \cdot (A e_1) \ \text{ in } Y, \tag{6.102b}$$

$$0 = \ \nabla_y \cdot \left(A \nabla_y \chi_2\right) + \nabla_y \cdot (A e_2) \ \text{ in } Y, \tag{6.102c}$$

where Y is the domain defining one period in y.

Now, the fact that $u_0 = u_0(x)$ together with (6.101) can be inserted into (6.98c) and we can complete the homogenisation process of (6.92). This process starts by rewriting (6.98c) as

$$\nabla_x \cdot \left(A \nabla_x u_0 + A \nabla_y u_1 - \nabla_x \cdot B\right) = \nabla_y \cdot \left(A \nabla_x u_1 + A \nabla_y u_2\right) \tag{6.103}$$

and realising that, due to periodicity

$$\nabla_x \cdot \left(\left(\int_Y A(x, y)\, dy \nabla_x u_0 + \int_Y A(x, y) \nabla_y u_1 \, dy - \nabla_x \cdot \int_Y B(x, y)\, dy \right) \right) = 0. \tag{6.104}$$

Making use of (6.101) the homogenised equation can be obtained

$$\nabla_x \cdot (A_0(x) \nabla_x u_0) - \nabla_x \cdot B_0(x) = 0, \tag{6.105}$$

where A_0 is a 2×2 matrix and B_0 is a 2×1 vector, i.e.

$$A_0(x) = \begin{pmatrix} a_{11} & a_{12} \\ a_{21} & a_{22} \end{pmatrix} \tag{6.106}$$

and

$$B_0(x) = \begin{pmatrix} b_1 \\ b_2 \end{pmatrix} \tag{6.107}$$

The coefficient functions of A_0 and B_0 read

$$a_{11}^0(x) = \int_Y A(x, y)\left(1 + \frac{\partial \chi_1}{\partial y_1}\right) dy, \tag{6.108}$$

$$a_{12}^0(x) = \int_Y A(x, y)\frac{\partial \chi_2}{\partial y_1} dy, \tag{6.109}$$

$$a_{21}^0(x) = \int_Y A(x, y)\frac{\partial \chi_1}{\partial y_2} dy = \int_Y A(x, y)\frac{\partial \chi_2}{\partial y_1} dy = a_{12}^0(x), \tag{6.110}$$

$$a_{22}^0(x) = \int_Y A(x, y)\left(1 + \frac{\partial \chi_2}{\partial y_2}\right) dy, \tag{6.111}$$

$$b_1^0(x) = \int_Y \left(B(x, y) - A(x, y)\frac{\partial \chi_0}{\partial y_1}\right), \tag{6.112}$$

and

$$b_2^0(x) = \int_Y \left(B(x, y) - A(x, y)\frac{\partial \chi_0}{\partial y_2}\right). \tag{6.113}$$

The homogenised equation, (6.105) is on the same form as the original one (6.95), where the coefficients matrices $A(x, y)$ and $B(x, y)$ are replaced by the homogenised counterparts $A_0(x)$ and $B_0(x)$.

Constant Bulk Modulus Fluid

Recall that an iso-viscous fluid with constant bulk modulus is characterised by a constant viscosity and a density of the form

$$\rho = \rho_a e^{(p-p_a)/\beta}. \tag{6.114}$$

By replacing pressure by density as the main dependent variable, the Reynolds equation can, in this case, be written as

$$\nabla \cdot \left(\frac{h^3 \beta}{12\mu_a} \nabla \rho \right) = \frac{\mathbf{u}_s}{2} \nabla (\rho h) . \tag{6.115}$$

More generally, this can be written, both in Cartesian and polar coordinates as

$$\nabla \cdot (A(x, y) \nabla u(x, y)) - \nabla \cdot (B(x, y) u(x, y)) = 0, \tag{6.116}$$

where, again, u is the dependent variable and A and B are known functions. Notice that u in this case represent fluid density and not pressure. Again, it is assumed that $A(x, y)$ and $B(x, y)$ are periodic in y and also that the expansion (6.96) holds. In this terminology (6.116) becomes

$$\begin{aligned} &\left(\nabla_x + \frac{1}{\varepsilon}\nabla_y\right) \cdot \left(A(x, y) \left(\nabla_x + \frac{1}{\varepsilon}\nabla_y\right) \left(u_0 + \varepsilon u_1 + \varepsilon^2 u_2 + \cdots\right)\right) \\ &\quad - \left(\nabla_x + \frac{1}{\varepsilon}\nabla_y\right) \cdot \left(B(x, y) \left(u_0 + \varepsilon u_1 + \varepsilon^2 u_2 + \cdots\right)\right) = 0. \end{aligned} \tag{6.117}$$

Following the procedure introduced in the previous section for incompressible and iso-viscous flow, this leads to the following set of determining equations:

$$\varepsilon^{-2}: \ 0 = \nabla_y \cdot \left(A \nabla_y u_0\right), \tag{6.118a}$$

$$\varepsilon^{-1}: \ 0 = \nabla_y \cdot \left(A \nabla_y u_1\right) + \nabla_y \cdot (A \nabla_x u_0) + \nabla_x \cdot \left(A \nabla_y u_0\right) - \nabla_y \cdot (B u_0), \tag{6.118b}$$

$$\begin{aligned} \varepsilon^{0}: \ 0 = {} & \nabla_y \cdot \left(A \nabla_y u_2\right) + \nabla_y \cdot (A \nabla_x u_1) + \nabla_x \cdot \left(A \nabla_y u_1\right) + \nabla_x \cdot (A \nabla_x u_0) + \\ & - \nabla_x \cdot (B u_0) - \nabla_y \cdot (B u_1). \end{aligned} \tag{6.118c}$$

Similarly to what we had before, (6.118a) implies that $u_0 = u_0(x)$ and (6.118b), therefore, simplifies to

$$\nabla_y \cdot \left(A \nabla_y u_1\right) + \nabla_y A \cdot \nabla_x u_0 - u_0 \nabla_y \cdot B = 0. \tag{6.119}$$

This implies that u_1 is on the form

$$u_1 = \chi_0 u_0 + \chi_1 \frac{\partial u_0}{\partial x_1} + \chi_2 \frac{\partial u_0}{\partial x_2}, \tag{6.120}$$

which means that χ_i is the solution to the same set of local problems as in the previous case (6.102). Following the same procedure as before leads to the homogenised equation

$$\nabla_x \cdot (A_0(x) \nabla_x u_0) - \nabla_x \cdot (B_0(x) u_0) = 0. \tag{6.121}$$

Remarkably, the coefficient functions of A_0 and B_0 are the same as for the incompressible fluid, thus given (6.108)–(6.113).

Ideal Gas

An ideal gas has a constant viscosity and a density of the form $\rho = \kappa p$. The Reynolds equation for these gases, in both cartesian and polar coordinates, admits the following generalisation:

$$\nabla \cdot (A(x)u(x)\nabla u(x)) - \nabla \cdot (B(x)u(x)), \tag{6.122}$$

where u is the dependent variable representing fluid pressure. For the subsequent homogenisation process, the auxiliary equation

$$\nabla \cdot (A(x, y)u(x, y)\nabla u(x, y)) - \nabla \cdot (B(x, y)u(x, y)) \tag{6.123}$$

is considered and it is assumed that the expansion (6.96) holds. Following the procedure introduced earlier for incompressible and iso-viscous flow, this leads to the following set of determining equations:

$$\varepsilon^{-2}: \; 0 = \nabla_y \cdot \left(Au_0\nabla_y u_0\right), \tag{6.124a}$$

$$\begin{aligned}\varepsilon^{-1}: \; 0 = {} & \nabla_y \cdot \left(Au_1\nabla_y u_0\right) + \nabla_y \cdot \left(Au_0\nabla_y u_1\right) + \nabla_y \cdot (Au_0\nabla_x u_0) + \\ & \nabla_x \cdot \left(Au_0\nabla_y u_0\right) + -\nabla_y \cdot (Bu_0),\end{aligned} \tag{6.124b}$$

$$\begin{aligned}\varepsilon^{0}: \; 0 = {} & \nabla_y \cdot \left(Au_0\nabla_x u_1 + Au_1\nabla_x u_0 + Au_0\nabla_y u_2 + Au_1\nabla_y u_1 + Au_2\nabla_y u_0\right) + \\ & \nabla_x \cdot \left(Au_0\nabla_x u_0 + Au_0\nabla_y u_1 + Au_1\nabla_y u_0\right) - \nabla_x \cdot (Bu_0) - \nabla_y \cdot (Bu_1)\end{aligned} \tag{6.124c}$$

Similarly to what we had before, (6.124a) implies that $u_0 = u_0(x)$ and (6.124b) simplifies to

$$u_0\nabla_y \cdot \left(A\nabla_x u_0 + A\nabla_y u_0\right) - u_0\nabla_y \cdot B = 0,$$

which, since $u_0 > 0$ (note that $u_0 = 0$ would imply a zero density), becomes

$$\nabla_y \cdot \left(A\nabla_x p_0 + A\nabla_y p_1\right) - \nabla_y \cdot B = 0, \tag{6.125}$$

i.e. the same as (6.100) with solution (6.101). Thus, the same (periodic) local problems (6.102) as for the two previous cases also apply here. The homogenised equation for this case finally becomes

$$\nabla_x \cdot (p_0(x)A_0(x)\nabla_x p_0(x)) - \nabla_x \cdot (p_0(x)B_0(x)) = 0, \tag{6.126}$$

Notably, the coefficient functions of A_0 and B_0 are the same as the ones for iso-viscous and incompressible case, as well as the one for the iso-viscous and constant bulk modulus compressible case, thus given (6.108)–(6.113).

Homogenised Coefficients and Patir and Cheng Flow Factors

The method proposed by Patir and Cheng (1978, 1979) considers a representative part of the surface roughness and model the complete surface as its periodic extension. This means that the film thickness can be expressed in the form of (6.88), which will be utilised in its stationary form in the following presentation of the Patir and Cheng averaging technique. Note that the x_1-direction must be chosen so that it is aligned with the direction of motion. Retaining as much as possible of the already introduced notation, the averaged Reynolds equation presented in Patir and Cheng (1978, 1979) is restated as

$$\nabla \cdot \left(\begin{pmatrix} \phi_1 & 0 \\ 0 & \phi_2 \end{pmatrix} \frac{h_g^3}{12\mu_a} \nabla p^{pc} \right) \tag{6.127}$$

$$= \nabla \cdot \left(\begin{pmatrix} u_s/2 \\ 0 \end{pmatrix} \left(\frac{1}{l_1 l_2 h_g} \int_Y h \, dy + \phi_0 \right) h_g \right) \quad \text{in } \Omega,$$

where the flow factors ϕ_i are given by

$$\phi_0 = -\frac{1}{l_1 l_2 h_g} \int_Y h^3 \frac{\partial v_0}{\partial y_1} dy, \tag{6.128}$$

$$\phi_1 = \frac{1}{l_2 h_g^3} \int_Y h^3 \frac{\partial v_1}{\partial y_1} dy, \tag{6.129}$$

and

$$\phi_2 = \frac{1}{l_1 h_g^3} \int_Y h^3 \frac{\partial v_2}{\partial y_2} dy. \tag{6.130}$$

The Patir and Cheng local problems, i.e. the so-called micro- bearing problems, determining v_i read

$$\nabla_y \cdot \left(h^3 \nabla_y v_0 \right) = \frac{\partial h}{\partial y_1} \quad \text{in } Y, \tag{6.131a}$$

$$v_0(x,0,y_2) = v_0(x,l_1,y_2) = 0, \quad \left.\frac{\partial v_0}{\partial y_2}\right|_{(x,y_1,0)} = \left.\frac{\partial v_0}{\partial y_2}\right|_{(x,y_1,l_2)} = 0,$$

$$\nabla_y \cdot \left(h^3 \nabla_y v_1\right) = 0 \text{ in } Y, \tag{6.131b}$$

$$v_1(x,0,y_2) = 0, \; v_1(x,l_1,y_2) = 1, \quad \left.\frac{\partial v_1}{\partial y_2}\right|_{(x,y_1,0)} = \left.\frac{\partial v_1}{\partial y_2}\right|_{(x,y_1,l_2)} = 0,$$

$$\nabla_y \cdot \left(h^3 \nabla_y v_2\right) = 0 \text{ in } Y \tag{6.131c}$$

$$\left.\frac{\partial v_2}{\partial y_1}\right|_{(x,0,y_2)} = \left.\frac{\partial v_2}{\partial y_1}\right|_{(x,l_1,y_2)} = 0, \quad v_2(x,y_1,0) = 0, \; v_2(x,y_1,l_2) = 1.$$

In Patir and Cheng (1978, 1979), these local solutions v_i are interpreted as local pressures.

To facilitate a comparison between Patir and Cheng flow factors and homogenised coefficients the results presented above are reformulated in the following, see also Almqvist et al. (2011). Indeed, another way of formulating (6.127) is

$$\nabla \cdot \left(A^{pc} \nabla p^{pc}\right) = \nabla \cdot B^{pc}, \tag{6.132}$$

where

$$A^{pc} = \frac{1}{12\mu_a} \begin{pmatrix} a_{11}^{pc} & a_{12}^{pc} \\ a_{12}^{pc} & a_{22}^{pc} \end{pmatrix},$$

$$a_{11}^{pc}(x) = \frac{1}{l_2} \int_Y h^3 \frac{\partial v_1}{\partial y_1} dy, \tag{6.133a}$$

$$a_{12}^{pc}(x) = a_{21}^{pc}(x) = 0, \tag{6.133b}$$

$$a_{22}^{pc}(x) = \frac{1}{l_1} \int_Y h^3 \frac{\partial v_2}{\partial y_2} dy, \tag{6.133c}$$

and

$$B^{pc} = \frac{u_s}{2} \begin{pmatrix} b_1^{pc}(x) \\ b_2^{pc}(x) \end{pmatrix} = \frac{u_s}{2} \begin{pmatrix} \frac{1}{l_1 l_2} \int_Y \left(h - h^3 \frac{\partial v_0}{\partial y_1}\right) dy \\ 0 \end{pmatrix}, \tag{6.134}$$

and where the functions v_i solve the local problems defined in (6.131).

An alternative way of presenting the homogenization results in section 'Homogenisation of the Reynolds Equation' is obtained by introducing the new dependent variable ψ_i;

$$\psi_i(y) = \frac{y_i + \chi_i(y)}{l_i}, i = 1, 2, \tag{6.135}$$

for which (6.102b) and (6.102c) become

$$\nabla_y \cdot \left(h^3 \nabla_y \psi_1\right) = 0 \text{ in } Y, \tag{6.136a}$$
$$\psi_1 (x, 0, y_2) + 1 = \psi_1 (x, l_1, y_2), \quad \psi_1 (x, y_1, 0) = \psi_1 (x, y_1, l_2),$$
$$\nabla_y \cdot \left(h^3 \nabla_y \psi_2\right) = 0 \text{ in } Y, \tag{6.136b}$$
$$\psi_2 (x, 0, y_2) = \psi_2 (x, l_1, y_2), \quad \psi_2 (x, y_1, 0) + 1 = \psi_2 (x, y_1, l_2).$$

Hence, (6.108)–(6.111) become

$$a_{11} (x) = \frac{1}{l_2} \int_Y h^3 \frac{\partial \psi_1}{\partial y_1} dy, \tag{6.137a}$$

$$a_{12} (x) = \frac{1}{l_1} \int_Y h^3 \frac{\partial \psi_2}{\partial y_1} dy, \tag{6.137b}$$

$$a_{21} (x) = \frac{1}{l_2} \int_Y h^3 \frac{\partial \psi_1}{\partial y_2} dy, \tag{6.137c}$$

$$a_{22} (x) = \frac{1}{l_1} \int_Y h^3 \frac{\partial \psi_2}{\partial y_2} dy. \tag{6.137d}$$

With this, the Patir and Cheng results and the homogenised ones are stated in an equal manner and they can now be easily compared. It is clear that the methods share quite a few features, and that the main differences are the boundary conditions for the local problems and that the off-diagonal terms of A^{pc} and the y_2-direction coefficient of B^{pc} are identical to zero while they are not for the homogenised model. The homogenisation method gives the correct flow factors for any kind of topography, provided it fulfils the Reynolds roughness assumption. The Patir and Cheng method will give the same result as the homogenised method for surface topographies that are symmetric in both the x- and y-directions. To elucidate on this, Figs. 6.8, 6.9, and 6.10, are presented. In Fig. 6.8, the solution of the local problem (6.131b), in the Patir and Cheng method for a bi-sinusoidal surface is depicted. The figure illustrates the effect of the Neumann boundary conditions, which impose the absence of flow over the upper and lower boundaries. This result should be compared to the solution of the local problem in (6.136a) in the homogenization method, for the same bi-sinusoidal surface patch. This is depicted in Fig. 6.9, illustrating how the periodic boundary conditions allows for flow around the protrusions of the bi-sinusoidal texture as well as to take place over the upper and lower boundaries. Figure 6.10 shows what the result becomes when replacing the bi-sinusoidal surface with a symmetric bi-cosinusoidal one. In this case, due to the horizontal symmetry, solving (6.136a) with periodic boundary conditions or (6.131b) with Neumann boundary conditions gives identical results.

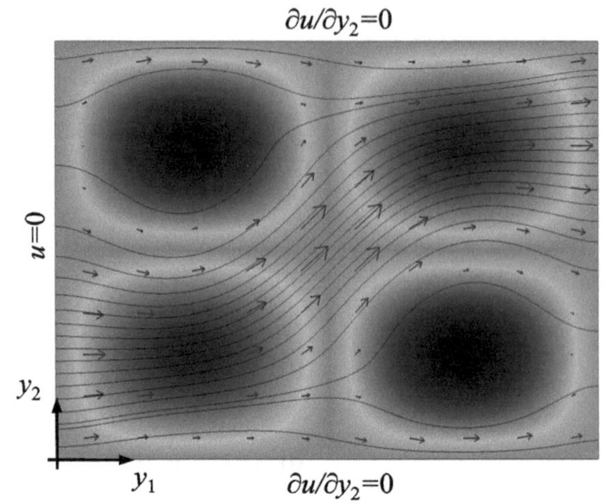

Fig. 6.8 Streamlines computed from the solution of the local problem in (6.131b) in the Patir and Cheng method. The contour map depicts the film thickness for a bi-sinusoidal surface roughness patch. Red for thinner and blue for thicker film. Arrows visualise fluid velocity

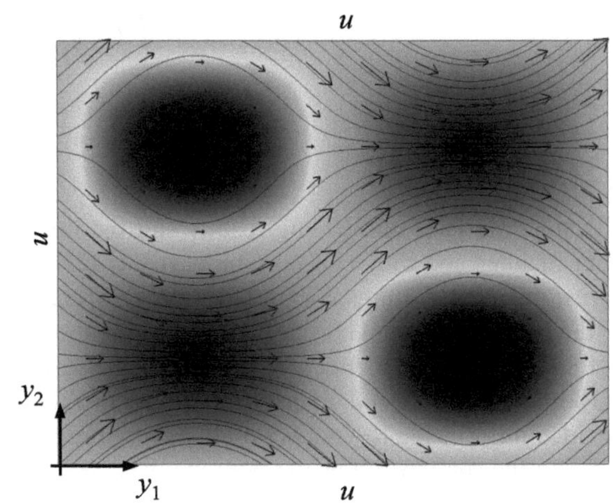

Fig. 6.9 Streamlines computed from the solution of the local problem in (6.136a) in the homogenisation method. The contour map depicts the film thickness for a bi-sinusoidal surface roughness patch. Red for thinner and blue for thicker film. Arrows visualise fluid velocity

Homogenised Flow Factors for Mixed Lubrication Conditions

In the previous sections, the expressions for coefficient functions a_{ij} and b_i in the homogenised matrix A_0 and vector B_0 were derived. Indeed, the explicit expressions (being the same for each of the three types of fluids considered) are given by (6.108)–(6.113). It is also clear that they are functions of the global coordinates (x_1, x_2) and this is also what couples the local- and the global-scale problems. Obviously, computing the coefficients for each node (x_{1i}, x_{2j}), in the discretised global domain require solving the set (6.102) of the local problems for all (i, j) belonging to the grid. Although this procedure could be used it is impractical for (at least) the following reasons: (i) it renders an unnecessarily large set of data, (ii) the values of the coefficient functions are computed only for the points (x_{1i}, x_{2j}). Thus if they

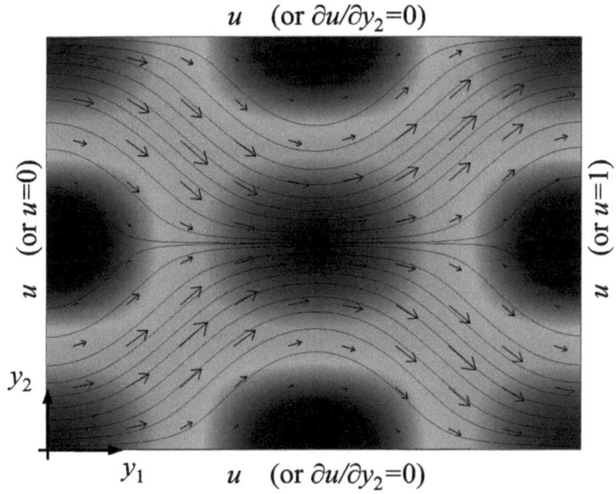

Fig. 6.10 The contour map depicts the film thickness for a symmetric bi-cosinusoidal surface roughness patch. Red for thinner and blue for thicker film. Streamlines obtained from (6.131b) and (6.136a) coincide. Arrows visualise fluid velocity

are not 'tabulated' together with the values of the discretised film thickness function h equated at exactly the same points (x_{1i}, x_{2j}), it would not be possible to use them together with another representation of the global-scale geometry. However, precisely as Patir and Cheng (1978) did, it is possible to render a more versatile set of coefficient functions by means of a simplistic parametrisation. In Patir (1978) a routine for generating randomly rough surfaces with Gaussian height distribution and with the possibility of specifying the aspect ratio (or lay) of the topography was presented. This routine was also used to generate the set of surfaces they used to generate the flow factors they presented. In connection to this, Pérez-Ràfols and Almqvist (2019) developed a routine that can be used to generate self-affine fractal surfaces with given height distribution and power spectrum. Let us now describe a procedure that can be applied in order to accomplish this.

We start by noting that on the local scale h_g is to be treated as a parameter representing the global-scale average interfacial separation. We denote this parameter α and let h_α replace h in (6.88). More precisely, for full-film conditions, we define h_α as

$$h_\alpha(y, \tau) = \alpha + h_r(y, \tau), \quad \alpha > \bar{h}_r \tag{6.138}$$

where

$$h_r := (h_u - h_l) - \min_{\forall(y,\tau)} (h_u - h_l)\,. \tag{6.139}$$

to ensure that $h_r \geq 0$ and $\bar{h}_r$ is the arithmetic mean of h_r.

For mixed lubrication conditions $\alpha \leq \bar{h}_r$ and the shape of the gap depend on the contact mechanics between the two rough surfaces being pressed together and before we proceed, we will very briefly explain how h (in (6.88)) is connected to the contact mechanics model. Indeed, let

$$h_d = h_r + u - g_{00} \tag{6.140}$$

describe the (local scale) gap that between the deformed surfaces h_u and h_l, that results due to the application of normal force F pressing them together. Note that $u = u(p_d)$, where p_d is the contact pressure, is the (local scale) displacement of h_r and g_{00} is the rigid body displacement. Then we can formulate the corresponding (local-scale contact mechanics) complementarity problem

$$h_d = 0, \quad p_d > 0, \tag{6.141a}$$
$$h_d > 0, \quad p_d = 0, \tag{6.141b}$$

subject to the force balance constraint

$$F - \int_Y p_d dy = 0. \tag{6.142}$$

Solving (6.141)–(6.142) for a given load F returns the dependent variables p_d, u and g_{00}. Recall that (6.140) describes the relation between h_d, u and g_{00}. Moreover, by solving (6.141)–(6.142) for a whole range of loads gives us the input required to solve the local problems (6.102) for a range of α-values. To this end, we supplement the α-parametrised film thickness equation defined for full-film conditions with an expression valid for mixed lubrication conditions. This is accomplished by forcing h to take exactly the values of h_d that was computed while solving the complementarity problem (6.141) for the specified range of loads F_k. This means that h for mixed lubrication conditions reads

$$h_\alpha(y, \tau) = \alpha + h_r(y, \tau) + u(y, \tau) + \epsilon, \quad \alpha = -g_{00} \tag{6.143}$$

where ϵ is a (small) auxiliary parameter that makes sure that $h_\alpha > 0$. Note now that there are two different specifications of the local scale film thickness, i.e. (i) the expression (6.138), for full-film conditions and (ii) the expression (6.143), for mixed lubrication conditions, which require solving the contact mechanics problem for a range of loads.

To summarise the, the procedure described above to obtain a widely applicable set of coefficient functions a_{ij} and b_i is composed of the following steps:

1. Specify a range of separations $\alpha > \bar{h}_r$ and a range of loads F.
2. Solve the complementarity contact mechanics problem (6.141) for the specified range of loads F.
3. Solve the local problems (6.102) for the auxiliary film thickness descriptions (6.138) and (6.143) for the specified α-values (including the values of g_{00} obtained from the contact mechanics model).
4. Compute the coefficient functions a_{ij} and b_i for all the local problems obtained in the previous steps.

It is important to realise that the procedure here outlined requires that the selected local-scale domain is representative of the considered surface roughness. This means that having selected any other domain of the same size, the computed coefficient functions a_{ij} and b_i would have been the same. In practice, this means that the coefficient functions should be computed for several such domains and the results compared. If these are sufficiently close, any of them (or possibly an average) can be used as representative of the whole surface roughness. If the results are far apart however, not even their average can be seen as representative for the surface roughness. Instead, a larger local-scale domain is to be chosen. A more thorough discussion of this issue in a similar context can be found in Pérez-Ràfols et al. (2016), where an alternative approach to be taken when the local-scale domain cannot be made larger is proposed.

Modelling Mixed Lubrication

A mixed lubrication model can be established by a combination of a model governing the hydrodynamic contribution and a model that accounts for contact mechanics. One of the first examples of such a model was presented by Patir and Cheng (1978, 1979). They derived an averaged form of the Reynolds equation, in which the surface roughness was accounted for by means of what we know today as 'flow factors'. Then, in order to simulate partial contact they include the effect of surface roughness by comparing the average interfacial separation $\bar{h}$ with the combined variance, $\sigma = \sqrt{{R_{q1}}^2 + {R_{q2}}^2}$, of any two digitalised surface roughness height descriptions, which full fill the assumptions for Reynolds equation to be valid on the local scale. Since the effect of roughness diminishes for large values of $\bar{h}/\sigma$ they focus their analysis to situations where $\bar{h}/\sigma < 3$ and their work has formed the starting point for a large number of similar contributions.

Obviously, Patir and Cheng did not include a contact mechanics model to account for local deformation of the contacting surfaces. This can, e.g. be achieved by incorporating an asperity-based contact model, which was also done by Rodhe et al. (1980). More precisely, they combined the averaged Reynolds equation that Patir and Cheng derived, with the model by Greenwood and Tripp (1970) and considered cavitation by means of the half-Sommerfeld boundary condition. Bolander et al. (2005) took this concept further by replacing the half-Sommerfeld condition with cavitation algorithm proposed by Elrod and Adams (1975).

A mixed lubrication model based on the homogenised Reynolds equation was later presented by Sahlin et al. (2010a, b). Indeed, this model combines half-space theory based contact mechanics of rough surfaces with the homogenised Reynolds equation in which the flow factors has been obtained in the way described in the last subsection of section 'Homogenisation of the Reynolds Equation'. In addition, to make the methodology even more versatile one can always generalise by transforming h_α into dimensionless form, i.e.

$$H_\alpha = h_\alpha / h_{ref}, \qquad (6.144)$$

where h_{ref} is an appropriate reference parameter. The local problems (6.102b) and (6.102c) are invariant under this transformation, meaning that the solutions χ_1 and χ_1 will be the same for any choice of h_{ref}. The local problem (6.102a) is, however, not invariant and becomes

$$0 = \nabla_y \cdot \left(\tilde{A}_\alpha \nabla_y \tilde{\chi}_0 \right) - \nabla_y \cdot \tilde{B}_\alpha \text{ in } Y, \tag{6.145}$$

where $\tilde{A}_\alpha$ and $\tilde{B}_\alpha$ indicate that they are transformed and where $\tilde{\chi}_0 = h_{ref}^2 \chi_0$. Moreover, the subscript also indicates that they are parametrised in α. This means that the homogenised matrix A_0 and vector B_0 transforms and their transformed correspondences should be computed from

$$\tilde{A}_0^\alpha = \iint_Y \begin{bmatrix} \tilde{a}_{11}^\alpha & \tilde{a}_{12}^\alpha \\ \tilde{a}_{21}^\alpha & \tilde{a}_{22}^\alpha \end{bmatrix} \left(e_i + \nabla_y \chi_i \right) dy \tag{6.146a}$$

$$\tilde{B}_0^\alpha = \iint_Y \tilde{B}_\alpha e_1 + \tilde{A}_\alpha \nabla_y \tilde{\chi}_0 dy \tag{6.146b}$$

Before the homogenised solution u_0 can be obtained, the α-parametrised homogenised matrix $\tilde{A}_0^\alpha$ and vector $\tilde{B}_0^\alpha$ must be mapped onto Ω. This is achieved by interpolating $\tilde{A}_0^{h_g(x_1,x_2)}$ and $\tilde{B}_0^{h_g(x_1,x_2)}$ for each point (x_1, x_2) in the global-scale grid point. Thus,

$$\nabla_x \cdot \left(\tilde{A}_0(x) \nabla_x \tilde{u}_0 \right) - \nabla_x \cdot \tilde{B}_0(x) = 0, \tag{6.147}$$

where $\tilde{u}_0 = h_{ref}^2 u_0$.

Summary

Lubrication theory including the derivation of the classical Reynolds equation, by means of scaling and asymptotic analysis of the Navier–Stokes equations has been presented to the reader. The concept of cavitation modelling, starting with the contribution from Elrod and Adams (1975) and ending with strictly formulated complementarity problems, i.e. Giacopini et al. (2010), Bertocchi et al. (2013), Almqvist et al. (2014), Almqvist and Wall (2016), was presented. Together with a complete description on a finite difference scheme, the reader should be equipped with the means to simulate hydrodynamic cavitation in various forms of lubricated contacts.

An effort to give the reader the basics of homogenisation and how it can be applied to efficiently average the effect of surface roughness is also provided. It shows the versatility and that the homogenised coefficients obtained for the flow of incompressible fluids as well as for compressible fluids, obeying the constant bulk modulus compressibility and even for ideal gases are the same. Thus, once obtained, they can be applied to study the influence of roughness in a wide range of applications. The relation between the homogenised coefficients and the Patir and Cheng flow factors is also presented. This shows why the homogenised coefficients are always representative and when and why the Patir and Cheng flow factors could be used. The concept of homogenised coefficients is also extended to incorporate partially lubricated situations.

Finally, a section suggesting how to model mixed lubrication situations occurring whenever the hydrodynamic action is not strong enough to generate the load carrying capacity to produce a fully separating film.

With this, we close the chapter and the authors thanks the reader for the attention.

References

Ahmed, S., Goodyer, C. E., & Jimack, P. K. (2012). An efficient preconditioned iterative solution of fully-coupled elastohydrodynamic lubrication problems. *Applied Numerical Mathematics, 62*(5), 649–663. https://doi.org/10.1016/j.apnum.2012.02.002.

Almqvist, A. (2011). Homogenization of the Reynolds equation governing hydrodynamic flow in a rotating device. *Journal of Tribology, 133*(2), 021705. https://doi.org/10.1115/1.4003650.

Almqvist, A., & Dasht, J. (2006). The homogenization process of the Reynolds equation describing compressible liquid flow. *Tribology International, 39*(9), 994–1002. ISSN 0301-679X. https://doi.org/10.1016/j.triboint.2005.09.036.

Almqvist, A., & Wall, P. (2016). Modelling cavitation in (elasto)hydrodynamic lubrication. In P. H. Darji (Eds.), *Advances in tribology*. Rijeka: IntechOpen. https://doi.org/10.5772/63533.

Almqvist, A., Essel, E. K., Persson, L. E., & Wall, P. (2007a). Homogenization of the unstationary incompressible Reynolds equation. *Tribology International, 40*(9), 1344–1350. ISSN 0301-679X. https://doi.org/10.1016/j.triboint.2007.02.021.

Almqvist, A., Sahlin, F., Larsson, R., & Glavatskih, S. (2007b). On the dry elasto-plastic contact of nominally flat surfaces. *Tribology International, 40*(4), 574–579. ISSN 0301-679X. https://doi.org/10.1016/j.triboint.2005.11.008.

Almqvist, A., Fabricius, J., Spencer, A., & Wall, P. (2011). Similarities and differences between the flow factor method by patir and cheng and homogenization. *Journal of Tribology, 133*(3), 031702. https://doi.org/10.1115/1.4004078.

Almqvist, A., Fabricius, J., & Wall, P. (2012). Homogenization of a Reynolds equation describing compressible flow. *Journal of Mathematical Analysis and Applications, 390*(2), 456–471. ISSN 0022-247X. https://doi.org/10.1016/j.jmaa.2012.02.005.

Almqvist, A., Spencer, A., & Wall, P. (2013). *Matlab routines solving a linear complementarity problem appearing in lubrication with cavitation*. http://www.mathworks.com/matlabcentral/fileexchange/41484.

Almqvist, A., Fabricius, J., Larsson, R., & Wall, P. (2014). A new approach for studying cavitation in lubrication. *Journal of Tribology, 136*(1), 011706-1–011706-6. https://doi.org/10.1115/1.4025875.

Ausas, R. F., Jai, M., & Buscaglia, G. C. (2009). A mass-conserving algorithm for dynamical lubrication problems with cavitation. *Journal of Tribology, 131*(3):031702. https://doi.org/10.1115/1.3142903.

Bayada, G., & Chambat, M. (1988). New models in the theory of the hydrodynamic lubrication of rough surfaces. *Journal of Tribology, Transactions of the ASME, 110*(3), 402–407. ISSN 0742-4787.

Bayada, G., Martin, S., & Vázquez, C. (2005a). An average flow model of the Reynolds roughness including a mass-flow preserving cavitation model. *Journal of Tribology, 127*(4), 793–802. https://doi.org/10.1115/1.2005307.

Bayada, G., Martin, S., & Vazquez, C. (2005b). Two-scale homogenization of a hydrodynamic Elrod-Adams model. *Asymptotic Analysis, 44*, 75–110.

Bayada, G., Cid, B., García, G., & Vázquez, C. (2013). A new more consistent Reynolds model for piezoviscous hydrodynamic lubrication problems in line contact devices. *Applied Mathematical Modelling, 37*(18–19), 8505–8517. https://doi.org/10.1016/j.apm.2013.03.072.

Bertocchi, L., Dini, D., Giacopini, M., Fowell, M. T., & Baldini, A. (2013). Fluid film lubrication in the presence of cavitation: A mass-conserving two-dimensional formulation for compressible, piezoviscous and non-Newtonian fluidsbaya. *Tribology International, 67*, 61–71. ISSN 0301-679X. https://doi.org/10.1016/j.triboint.2013.05.018.

Bolander, N. W., Steenwyk, B. D., Sadeghi, F., & Gerber, G. R. (2005). Lubrication regime transitions at the piston ring-cylinder liner interface. *Proceedings of the Institution of Mechanical Engineers, Part J: Journal of Engineering Tribology, 219*(1), 19–31. https://doi.org/10.1243/135065005X9664.

Boukrouche, M., & Bayada, G. (1993). Mathematical model. Existence and uniqueness of cavitation problems in porous journal bearing. *Nonlinear Analysis: Theory, Methods & Applications, 20*(8), 895–920. https://doi.org/10.1016/0362-546X(93)90084-6.

Buresti, G. (2015). A note on stokes' hypothesis. *Acta Mechanica, 226*(10), 3555–3559. ISSN 1619-6937. https://doi.org/10.1007/s00707-015-1380-9.

Cottle, R. W., Pang, J. S., & Stone, R. E. (2009). *The linear complementarity problem*. Report No. 60. Philadelphia, PA: SIAM.

Dowson, D., & Higginson, G. R. (1966). *Elasto-hydrodynamic lubrication: The fundamentals of roller or gear lubrication*. Oxford: Pergamon Press.

Elrod, H. G. (1981). A cavitation algorithm. *Journal of Tribology, 103*, 350–354.

Elrod, H. G., & Adams, M. L. (1975). A computer program for cavitation and starvation problems. In D. Dowson, M. Godet, & C. M. Taylor (Eds.), *Cavitation and related phenomena in lubrication* (pp. 37–43). London: Mechanical Engineering Publications.

Evans, H. P., Elcoate, C. D., Hughes, T. G., & Snidle, R. W. (2001). Transient elastohydrodynamic analysis of rough surfaces using a novel coupled differential deflection method. *Proceedings of the Institution of Mechanical Engineers Part J, 215*, 319–337.

Evans, L. C. (2010). Graduate studies in mathematics providence. In *Partial differential equations* (2nd ed., Vol. 19). Rhode Island: American Mathematical Society.

Fabricius, J., Koroleva, Y. O., & Wall, P. (2014). A rigorous derivation of the time-dependent reynolds equation. *Asymptotic Analysis, 84*, 103–121.

Floberg, L. (1960). *The two-groove journal bearing, considering cavitation*. Technical Report 231, Institute of Machine Elements, Chalmers University of Technology, Gothenburg, Sweden.

Floberg, L. (1961). *Lubrication of two cylinder surfaces, considering cavitation*. Technical Report 232, Institute of Machine Elements, Chalmers University of Technology, Gothenburg, Sweden.

Fowell, M., Olver, A. V., Gosman, A. D., Spikes, H. A., & Pegg, I. (2007). Entrainment and inlet suction: Two mechanisms of hydrodynamic lubrication in textured bearings. *Journal of Tribology, 129*(2), 336–347. https://doi.org/10.1115/1.2540089.

Giacopini, M., Fowell, M. T., Dini, D., & Strozzi, A. (2010). A mass-conserving complementarity formulation to study lubricant films in the presence of cavitation. *Journal of Tribology, 132*(4), 041702. https://doi.org/10.1115/1.4002215.

Greenwood, J. A., & Tripp, J. H. (1970). The contact of two nominally flat rough surfaces. *Proceedings of the Institution of Mechanical Engineers, 185*(48), 625–633. ISSN 0020-3483.

Gustafsson, I., Rajagopal, K. R., Stenberg, R., & Videman, J. (2015). Nonlinear Reynolds equation for hydrodynamic lubrication. *Applied Mathematical Modelling, 39*(17), 5299–5309. ISSN 0307-904X. https://doi.org/10.1016/j.apm.2015.03.028.

Habchi, W. (2018). *Finite element modeling of elastohydrodynamic lubrication problems*. https://doi.org/10.1002/9781119225133.

Habchi, W. (2019). A schur-complement model-order-reduction technique for the finite element solution of transient elastohydrodynamic lubrication problems. *Advances in Engineering Software, 127*, 28–37. https://doi.org/10.1016/j.advengsoft.2018.10.007.

Holmes, M. A. J. (2002). *Transient analysis of the point contact elastohydrodynamic lubrication problem using coupled solution methods*. Ph.D. thesis, Cardiff University.

Holmes, M. J. A., Evans, H. P., Hughes, T. G., & Snidle, R. W. (2003a). Transient elastohydrodynamic point contact analysis using a new coupled differential deflection method part 1: Theory and validation. *Proceedings of the Institution of Mechanical Engineers, Part J: Journal of Engineering Tribology, 217*(4), 289–304. https://doi.org/10.1243/135065003768618641.

Holmes, M. J. A., Evans, H. P., Hughes, T. G., & Snidle, R. W. (2003b). Transient elastohydrodynamic point contact analysis using a new coupled differential deflection method part 2: Results. *Proceedings of the Institution of Mechanical Engineers, Part J: Journal of Engineering Tribology, 217*(4), 305–322. https://doi.org/10.1243/135065003768618650.

Hooke, C. J., & Li, K. Y. (2006). Rapid calculation of the pressures and clearances in rough, elastohydrodynamically lubricated contacts under pure rolling. part 1: Low amplitude, sinusoidal roughness. *Proceedings of the Institution of Mechanical Engineers, Part C: Journal of Mechanical Engineering Science, 220*(6), 901–913. https://doi.org/10.1243/09544062C03405.

Jakobsson, B., & Floberg, L. (1957). *The finite journal bearing, considering vaporization*. Technical Report 190, Institute of Machine Elements, Chalmers University of Technology, Gothenburg, Sweden.

Liu, S., Wang, Q., & Liu, G. (2000). A versatile method of discrete convolution and FFT (DC-FFT) for contact analyses. *Wear, 243*(1–2), 101–111. ISSN 0043-1648. https://doi.org/10.1016/S0043-1648(00)00427-0.

Liu, S., Hua, D., Chen, W. W., & Wang, Q. J. (2007). Tribological modeling: Application of fast Fourier transform. *Tribology International, 40*(8), 1284–1293. ISSN 0301-679X. https://doi.org/10.1016/j.triboint.2007.02.004.

Lukkassen, D., Meidell, A., & Wall, P. (2009). Homogenization of some variational problems connected to the theory of lubrication. *International Journal of Engineering Science*, *47*(1), 153–162.

Lukkassen, D., Nguetseng, G., & Wall, P. (2002). Two-scale convergence. *International Journal of Pure and Applied Mathematics*, *2*(1), 33–81.

Olsson, K. O. (1965). *Cavitation in dynamically loaded bearings*. Technical Report 308, Institute of Machine Elements, Chalmers University of Technology, Gothenburg, Sweden.

Olver, A. V., Fowell, M. T., Spikes, H. A., & Pegg, I. G. (2006). 'inlet suction', a load support mechanism in non-convergent, pocketed, hydrodynamic bearings. *Proceedings of the Institution of Mechanical Engineers, Part J: Journal of Engineering Tribology, 220*(2), 105–108. 10.1243/13506501JET168. https://doi.org/10.1243/13506501JET168.

Patir, N. (1978). A numerical procedure for random generation of rough surfaces. *Wear, 47*(2), 263–277. ISSN 0043-1648. https://doi.org/10.1016/0043-1648(78)90157-6.

Patir, N., & Cheng, H. S. (1978). An average flow model for determining effects of three-dimensional roughness on partial hydrodynamic lubrication. *Journal of Tribology, Transactions of the ASME, 100*, 12–17.

Patir, N., & Cheng, H. S. (1979). Application of average flow model to lubrication between rough sliding surfaces. *Journal of Tribology, Transactions of the ASME, 101*, 220–230.

Pérez-Ràfols, E., & Almqvist, A. (2019). Generating randomly rough surfaces with given height probability distribution and power spectrum. *Tribology International, 131*, 591–604. ISSN 0301-679X. https://doi.org/10.1016/j.triboint.2018.11.020.

Pérez-Ràfols, E., Larsson, R., Lundström, T. S., Wall, P., & Almqvist, A. (2016). A stochastic two-scale model for pressure-driven flow between rough surfaces. *Proceedings of the Royal Society. Mathematical, Physical and Engineering Sciences, 472*(2190), 20160069. https://doi.org/10.1098/rspa.2016.0069.

Persson, B. N. J. (2010). Fluid dynamics at the interface between contacting elastic solids with randomly rough surfaces. *Journal of Physics Condensed Matter, 22*(26). https://doi.org/10.1088/0953-8984/22/26/265004.

Rajagopal, K. R., & Szeri, A. Z. (2003). On an inconsistency in the derivation of the equations of elastohydrodynamic lubrication. *Proceedings of the Royal Society of London. Series A: Mathematical, Physical and Engineering Sciences, 459*, 2771–2786.

Reynolds, O. (1886). On the theory of lubrication and its application to Mr. Beauchamps tower's experiments, including an experimental determination of the viscosity of olive oil. *Philosophical Transactions of the Royal Society of London A, 177*, 157–234.

Rodhe, S. M., Whitaker, K. W., & McAllister, G. T. (1980). A mixed friction model for dynamically loaded contacts with application to piston ring lubrication. In *Surface roughness effects in hydrodynamic and mixed lubrication. ASME Winter Annual Meeting, Chicago* (pp. 19–50).

Sahlin, F., Almqvist, A., Larsson, R., & Glavatskih, S. (2007). A cavitation algorithm for arbitrary lubricant compressibility. *Tribology International, 40*(8), 1294–1300. ISSN 0301-679X. https://doi.org/10.1016/j.triboint.2007.02.009.

Sahlin, F., Larsson, R., Marklund, P., Lugt, P. M., & Almqvist, A. (2010a). A mixed lubrication model incorporating measured surface topography. part 1: theory of flow factors. *Proceedings of the Institution of Mechanical Engineers, Part J: Journal of Engineering Tribology, 224*(4), 335–351. https://doi.org/10.1243/13506501JET658.

Sahlin, F., Larsson, R., Marklund, P., Lugt, P. M., & Almqvist, A. (2010b). A mixed lubrication model incorporating measured surface topography. Part 2: Roughness treatment, model validation, and simulation. *Proceedings of the Institution of Mechanical Engineers, Part J: Journal of Engineering Tribology, 224*(4), 353–365. https://doi.org/10.1243/13506501JET659.

Scaraggi, M., & Carbone, G. (2012). A two-scale approach for lubricated soft-contact modeling: An application to lip-seal geometry. *Advances in Tribology*. https://doi.org/10.1155/2012/412190.

Shirzadegan, M., Almqvist, A., & Larsson, R. (2016). Fully coupled ehl model for simulation of finite length line cam-roller follower contacts. *Tribology International, 103*, 584–598. https://doi.org/10.1016/j.triboint.2016.08.017.

Söderfjäll, M. (2017). *Friction in piston ring—Cylinder liner contacts*. Ph.D. thesis, Luleå University of Technology, Machine Elements.

Tian, X., & Bhushan, B. (1996). A numerical three-dimensional model for the contact of rough surfaces by variational principle. *Journal of Tribology, 118*(1), 33–42. ISSN 07424787. https://doi.org/10.1115/1.2837089.

Venner, C. H., & Lubrecht, A. A. (2000). Multilevel methods in lubrication. *Tribology Series, 37*.

Vijayaraghavan, D., & Keith Jr, T. G. (1989). Development and evaluation of a cavitation algorithm. *STLE Tribology Transactions, 32*(2), 225–233.

Vijayaraghavan, D., & Keith Jr, T. G. (1990). An efficient, robust, and time accurate numerical scheme applied to a cavitation algorithm. *Journal of Tribology, Transactions of the ASME, 112*(1), 44–51. ISSN 0742-4787.

Wall, P. (2007). Homogenization of Reynolds equation by two-scale convergence. *Chinese Annals of Mathematics—Series B, 28*(3), 363–374.

Wang, W. Z., Wang, H., Liu, Y. C., Hu, Y. Z., & Zhu, D. (2003). A comparative study of the methods for calculation of surface elastic deformation. *Proceedings of the Institution of Mechanical Engineers, Part J: Journal of Engineering Tribology, 217*(2), 145–154. ISSN 1350-6501. https://doi.org/10.1243/13506500360603570.

Woods, C. M., & Brewe, D. E. (1989). The solution of the Elrod algorithm for a dynamically loaded journal bearing using multigrid techniques. *Journal of Tribology, 111*(2), 302–308.

Zhu, D., Liu, Y., & Wang, Q. (2015). On the numerical accuracy of rough surface ehl solution. *Tribology and Lubrication Technology, 71*(1), 40–55. https://doi.org/10.1080/10402004.2014.886349.

Chapter 7
Contact Mechanics of Rubber and Soft Matter

Carmine Putignano and Daniele Dini

Abstract This chapter reviews recent advances made in the treatment of contact problems involving soft materials often characterized by non-linearly elastic material properties, such as rubber and soft biological tissues. Starting from the fundamental formulation developed to solve viscoelastic contact mechanics, the treatment of complex problems involving surface roughness, layered materials, and reciprocating contacts in dry contacts is presented in increased order of complexity. The reader is then introduced to the study of lubricated contacts, with a discussion of the interplay between viscoelastic effects in the solids and the viscosity marking the lubricant behavior. Experimental validations that cover various aspects of the work are also presented.

Introduction

Rubber and rubber-based composites are receiving widespread attention as engineering solutions that combine good mechanical and chemical properties in terms of resilience, elasticity, and durability. Tires, belts, rollers, and seals are only examples of mechanical devices where these materials are considered as the first choice. The viscoelastic mechanical response of such materials must be properly considered during the optimization process of engineering components, whose design needs, inter alia, the full comprehension of phenomena involving viscoelastic energy dissipation. Examples are: (i) rolling contacts (Hunter 1961; Persson 2010; Panek and Kalker 1980; Harrass et al. 2010; Dumitru 2009; Yoneyama et al. 2010) (ii) sliding contacts (Carbone and Mangialardi 2004; Persson 2001, 2006b; Grosch 1963), (iii)

C. Putignano
Department of Mechanics Mathematics and Management, Polytechnic University of Bari, Bari, Italy
e-mail: carmine.putignano@poliba.it

D. Dini (✉)
Tribology Group, Department of Mechanical Engineering, Imperial College London, London, UK
e-mail: d.dini@imperial.ac.uk

M. Paggi and D. Hills (eds.), *Modeling and Simulation of Tribological Problems in Technology*, CISM International Centre for Mechanical Sciences 593,
https://doi.org/10.1007/978-3-030-20377-1_7

crack propagation (Carbone and Persson 2005a, b; Persson et al. 2005; Persson and Brener 2005; D'Amico et al. 2012), (iv) seals (Bottiglione et al. 2009a, b; Lorenz and Persson 2010a, b, c), and (v) adhesives and biomimetic adhesives (Carbone et al. 2011; Carbone and Pierro 2012a, b; Martina et al. 2012). In this Chapter, our aim is to assess the role played by a non-purely elastic behavior in determining the peculiarities of soft contact mechanics. Particular attention is paid to linear viscoelasticity, which is the rheological constitutive model commonly adopted when dealing with soft matter.

With regard to these issues, a very large number of scientific contributions have been dedicated to develop theories (Persson 2001, 2006b, 2010; Hunter 1961; Panek and Kalker 1980) and numerical methodologies (Le Tallec and Rahler 1994; Vollebregt 2009; Padovan and Paramadilok 1984; Padovan 1987; Padovan et al. 1992; Nackenhorst 2004; Nasdala et al. 1998) to investigate rolling and sliding contacts of viscoelastic materials. In this book, we provide a review of the main numerical and experimental methodologies to investigate viscoelastic contacts, and in particular to predict the viscoelastic contribution to the friction coefficient and its dependence on sliding/rolling velocity. Such a viscoelastic contribution to friction has been pioneeringly treated in an analytical manner by Hunter (1961) for the case of two-dimensional (2D) contacts, specifically for a rigid cylinder in contact with a viscoelastic half-space. Hunter's approach has been then extended to the three-dimensional case by enforcing the so-called line contact approximation (Panek and Kalker 1980). However, in many cases, e.g., a sphere rolling or sliding on a viscoelastic foundation, this assumption represents a very strong approximation, and cannot be employed to fully assess contact problems.

Moreover, it should be noted that these fully analytical models present a further limitation as they are only able to handle ideal viscoelastic materials with one single relaxation time, i.e., materials described by the so-called standard viscoelastic model (see Fig. 7.1a). Real materials, instead, present a very wide spectrum of relaxation times, and the most general form of their linear viscoelastic response is the Maxwell–Wiechert model (see the next section). As a matter of fact, all the classes of theories, which are able to treat only idealised one relaxation time material, cannot be employed to analyze and optimize the design of real engineering components. Moving from this context, in the last decade, Persson (2001, 2006b, 2010) has proposed a new theory of contact mechanics to calculate, among the other quantities, the contact area and the sliding and rolling friction of a linear viscoelastic materials moving on rigid smooth or rough surface. His calculations are based on some reasonable assumptions which allow to strongly simplify the problem, but these assumptions needed to be tested against experiments or compared to reliable numerical calculations (see also Carbone et al. 2009; Putignano et al. 2012a, b; Campana et al. 2008) for a final verification and tuning of his models. Among the numerical approaches, we have finite element methods (FEM) (Le Tallec and Rahler 1994; Padovan and Paramadilok 1984; Padovan 1987; Padovan et al. 1992; Nackenhorst 2004; Nasdala et al. 1998) that are able to treat real viscoelastic materials, but have been conceived for structural modeling (e.g., tire modeling) and are not accurate enough when exploited to calculate the pressure, displacement, strain distributions and friction at the inter-

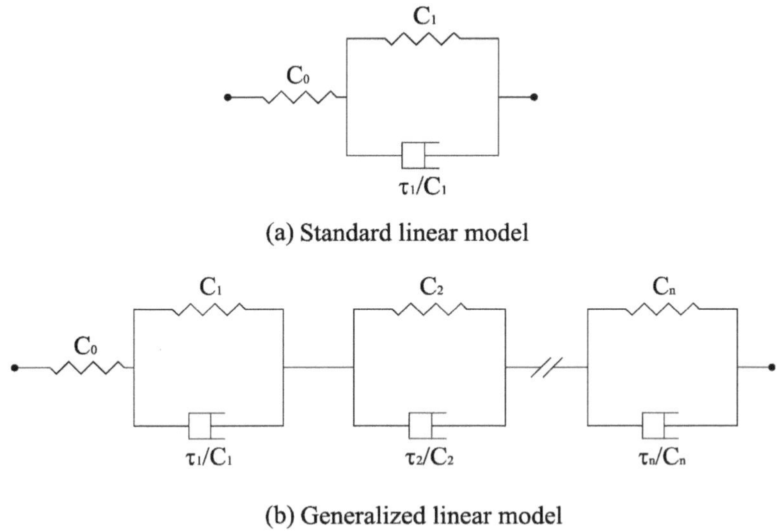

Fig. 7.1 The standard linear viscoelastic solid constituted by a spring in series with a Voigt element (the latter consists of a Hookean spring in parallel with a Newtonian dashpot) (**a**); the generalized linear viscoelastic model (**b**)

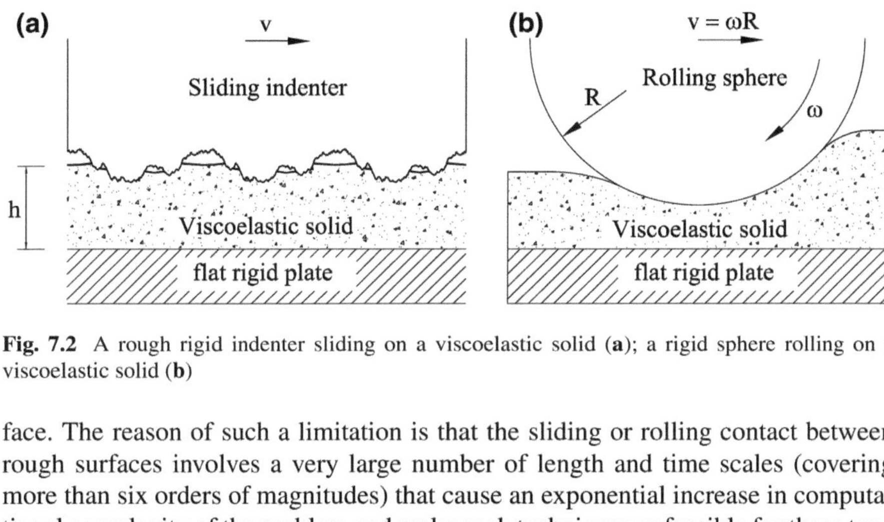

Fig. 7.2 A rough rigid indenter sliding on a viscoelastic solid (**a**); a rigid sphere rolling on a viscoelastic solid (**b**)

face. The reason of such a limitation is that the sliding or rolling contact between rough surfaces involves a very large number of length and time scales (covering more than six orders of magnitudes) that cause an exponential increase in computational complexity of the problem and make such techniques unfeasible for these type of investigations. In this book, we focus on some boundary element formulations which, as we will see in detail in the next Sections, reduce the full viscoelastic problem formulated in the space and the time domains to an equivalent one to be solved only in the space domain. This enables, for the generic contact problem, the study of complex geometrical domains as it is indeed the case of rough surfaces. Ultimately, these formulations allow the numerical study of the rolling/sliding contact between a rigid (smooth or rough) indenter over a linear viscoelastic layer (see Fig. 7.2).

Time-Dependent Material Effects: Viscoelastic Rheological Models

Here, we briefly recall the main point of linear viscoelastic theory. Real viscoelastic materials present a very wide spectrum of relaxation times, and the most general form of their linear viscoelastic response is

$$\varepsilon(t) = \int_{-\infty}^{t} d\tau \mathcal{J}(t-\tau)\dot{\sigma}(\tau) \tag{7.1}$$

where the $\varepsilon(t)$ is the time-dependent strain, $\dot{\sigma}(t)$ is the stress [the symbol "·" stands for the time derivative], and the function $\mathcal{J}(t)$ is the creep function [we recall that $\mathcal{J}(t)$ must satisfy causality, i.e., $\mathcal{J}(t<0)=0$]. One can show that the most general form of $\mathcal{J}(t)$ is (Christensen 1982; Ferry 1980):

$$\mathcal{J}(t) = \mathcal{H}(t)\left[\frac{1}{E_0} - \int_0^{+\infty} d\tau \mathcal{C}(\tau)\exp(-t/\tau)\right] \tag{7.2}$$

where $\mathcal{H}(t)$ is the Heaviside step function, the real quantity E_0 is the elastic modulus of the material at zero frequency, $\mathcal{C}(\tau)$ is a positive function usually referred to as the creep (or retardation) spectrum (Christensen 1982), and τ is the relaxation time continuously distributed on the real axis. We recall that $J(t=+\infty)=E_0^{-1}$ and $J(t=0)=E_0^{-1}-\int_0^{+\infty} d\tau \mathcal{C}(\tau)=E_\infty^{-1}$, where the real quantity E_∞ is the high-frequency elastic modulus of the material. Usually, a discrete version of Eq. (7.2) is employed to characterize linear viscoelastic solids, i.e., one writes $\mathcal{C}(\tau)=\sum_k C_k\delta(\tau-\tau_k)$ to get

$$\mathcal{J}(t) = \mathcal{H}(t)\left[\frac{1}{E_0} - \sum_{k=1}^{n} C_k \exp(-t/\tau_k)\right] \tag{7.3}$$

This latter representation of $\mathcal{J}(t)$ corresponds to the general linear viscoelastic model represented in Fig. 7.1b, where the quantities C_k represent the elastic compliances of the springs. Equivalently, we can introduce the viscoelastic moduli E_k and write:

$$\mathcal{J}(t) = \mathcal{H}(t)\left[\frac{1}{E_0} - \sum_{k=1}^{n} \frac{1}{E_k} \exp(-t/\tau_k)\right] \tag{7.4}$$

Now, taking the Fourier transform of Eq. (7.1) one obtains $\varepsilon(\omega)=\sigma(\omega)/E(\omega)$ with $E(\omega)=[i\omega J(\omega)]^{-1}$ being the viscoelastic modulus of the material. We have defined $\mathcal{J}(\omega)=\int dt \mathcal{J}(t)\exp(-i\omega t)$, $\sigma(\omega)=\int dt\sigma(t)\exp(-i\omega t)$ and $\varepsilon(\omega)=\int dt\varepsilon(t)\exp(-i\omega t)$. Using Eq. (7.2) one can also show that

$$\frac{1}{E(\omega)} = \frac{1}{E_0} - \int_0^\infty d\tau \frac{i\omega\tau C(\tau)}{1+i\omega\tau} = \frac{1}{E_\infty} + \int_0^\infty d\tau \frac{C(\tau)}{1+i\omega\tau} \tag{7.5}$$

and using (7.5) one can easily prove the *sum rule*:

$$\frac{1}{E_0} - \frac{1}{E_\infty} = \frac{2}{\pi}\int_0^\infty d\omega\, \frac{1}{\omega}\mathrm{Im}\frac{1}{E(\omega)} \tag{7.6}$$

The general structure of the viscoelastic modulus $E(\omega)$ is shown in Fig. 7.3. At "low" frequencies the material is in the "rubbery" region where $E_1 = \mathrm{Re}E(\omega)$ is relatively small and approximately constant, and viscoelastic dissipations related to the imaginary part $E_2 = \mathrm{Im}E(\omega)$ of the viscoelastic modulus becomes negligible. At very high frequencies, the material is elastically very stiff (brittle-like). In this "glassy" region $E_1(\omega)$ is again nearly constant but much larger (generally by 3–4 orders of magnitude) than in the rubbery region. In the intermediate frequency range (the so-called transition region), the loss tangent $\mathrm{Im}E(\omega)/\mathrm{Re}E(\omega)$ is very large (see Fig. 7.3b), and it is mainly this region which determines the energy dissipation during sliding or rolling motion.

We observe that, as we see also in next Sections, for real viscoelastic solids the spectrum of relaxation times may cover more than 8–10 decades so that the number n of Voigt element to be arranged in series as in Fig. 7.1b to correctly describe the entire response of the material may be relatively large and usually, 10–30 elements, i.e., 10–30 relaxation times, are needed for a full characterization of the constitutive

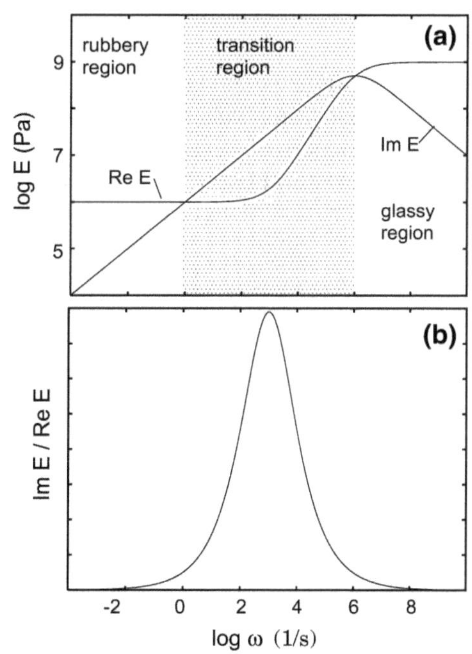

Fig. 7.3 The real $E_1 = \mathrm{Re}[E(\omega)]$ and the imaginary $E_2 = \mathrm{Im}[E(\omega)]$ parts of the viscoelastic modulus $E(\omega)$ of a typical rubber-like material (**a**); the loss tangent $E_2(\omega)/E_1(\omega)$ (**b**)

equation. Therefore, for any approach to viscoelastic problems, it is of outstanding importance to try to account for real multi-relaxation-time materials.

Viscoelastic Steady-State Sliding and Rolling Contact Mechanics: Numerical Formulations and Experiments

Boundary Element Method: Mathematical Formulation

By following the Green function approach developed in (Carbone et al. 2009; Putignano et al. 2012a, b; Carbone and Mangialardi 2008b) and recalling the translational invariance and the elastic–viscoelastic correspondence principle (Christensen 1982), we may formulate the general linear viscoelastic contact problem between a rigid indenter and a viscoelastic slab as

$$u(\mathbf{x}, t) = \int_{-\infty}^{t} d\tau \int d^2x \mathcal{J}(t-\tau) \mathcal{G}(\mathbf{x}-\mathbf{x}') \dot{\sigma}(\mathbf{x}', \tau) \tag{7.7}$$

where $\mathbf{x}$ is the in-plane position vector, t is the time, $u(\mathbf{x}, t)$ is the normal surface displacement of the viscoelastic solid, $\sigma(\mathbf{x}, t)$ is the normal interfacial stress, $\mathcal{J}(t)$ is given in Eq. (7.2) and the quantity $\mathcal{G}(\mathbf{x})$ is (see Carbone and Mangialardi 2004, 2008b)

$$\mathcal{G}(\mathbf{x}) = -\frac{2(1-\nu^2)}{(2\pi)^2} \int d^2q \frac{\mathcal{S}(qh)}{q} e^{i\mathbf{q}\cdot\mathbf{x}} \tag{7.8}$$

where $q = |\mathbf{q}|$ is the modulus of the wave vector $\mathbf{q} = (q_x, q_y)$, $r = |\mathbf{x}|$, and $\mathcal{S}(qh)$ is a correction factor which accounts for different constraint or boundary conditions. For an elastic slab of thickness h sandwiched between a flat rigid plate and a rigid body, as shown in Fig. 7.2 the correction factor $\mathcal{S}(qh)$ (Carbone and Mangialardi 2004; Carbone and Mangialardi 2008b) is:

$$\mathcal{S}(qh) = \frac{(3-4\nu)\sinh(2qh) - 2qh}{5 + 2(qh)^2 - 4\nu(3-2\nu) + (3-4\nu)\cosh(2qh)} \tag{7.9}$$

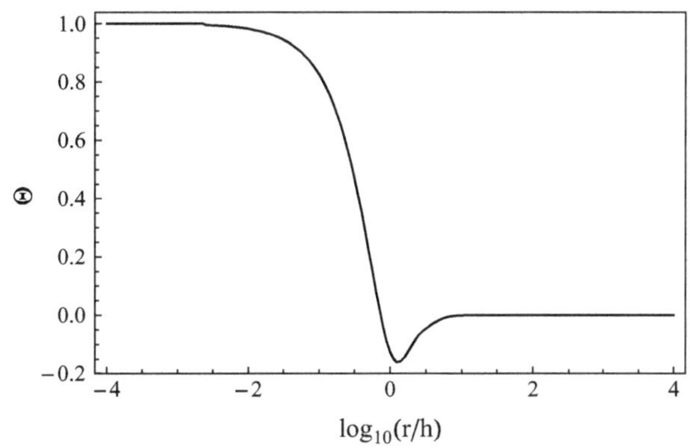

Fig. 7.4 The correction factor $\Theta(r/h)$ as a function of the ratio r/h in a log-linear diagram. As expected, for relatively small values of r/h, the correction factor is equal to 1, whereas for large values of r/h, it vanishes

For a semi-infinite solid (i.e., $h \to \infty$) the corrective factor is $\mathcal{S}(|\mathbf{q}|\,h) = 1$, and one recovers the classical Boussinesq solution

$$\mathcal{G}_{h\to\infty}(\mathbf{x}) = -\frac{1-\nu^2}{\pi r} \tag{7.10}$$

Equation (7.8) can be rephrased as

$$\mathcal{G}(\mathbf{x}) = -\frac{1-\nu^2}{\pi}\int_0^{+\infty} dq\mathcal{S}(qh)\,J_0(qr) = -\frac{1-\nu^2}{\pi r}\Theta(r/h) \tag{7.11}$$

with

$$\Theta(r/h) = \int_0^{+\infty} dw\mathcal{S}(wh/r)\,J_0(w) \tag{7.12}$$

where $J_0(w)$ is the zeroth-order Bessel function. Figure 7.4 shows the quantity $\Theta(r/h)$ as a function of the ratio r/h in a log-linear diagram. It is worth noticing that $\Theta(r/h)$ is a very well behaved function which, as expected, approaches the unit value at relatively low values of r/h, and rapidly vanishes as $r/h \to \infty$.

Now, let us observe that tackling directly Eq. (7.8) may be extremely complicated as it requires, for a full solution, to discretize both the time and the space domain. This may be often unfeasible as, for example, in the case of rough interfaces with a large number of space and time scales. However, given the steady-state properties of the system, and, thus, observing that the sliding/rolling motion occurs at constant velocity $\mathbf{v}$, we can simplify Eq. (7.8). Indeed, we have $\sigma(\mathbf{x}, t) = \sigma(\mathbf{x} - \mathbf{v}t)$ and $u(\mathbf{x}, t) = u(\mathbf{x} - \mathbf{v}t)$. Therefore, applying the transformation rule $\mathbf{X} = \mathbf{x} - \mathbf{v}t$, and recalling linearity and translational invariance one can also write Eq. (7.8) as

$$u(\mathbf{X}, \mathbf{v}) = \int d^2X'\mathcal{G}\left(\mathbf{X} - \mathbf{X}', \mathbf{v}\right)\sigma\left(\mathbf{X}'\right) \tag{7.13}$$

where the velocity-dependent function $\mathcal{G}(\mathbf{X}, \mathbf{v})$ is a "viscoelastic" Green's function, which can be easily determined assuming that the stress distribution at the interface is represented by Dirac delta moving at constant velocity $\mathbf{v}$. So, let us assume $\sigma(\mathbf{x}, t) = \delta(\mathbf{x} - \mathbf{v}t)$, substituting into Eq. (7.13) recalling that $\mathbf{X} = \mathbf{x} - \mathbf{v}t$ we obtain

$$\mathcal{G}(\mathbf{X}, \mathbf{v}) = \mathcal{J}(0)\, \mathcal{G}(\mathbf{X}) + \int_{0^+}^{\infty} dt \mathcal{G}(\mathbf{X} + \mathbf{v}t)\, \dot{\mathcal{J}}(t) \tag{7.14}$$

By recalling that $\mathcal{J}(0) = 1/E_\infty$, we may rephrase the equation above as

$$\begin{aligned} \mathcal{G}(\mathbf{X}, \mathbf{v}) = -\frac{1-\nu^2}{\pi} \Bigg\{ & \frac{1}{E_\infty} \frac{1}{|\mathbf{X}|} \Theta\left(\frac{|\mathbf{X}|}{h}\right) \\ & + \int_0^{+\infty} d\tau C(\tau) \int_{0^+}^{+\infty} dz \frac{1}{|\mathbf{X} + \mathbf{v}\tau z|} \Theta\left(\frac{|\mathbf{X} + \mathbf{v}\tau z|}{h}\right) \exp(-z) \Bigg\} \end{aligned} \tag{7.15}$$

It is worth noticing that in the limit case of very small sliding velocities, i.e., $|\mathbf{v}| \to 0$ we recover the classical elastic solution

$$\mathcal{G}(\mathbf{X}, \mathbf{0}) = -\frac{1-\nu^2}{\pi E_0} \frac{1}{|\mathbf{X}|} \Theta\left(\frac{|\mathbf{X}|}{h}\right) \tag{7.16}$$

where E_0 is the zero frequency elastic modulus of the material. On the other hand, for $|\mathbf{v}| \to \infty$ we obtain

$$\mathcal{G}(\mathbf{X}, \infty) = \frac{1-\nu^2}{\pi E_\infty} \frac{1}{|\mathbf{X}|} \Theta\left(\frac{|\mathbf{X}|}{h}\right) \tag{7.17}$$

which again is the standard elastic solution but for a stiffer material of Young's modulus E_∞. In the midrange of velocities, the viscoelastic response appears as a retardation effect embedded in the term $\mathbf{v}\tau z$. However, we observe that the main analytical properties of $\mathcal{G}(\mathbf{X}, \mathbf{v})$ remain exactly the same as in the case of elastic materials: in particular, its singular behavior is always of the type r^{-1}. This allows to tackle the problem of inverting Eq. (7.13) by following the same approach already developed by the authors for elastic materials (Putignano et al. 2012a, b). Here we briefly summarize the numerical strategy we exploit to solve the problem. First we discretize the contact domain in N square cells (we actually use an adaptive mesh (Putignano et al. 2012a, b); then, assuming that in each square cell the normal stress

σ is constant and equal to $\sigma_k = \sigma(\mathbf{X}_k)$ where $\mathbf{X}_k$ is the position vector of the center of the square cell D_k, the normal displacement $u_i = u(\mathbf{X}_i)$ at the center of the ith square cell is

$$u_i = -\frac{1-\nu^2}{\pi} \sum_{k=1}^{N} \sigma_k \left\{ \Theta\left(\frac{|\mathbf{X}_i - \mathbf{X}_k'|}{h}\right) \frac{1}{E_\infty} \int_{D_k} d^2X' \frac{1}{|\mathbf{X}_i - \mathbf{X}'|} \right. \tag{7.18}$$
$$\left. + \int_0^{+\infty} d\tau C(\tau) \int_{0^+}^{+\infty} dz \exp(-z)\, \Theta\left(\frac{|\mathbf{X}_i + \mathbf{v}\tau z - \mathbf{X}_k'|}{h}\right) \int_{D_k} d^2X' \frac{1}{|\mathbf{X}_i + \mathbf{v}\tau z - \mathbf{X}'|} \right\}$$

In Eq. (7.18) the term $\int_{D_k} d^2X' |\mathbf{X}_i - \mathbf{X}'|^{-1}$ and the quantity $\int_{D_k} d^2X' |\mathbf{X}_i + \mathbf{v}\tau z - \mathbf{X}'|^{-1}$ can be easily calculated by exploiting Love's solution for elastic materials (Johnson 1985), as shown in (Putignano et al. 2012b). Proceeding in this way the problem is converted into a system of linear equations of the type

$$u_i = L_{ik}(\mathbf{v})\, \sigma_k \tag{7.19}$$

where the response matrix $L_{ik}(\mathbf{v})$ parametrically depends on the velocity $\mathbf{v}$. Equation (7.19) can be easily solved, together with the determination of the real contact area, by employing the iterative scheme based on a nonuniform adaptive mesh already presented by the authors for the case of elastic materials (Putignano et al. 2012a, b).

As mentioned before, one of the aims of our study is related to the calculation of the viscoelastic contribution to friction. To this end, we observe that the friction force is easily determined recalling that the energy per unit time W provided by the external tangential applied force F_T must balance the energy per unit time dissipated as a consequence of viscoelastic response of the material. Hence, in steady-state conditions, we write

$$W = F_T v = -\int d^2x \sigma(\mathbf{x} - \mathbf{v}t) \frac{\partial u(\mathbf{x} - \mathbf{v}t)}{\partial t} = \int_\Omega d^2X \sigma(\mathbf{X}) \mathbf{v} \cdot \nabla u(\mathbf{X}) \tag{7.20}$$

where Ω is the contact domain. Because of isotropy, we may assume without any loss of generality $\mathbf{v} = v\mathbf{i}$, where $\mathbf{i}$ is the unit vector of the X-axis, and write the final relation as

$$F_T = \int_\Omega d^2X \sigma(\mathbf{X}) \frac{\partial u}{\partial X} \tag{7.21}$$

The friction coefficient is then calculated as $\mu = F_T / F_N$ where F_N is the external applied load.

Numerical Results

Now, to show the main feature of viscoelastic contacts and to compare our numerical predictions with available analytical solutions, let us focus on the rolling contact of a rigid sphere of radius $R = 1$ cm moving at constant velocity on a viscoelastic half-space ($h \rightarrow +\infty$) characterized by only one relaxation time. Thus, we write $\mathcal{C}(t) = 1/E_1 \delta(t - \tau)$ to obtain:

$$\mathcal{J}(t) = \mathcal{H}(t)\left[\frac{1}{E_0} - \frac{1}{E_1}\exp(-t/\tau)\right] = \mathcal{H}(t)\left\{\frac{1}{E_\infty} + \frac{1}{E_1}\left[1 - \exp(-t/\tau)\right]\right\} \tag{7.22}$$

Specifically, let us employ the following numerical values: $E_\infty = 10^7$ Pa, $E_\infty/E_0 = 10$, and $\tau = 0.01$ s. In Fig. 7.5 for a fixed value of the indentation (i.e., rigid displacement) $\delta = 0.1$ mm of the sphere into the half-space, and for increasing values of the dimensionless velocity $v\tau/a_0$ with $a_0 = \sqrt{R\delta}$ being the classical Hertzian value of the contact radius, the interfacial pressure $p(\mathbf{X}) = -\sigma(\mathbf{X})$ distribution along plane of symmetry $Y = 0$ and the shape of the contact area are shown.

It is interesting to point out that, given the value of the penetration $\delta = 0.1$ mm, coherently with the physical nature of the problem, the Hertzian solution is recovered both at very low speed, when the material enters the so-called rubbery region and behaves as a soft elastic solid, and at very high speed when the material enters the glassy region and is again elastic but much stiffer. Recalling that we have fixed the penetration δ, the aforementioned limiting cases are characterized by the same displacement distributions and contact areas but different interfacial pressure values. In particular, given a ratio $E_\infty/E_0 = 10$, the value of interfacial pressure in the limiting case of very high rolling speed is expected to be 10 times larger than in the case of zero rolling velocity. This is, indeed, confirmed in Fig. 7.5. For intermediate values of the dimensionless speed $v\tau/a_0$, the pressure distribution shows an asymmetric profile with a peak closer to the leading edge, and a pressure center (red spot in Fig. 7.5) displaced toward the leading edge of the contact. Moreover, the contact area shrinks in the direction of the rolling speed. For low-speed values, these effects increase in magnitude as the speed v increases. However, when the excitation frequency $f_c \approx v/(2a_0)$ approaches the value τ^{-1}, viscoelastic effects must reach their maximum level (see Fig. 7.5 for $v\tau/a_0 = 1.7$) and start to decrease as v is further increased. Figure 7.6 shows the 3D interfacial pressure distributions at $v\tau/a_0 = 1.7$, i.e., at the maximum of contact area shrinkage.

Figure 7.7 compares the viscoelastic displacement (solid line) with the elastic one (dashed line) and shows that the former presents, as expected, a pronounced "detachment" at the trailing edge of the contact.

Now, using Eq. (7.21), we can calculate the rolling friction coefficient $\mu = F_T/F_N$ as a function of $v\tau/a_0$. Figure 7.8 reports the results of our calculations at fixed $\delta = 0.1$ mm. As expected the friction coefficient shows a bell-shaped curve (in agreement with Hunter 1961; Persson 2010; Panek and Kalker 1980), which confirms the strong dependence of the friction coefficient on the viscoelastic response spectrum

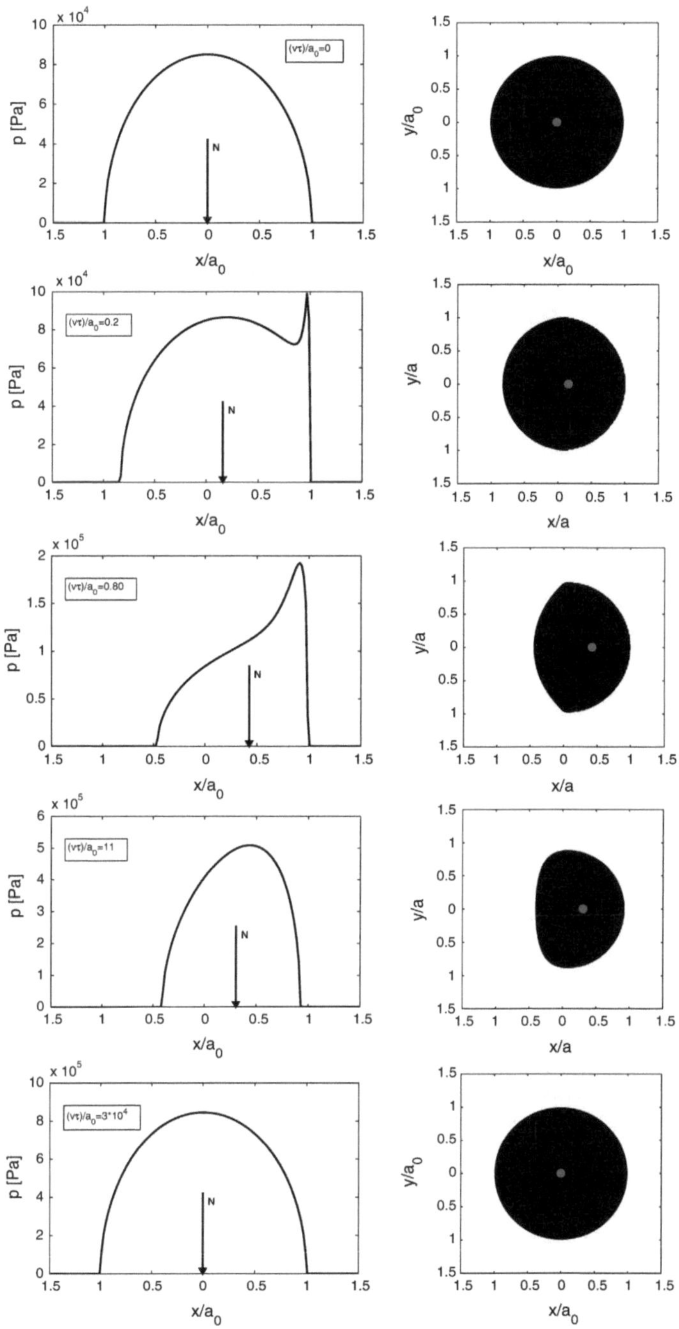

Fig. 7.5 The interfacial pressure distribution $p(X, Y = 0)$ (on the left), and contact area (on the right), for a viscoelastic material with one relaxation time $\tau = 0.01$ s, and $E_\infty = 10^7$ Pa, $E_\infty / E_0 = 10$. Results are reported for penetration $\delta = 0.1$ mm and different values of the dimensionless speed $v\tau / a_0$. The red spot in the figure represents the center of pressure distribution

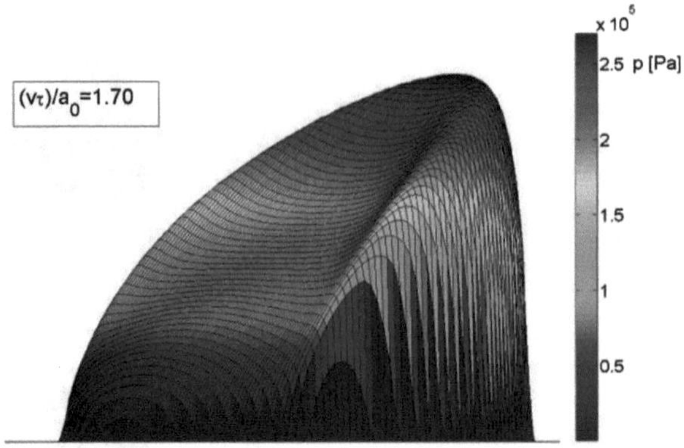

Fig. 7.6 The interfacial pressure distribution for a viscoelastic material with $E_\infty = 10^7$ Pa, $E_\infty/E_0 = 10$, and one single relaxation time $\tau = 0.01$ s. Calculations are reported at fixed penetration $\delta = 0.1$ mm. The dimensionless speed is $v\tau/a_0 = 1.7$

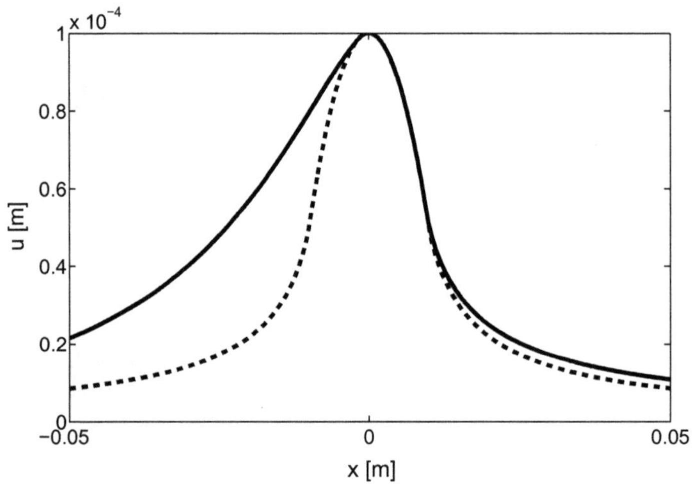

Fig. 7.7 The interfacial displacement distributions for a viscoelastic material with one relaxation time $\tau = 0.01$ s, $E_\infty = 10^7$ Pa, and $E_\infty/E_0 = 10$ (solid line), the corresponding elastic—Hertzian— solution (dashed line). Results are shown for $\delta = 0.1$ mm and $v\tau_0/a_0 = 1.7$. Observe the strong shrinkage of the contact area at the trailing edge

of the materials. Indeed, dimensional arguments show that μ must depend on the viscoelastic modulus of the material through the relation

$$\mu \approx \frac{\delta}{\sqrt{A(\omega)}} \frac{\operatorname{Im}[E(\omega)]}{|E(\omega)|} = -\frac{\delta}{\sqrt{A(\omega)}} \frac{\operatorname{Im}[1/E(\omega)]}{|1/E(\omega)|} \tag{7.23}$$

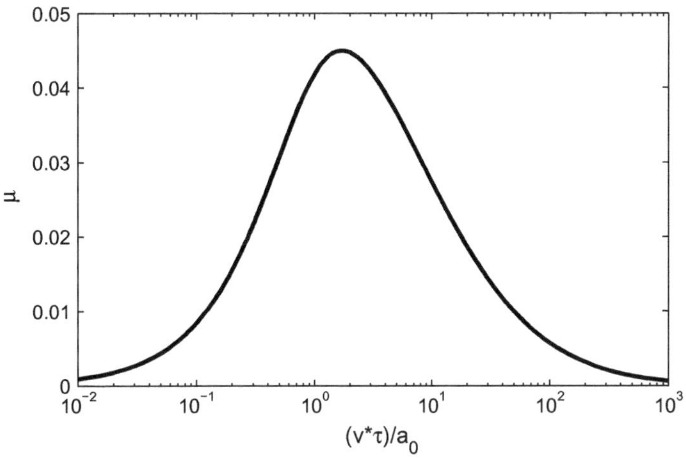

Fig. 7.8 The friction coefficient $\mu = F_T/F_N$ as a function of the dimensionless speed $v\tau/a_0$, for a viscoelastic material with one relaxation time $\tau = 0.01$ s, $E_\infty = 10^7$ Pa, and $E_\infty/E_0 = 10$. Results are shown for fixed penetration $\delta = 0.1$ mm

Now, considering that the linear size $\sqrt{A}$ of the contact area A is always of the same order of magnitude of a_0, one concludes that the maximum value of friction is obtained when the radian frequency $\omega \approx \pi v/a_0$ takes a value close to the one which maximizes $\mathrm{Im}\,[E\,(\omega)]\,/\,|E\,(\omega)|$. This indeed occurs when $\omega\tau = \sqrt{E_\infty/E_0}$, i.e., $v\tau/a_0 = \pi^{-1}\sqrt{E_\infty/E_0}$. In the calculations presented in this book, we have chosen $E_\infty/E_0 = 10$, so that one obtains that the maximum friction value is obtained when $v\tau/a_0 \approx 1$ in very good agreement with Fig. 7.8. Moreover, from Eq. (7.23), one should expect that at very low and very high speeds when the material enters the rubbery and glassy regions, respectively, the friction coefficient μ must disappear as, indeed, shown in Fig. 7.8.

One of the main advantages of the numerical boundary approach presented above is the possibility of managing a wide spectrum for the surface roughness. As proposed in Carbone and Putignano (2014), let us focus on the paradigmatic case of the contact of a rigid rough fractal surface sliding over a viscoelastic half-space ($h \to +\infty$) characterized by one relaxation time. In particular, we employ for the viscoelastic material the values of $E_\infty = 10^7$ Pa, $E_\infty/E_0 = 3$, and $\tau = 0.01$ s. As for the rough surface, here self-affine fractal surfaces, numerically generated by means of the spectral method described in Putignano et al. (2012b), are employed. These surfaces have spectral components in the range $q_0 < q < q_1$, where $q_0 = 2\pi/L$, the side of the square computational cell is $L = 0.01$ m, $q_1 = Nq_0$ and N number of scales (or wavelengths). In particular, the results shown in this section are obtained with $N = 64$.

In Fig. 7.9, we analyze the viscoelastic friction as a function of the dimensionless speed $\xi = v\tau_0/L$ for a fixed normal load $F_N = 0.30$ N. As expected, we have again a bell-shaped curve that vanishes for very low and very high speeds, i.e., when the

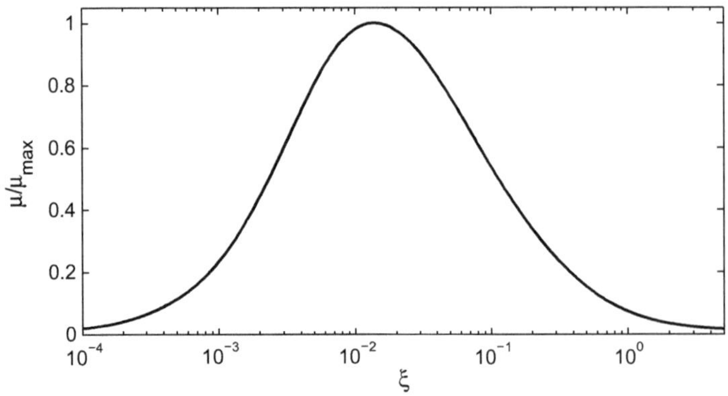

Fig. 7.9 Viscoelastic friction coefficient as a function of the dimensionless sliding speed ξ for a constant normal load $P = 0.30$ N

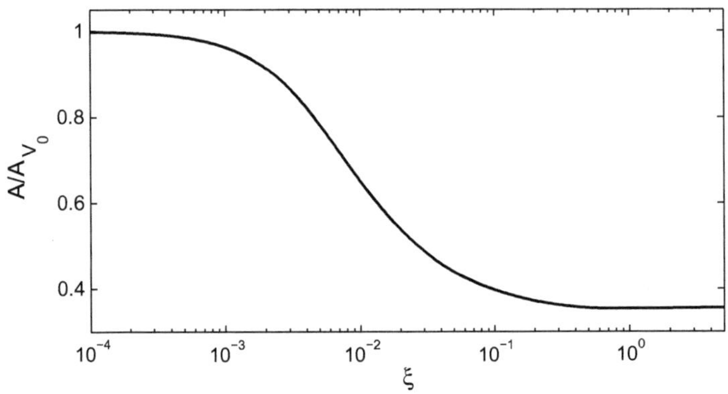

Fig. 7.10 Dimensionless contact area A/A_{v0} as a function of the dimensionless sliding speed ξ for a constant normal load $P = 0.30$ N

solid behaves as an elastic material. The maximum of the curve corresponds, instead, to maximum of the imaginary modulus $\mathrm{Im}\,[(E(\omega)]$.

We can, now, focus on other aspects dealing with the sliding contacts of rough surfaces and, in particular, with the contact area. Specifically, our analysis starts from a quantitative analysis of the real contact area A. In Fig. 7.10, for a nominal contact pressure $\sigma_0 = F_N/L^2 = 3$ kPa, we study, as a function of the dimensionless speed ξ, the ratio A/A_{v_0} between the real contact area A and the real contact area in stationary conditions $A_{v_0} = A(\xi = 0)$. As expected, due to the viscoelastic stiffening, the contact area decreases as the dimensionless sliding velocity ξ is increased. Interestingly, if we focus our attention on the ratio $A\,(\xi)\,/A_{v_0}$, we expect that this quantity approaches the value $E_0/E_\infty = 1/3$ at large sliding velocity. This is clearly shown in Fig. 7.10 and is a consequence of the direct proportionality of the intimate contact area on the ratio between the nominal applied pressure and the composite

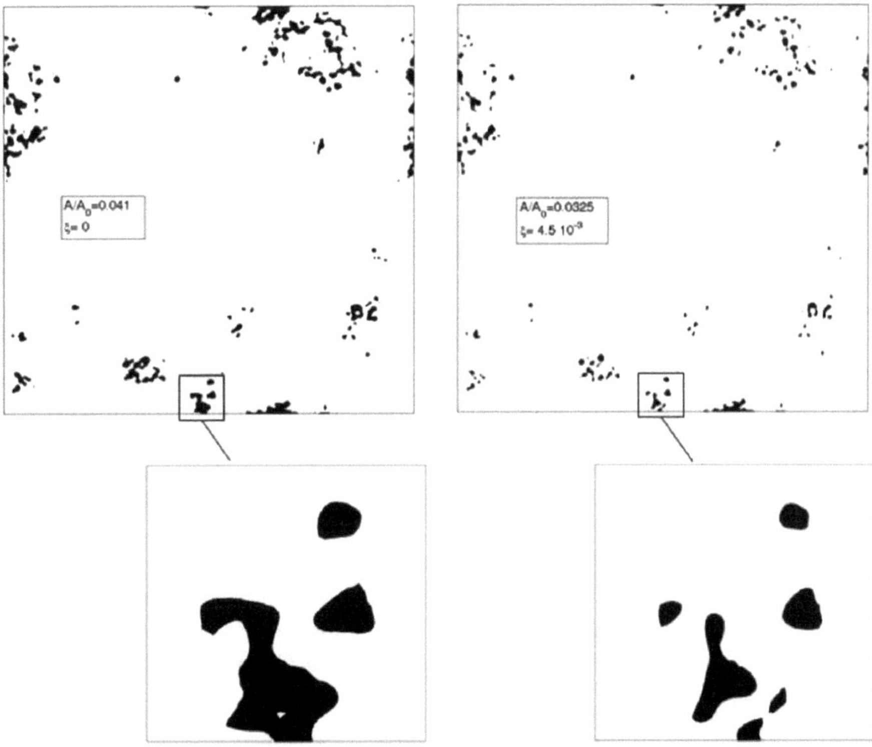

Fig. 7.11 Real contact area for $\xi = 0$ (*on the left*) and $\xi = 4.5\ 10^{-3}$ (*on the right*), given a constant normal load $P = 0.30$ N. The same region is extracted to enlighten anisotropic effects

elastic modulus (see Persson 2001; Greenwood and Williamson 1966; Putignano et al. 2012a), i.e.,

$$\frac{A}{A_0} = \frac{\kappa}{\sqrt{\langle \nabla h \rangle^2}} \frac{\sigma_0}{E^*} \tag{7.24}$$

where A_0 is the nominal contact area, E^* is the composite Young's modulus $E^* = E/\left(1 - \nu^2\right)$ and κ is a constant proportionality coefficient proved to be very close to 2 (Putignano et al. 2012a).

For intermediate values of ξ, viscoelastic effects have a prominent importance. Figure 7.11 shows that, besides the decrease in contact area, increasing ξ leads a marked shrinkage of the contact area at the trailing edge. This is particularly evident in the magnified views (Fig. 7.11). Therefore, despite the isotropy of the rigid randomly rough surface, the interfacial displacement field of the viscoelastic solid (i.e., the shape of the deformed surfaces) will show a certain degree of anisotropy.

To quantify the degree of anisotropy of the deformed surface in a certain range of wave vectors $\zeta_1 q_0 < |\mathbf{q}| < \zeta_2 q_0$, one can use the (symmetric) anisotropy tensor defined as

$$\mathbf{M}(\zeta_1, \zeta_2) = \int_{\zeta_1 q_0 < |\mathbf{q}| < \zeta_2 q_0} d^2 q \mathbf{q} \otimes \mathbf{q} C(\mathbf{q}) \tag{7.25}$$

where $C(\mathbf{q}) = (2\pi)^{-2} \int d^2 x \langle u(\mathbf{0}; \zeta_1, \zeta_2) u(\mathbf{x}; \zeta_1, \zeta_2) \rangle \exp(-i\mathbf{q} \cdot \mathbf{x})$ is the power spectral density of the filtered deformed surface $u(\mathbf{x}; \zeta_1, \zeta_2)$ (the symbol $\langle\rangle$ stands for the ensemble average). Observe that the quantity $M_{ij} = \int_{\zeta_1 q_0 < |\mathbf{q}| < \zeta_2 q_0} d^2 q q_i q_j C(q)$, with i and $j = 1, 2$, corresponds to the second-order moments of the power spectral density of the filtered surface, i.e., $M_{11} = m_{20} = \langle u_x^2 \rangle$, $M_{22} = m_{20} = \langle u_y^2 \rangle$, $M_{12} = m_{11} = \langle u_x u_y \rangle$, where $u_x = \partial u / \partial x$, $u_y = \partial u / \partial y$. Incidentally, we observe that if the range $[\zeta_1, \zeta_2]$ is too wide, in order to balance the dominant contribution of large wave vectors, it would be preferable to define the anisotropic tensor as

$$\mathbf{M}(\zeta_1, \zeta_2) = \int_{\zeta_1 q_0 < |\mathbf{q}| < \zeta_2 q_0} d^2 q \frac{\mathbf{q} \otimes \mathbf{q}}{|\mathbf{q}|^2} C(\mathbf{q}) \tag{7.26}$$

However, in our case, the two definitions do not lead to significantly qualitative differences and we will use formulation (7.25). Furthermore, associated to the symmetric tensor $\mathbf{M}$ one can conveniently use the quadratic form $Q(\mathbf{x}) = M_{ij} x_i x_j$. Assuming $x = r \cos\theta$ and $y = r \sin\theta$, one easily obtains

$$Q(\mathbf{x}) = r^2 |\nabla h \cdot \mathbf{e}(\theta)|^2 = r^2 m_2(\theta)$$

where $\mathbf{e}(\theta)$ is the unit vector $(\cos\theta, \sin\theta)$ and

$$m_2(\theta) = m_{20} \cos^2(\theta) + 2m_{11} \sin(\theta)\cos(\theta) + m_{02} \sin^2(\theta) \tag{7.27}$$

is simply the average square slope of the a profile obtained by cutting the deformed surface $u(\mathbf{x}; \zeta_1, \zeta_2)$ along the direction θ. The quantity $m_2(\theta)$ can be represented in a polar diagram: a circumference means isotropy; otherwise, the surface is anisotropic.

To characterize the anisotropy of the deformed surface, one can define the degree of anisotropy as the ratio $\gamma = m_{2\min}/m_{2\max}$ between the minimum $m_{2\min}$ and the maximum $m_{2\max}$ eigenvalues of the tensor $\mathbf{M}$, and the principal directions of anisotropy through the eigenvectors of the tensor $\mathbf{M}$, e.g., by the value of the angle θ_P which maximize $m_2(\theta)$, i.e., $m_2(\theta_P) = m_{2\max}$. In this analysis, we use $\zeta_1 = 1$ and $\zeta_2 = N = 64$ to obtain the polar diagram shown in Fig. 7.12. We notice that, unlike what occurs at zero sliding speed, where the deformed surface is almost perfectly isotropic (a circular polar diagram is shown in Fig. 7.12 with $\gamma = 0.93$, very close to 1), at nonzero sliding velocity, e.g., for ξ equal to 1.17×10^{-2}, which correspond to the maximum degree of anisotropy, $\gamma = 0.70$. Interestingly, the speed $\xi = 1.17 \times 10^{-2}$, for which we find the greatest anisotropy, is the value where the maximum viscoelastic friction is found. This confirms that anisotropy and friction are strictly connected and both related to the loss modulus $\mathrm{Im}[(E(\omega)]$, i.e., to the

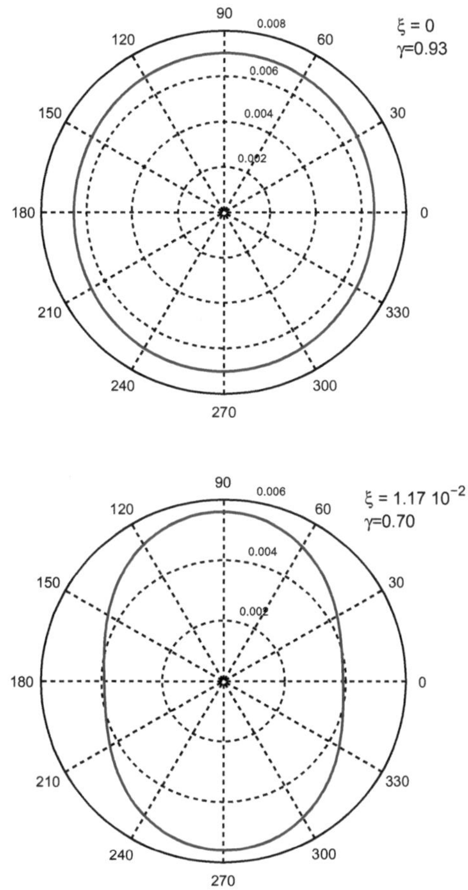

Fig. 7.12 Polar plots of $m_2(\theta)$ for $\xi = 0$ and $\xi = 1.17\ 10^{-2}$, given a constant normal load $P = 0.30$ N

parameter governing the viscoelastic friction. Furthermore, the principal direction θ_P is almost perfectly parallel to the sliding speed (i.e., to x-axis). This is fully consistent with results shown in Fig. 7.11, where the contact area looks stretched along the y-axis.

Beyond the Half-Space Assumption: Effects of Contacting Layers of Finite Thickness

When contact areas have a characteristic length comparable with the bodies' thickness, the half-space assumption adopted in the previous analysis has to be removed. Indeed, as shown in the literature (see, e.g., Carbone and Putignano 2014; Putignano et al. 2015) there occurs a significant modification in the contact stiffness and,

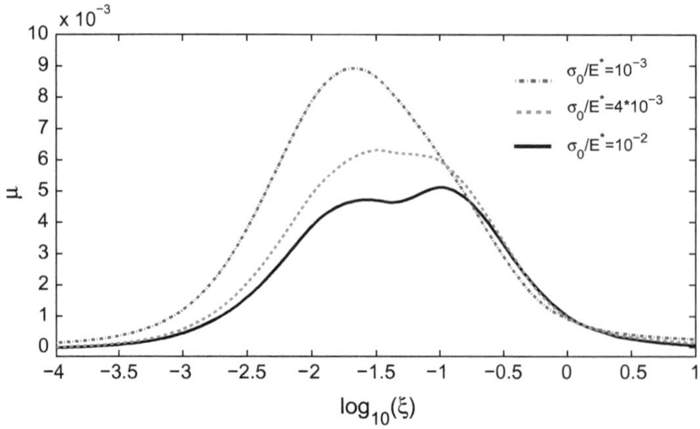

Fig. 7.13 Viscoelastic friction coefficient μ as a function of the dimensionless sliding speed $\xi = v\tau/L$ for the constant thickness $h/L = 0.06$ and different values of the normal load σ_0/E^*: $\sigma_0/E^* = 10^{-2}$ (black curve), $\sigma_0/E^* = 4\,10^{-3}$ (green dotted curve), and $\sigma_0/E^* = 10^{-2}$ (red dotted curve). Note that h and L denote the thickness of the thin layer and the size of the computational domain, respectively. (color figure online)

therefore, in all the quantities marking the contact solution, including contact areas, separation, and friction. The nature of these changes may be different and depends on the boundary conditions imposed on the system: for example, a rigid foundation supporting the deformable layer entails a system stiffening, whereas a free layer supported by a constant pressure has a larger compliance in comparison with a half-space. In the case of thin layers, what is independent on the several boundary conditions is the reduction in the amount of material potentially capable of dissipating energy due to viscoelastic hysteresis. Fig. 7.13 from example shows the friction coefficient as a function of the dimensionless speed for a fixed thickness value and for different normal loads. We observe that the friction coefficient curve has the usual bell-shaped behavior only for the lowest normal load, whereas for largest load value a completely different trend, with two peaks, is observed. It is also surprising to observe that the friction coefficient decreases with the normal load as more dissipation is expected when the deformed volume increases due to viscoelastic losses. To fully understand such an effect, we should focus on the physics governing viscoelastic friction: this is due to the dissipation happening in the material volume deformed during the sliding motion. Now, when the contact layer has a finite thickness, the amount of material, which can be deformed and, consequently, can dissipate, is finite: once the region available for the dissipation to take place is saturated, no further increase in the friction force can be obtained by increasing the normal force. This is clarified in Fig. 7.14, where a schematic shows that in the layered case the region capable of dissipating is saturated and the saturation depends on the relative sliding speed for the viscoelastic case.

In order to explore this behavior, it is necessary to introduce a characteristic length, l_{eq}, which qualitatively captures the extent of the dissipative region; this allows a direct comparison with the other key length scale of the problem, namely

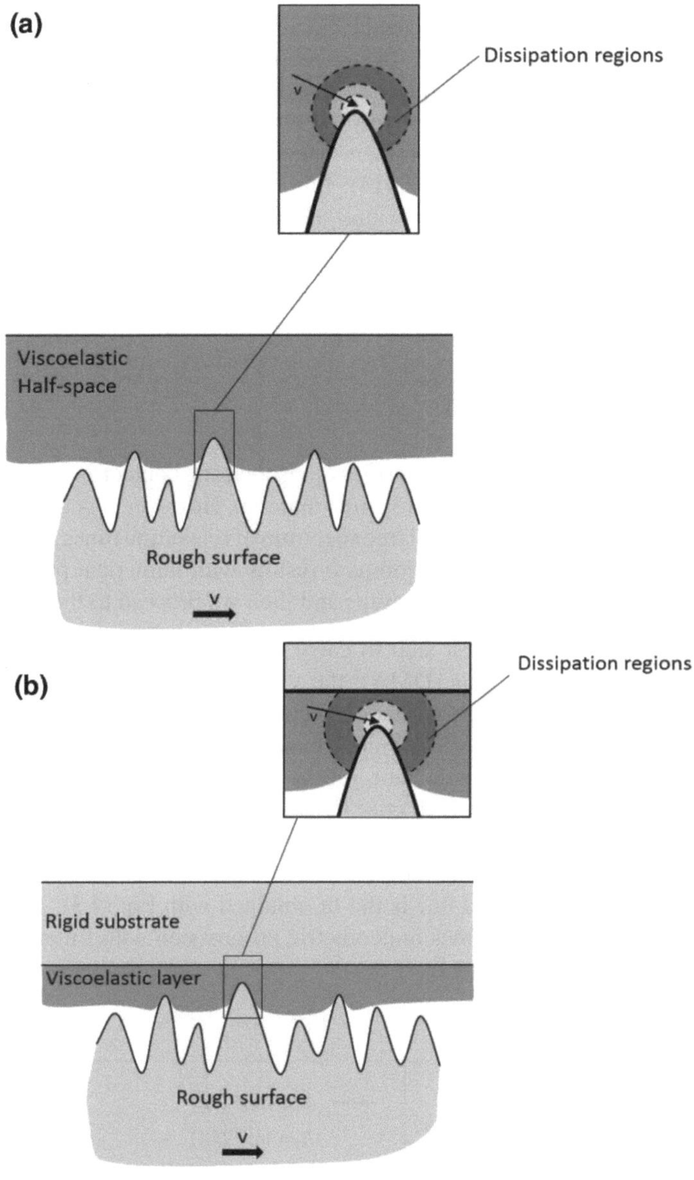

Fig. 7.14 Schematic of the viscoelastic friction for half-space and for a thin layer

the thickness, h, of the viscoelastic layer. This characteristic length should be ideally defined so that it can capture the saturation of the material's capability to dissipate. In particular, if l_{eq} is thought as a mono-dimensional measure of the volume over which dissipation takes place, when the ratio l_{eq}/h is smaller than 1 there is still material capable of dissipating; conversely, as soon as l_{eq}/h becomes greater than 1, the dissipative region gets saturated. In contact problems characterized by regular or smooth surfaces, it would be straightforward to define l_{eq} as the contact characteristic wavelength or contact width, respectively; however, it is less easy to define a unique wavelength for a specific contact problem when dealing with multiscale rough surfaces. In Carbone and Putignano (2014); Putignano et al. 2015 l_{eq} is defined as $l_{eq} = \sqrt{A_m}$, where A_m is the mean value of each individual contact cluster. Such a dimensionless quantity can be employed as a measure of the influence of finite thickness on rough contact mechanics.

Experimental Testing

The last Section has shown the peculiarities of viscoelastic contact and, to this aim, we have focused on an ideal one relaxation time material. However, it is well-known that real materials are characterized by a large spectrum of relaxation times. In this section, we study viscoelastic rubbers and compare results with numerical predictions. We present the particular experimental setups and then we proceed to the results.

Materials To this aim, the first step to carry out is to characterize, by means of a dynamic mechanical analysis (DMA), the viscoelastic materials experimentally tested. Such a mechanical characterization is carried out over a set range of frequencies and temperatures. Let us provide some examples.

The first one is a styrene butadiene rubber copolymer (SBR), whose dynamic analysis can be found in Putignano et al. (2013). Therefore, the real part $E_1(\omega) = \mathrm{Re}[E(\omega)]$ and the imaginary part $E_2(\omega) = \mathrm{Im}[E(\omega)]$ of the measured viscoelastic modulus $E(\omega)$ at environmental temperature of 12 °C are shown in Fig. 7.15 (see points in the figure). The solid line is the fit obtained with Eq. (7.3). The best fit is obtained with 34 relaxation times in geometric progression with Euler's number as common ratio, i.e., $\tau_{k+1}/\tau_k = e$.

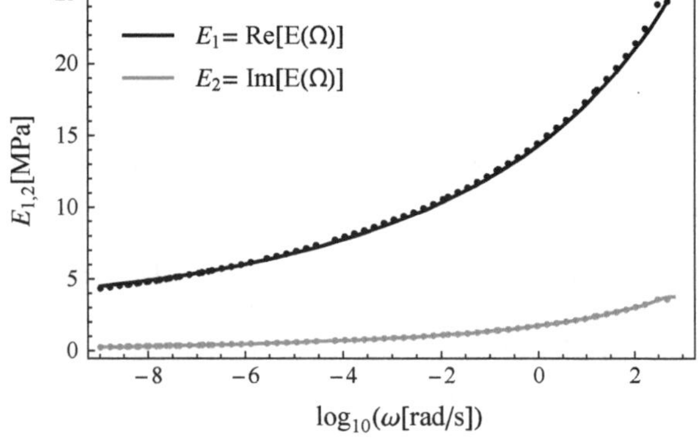

Fig. 7.15 The real part $E_1 = \mathrm{Re}[E(\omega)]$ (black line) and the imaginary part $E_2 = \mathrm{Im}$ (blue line) of the viscoelastic modulus $E(\omega)$ of the SBR rubber samples at the room temperature of 12 °C. Points represent the measured values, and the solid lines represent the fit with Eq. (7.3)

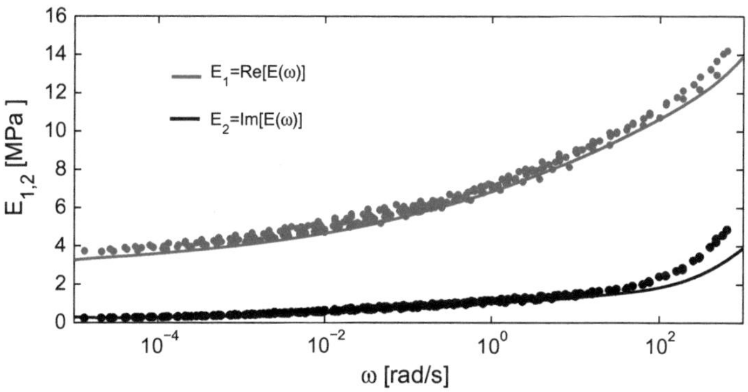

Fig. 7.16 The real part $E_1 = \mathrm{Re}\,[E\,(\omega)]$ (red line) and the imaginary part $E_2 = \mathrm{Im}$ (blue line) of the complex modulus $E\,(\omega)$ at 30°C. Points represent the measured values, and the solid lines represent the fit obtained by using Eq. (7.3) (color figure online)

The second material is a nitrile rubber (provided by The Precision Plastic Ball Co Ltd.). In this case, we have directly carried out by means of a Q800 Dynamic Mechanical Analyzer (DMA) manufactured by TA Instruments. In detail, the test was carried on rubber strips with cross section 2 mm × 5.5 mm and length 30 mm. The preload at which the measurements are carried out is 7.5 N. The range of frequency is from 0.01 to 25 Hz and temperature varies from 0 to 100 °C with 10 °C step. In order to extend the data to the very low frequencies needed to represent the test conditions, as described in Christensen (1982), Ferry (1980), the well-known WLS relations were employed to shift the data measured at different temperatures and obtain the viscoelastic spectrum of the rubber, shown in Fig. 7.16. Here, the experimental data, obtained in such a way, are fitted with Eq. (7.3). We observe that, as suggested in literature Kessler (2004), the discrepancy between measured and fitted data at high frequency can possibly be attributed to slippage occurring in the sample grip of the DMA equipment.

A final example deals with a thermoplastic and, in particular, with the poly(methyl methacrylate) (PMMA). This test is again carried out in tensile conditions to obtain data over a set interval of frequencies and temperatures. The characterization is performed on rubber strips, having a cross section of 2 mm × 5.5 mm and a length of 35 mm, over a frequency interval between 0.1 and 10 Hz and temperature varying from −50 to 150 °C with 10 °C step. Thus, as we have seen, it is possible to build up a master curve for the complex viscoelastic modulus over a much extended frequency range. This is what is shown in Fig. 7.17, where the experimental data are plotted in blue—the squares and the circles refer, respectively, to the imaginary and the real part of the viscoelastic modulus. In the same figure, we fit the experimental data using the discretized version of Eq. (7.2) with a spectrum of relaxation times being assumed to be in geometric progression with ratio equal to $\tau_{k+1}/\tau_k = 1.5\,e$. A good agreement is found over the entire frequency range between the experiment and the theoretical fitting.

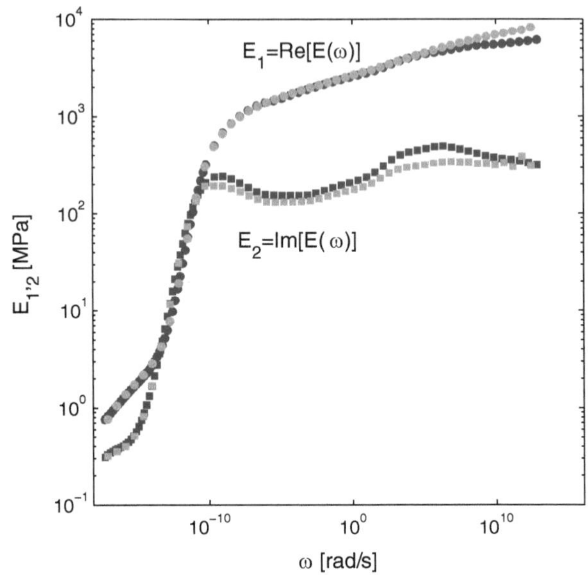

Fig. 7.17 The real part $E_1 = \text{Re}\,[E\,(\omega)]$ (squares) and the imaginary part $E_2 = \text{Im}$ (circles) of the complex modulus $E\,(\omega)$ at 30 °C. The blue color refers to the measured values, and the red is used for the fit obtained by using Eq. (7.3) (color figure online)

Penetration and contact area Among the main quantities to measure experimentally, we have penetration and contact area. To this end, in Putignano et al. (2013), a conventional EHD ball and disk rig (PCS Instruments Ltd., Acton, UK) is employed: on this rig, a 19.05 mm diameter nitrile rubber ball (provided by The Precision Plastic Ball Co Ltd.), is loaded against a glass disk, as shown in Fig. 7.18. The ball specimen has no shaft and is located on a roller carriage, so that when loaded, the disk rotation drives under pure rolling contact conditions.

Additional components have been incorporated in the conventional EHL rig setup to enable the accurate measurement of the displacements. In particular, a T-shaped stem is attached to the ball loading system supporting the ball roller carriage. This enables a laser displacement sensor (LK-G32 produced by Keyence Ltd.) to measure the vertical displacement corresponding to the penetration of the ball specimen. This approach is reasonable since the loading system is considerably much stiffer than the rubber ball specimen. Furthermore, the high resolution of the displacement sensor (0.05 μm) enables the small variations in penetration that occur due to changes in rolling speed to be measured.

Contact area measurements are obtained from images taken with a camera located above the glass disk (see Fig. 7.18). Due to the very low modulus of the rubber ball, contact area diameters are likely to be sizeable—i.e., several millimeters large—even under low loads. To capture such areas, an SLR camera (a Canon EOS 500D) and lens (a 7 Tamron AF 28–75 mm f/2.8 XR Di LD) are used.

Once the images are collected, a suitable MATLAB algorithm is employed to digitize the contact area and extract its dimensions as they evolved due to viscoelastic effects. After importing the black-and-white snapshots, each picture, managed

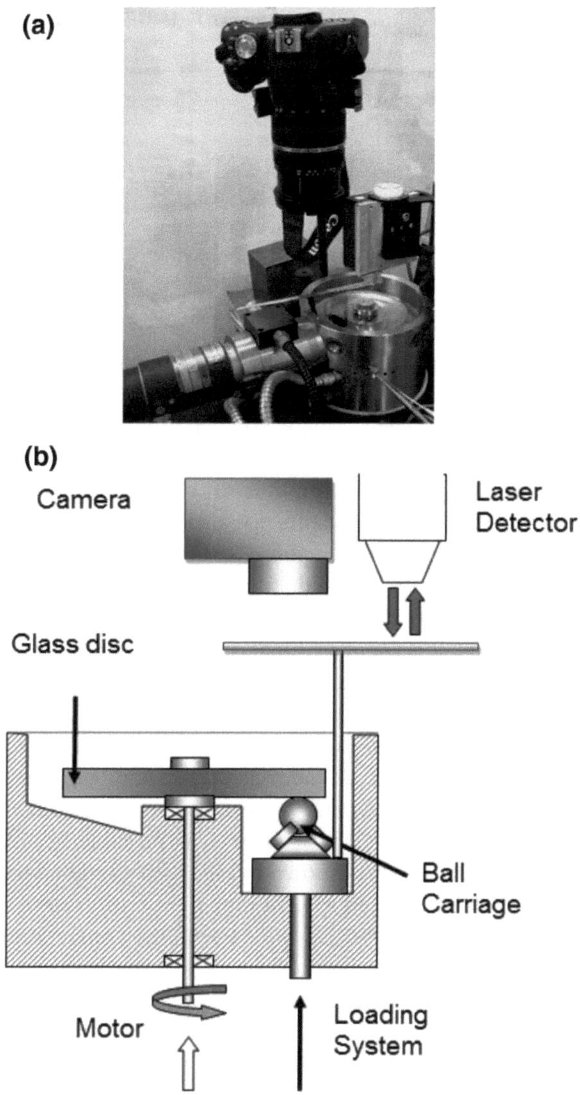

Fig. 7.18 EHL test rig: **a** photograph of the experimental configuration; **b** schematic of the original setup and additional features used to accurately measure surface displacements

as a matrix in the MATLAB environment, is Fourier-transformed and undergoes a filtering process, aimed at detecting the contact borders. The transition zone (i.e., the border between contacting and non-contacting regions) could then be found. Finally, the contact area was calculated knowing pixel size of the images (the latter being obtained from an image of a specimen of known dimensions, positioned at the contact interface).

Now, once we have defined the experimental setup, we can carry out the measures. The first part of our experimental investigation involves measuring the variation in penetration as a function of the rolling speed. As in previous researches (Carbone and Putignano 2013), we observe that an increase in rolling speed causes a signifi-

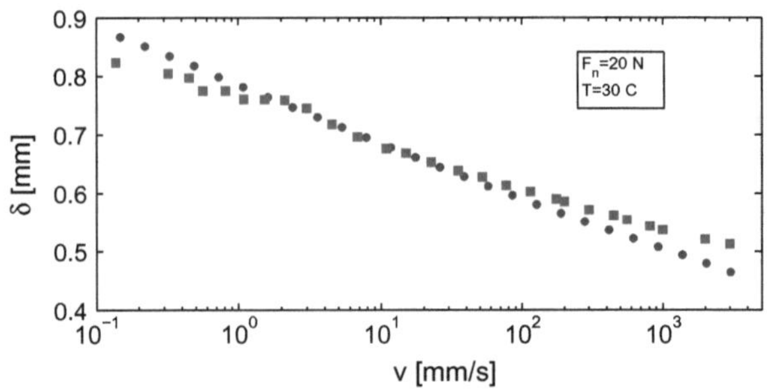

Fig. 7.19 Contact penetration versus rolling speed for a constant normal load of $F_N = 20$ N and a temperature $T = 30\,°\mathrm{C}$. Red squares refer to experimental results, and blue circles to numerical predictions (color figure online)

cant stiffening of the viscoelastic specimen. Therefore, if the applied normal load is kept constant (e.g., equal to $F_N = 20$ N), the penetration should show a decreasing trend with the speed. This is confirmed in Fig. 7.19, which shows the measured and predicted penetrations with varying speed. Here, each experimental data point (red square in Fig. 7.19) represents an average of 15 measurements, with a scatter of less than 5%.[1] These results not only confirm the expected trend but also show very good agreement with the numerical predictions (blue circle in Fig. 7.19), over a very large speed range. Indeed, such a strong correlation is observed up to 100 mm/s; however, for greater speeds, experimental values tend to diverge from the numerical predictions. It is suggested that this divergence may be due to thermal heating effects that occur at high speed and cause a softening of the polymer and hence higher than predicted penetration values. Additionally, at such speeds, wear of the polymer ball may be affecting displacement measurements.

The importance of thermal effects can be confirmed by performing thermal microscopy in the contact. As described in Putignano et al. (2014), by means of such an experimental procedure, it is possible to obtain a detailed thermal map of the contact region. As shown in Fig. 7.20, obtained for a normal load $P = 2$ N and a constant rolling speed $v = 54$ mm/s, the temperature in the contact is higher than the room temperature ($T_{room} = 30$ C). In the contact region, the presence of hot spots, which are due to the surface roughness and correspond to asperities in the contact. This phenomenon of frictional heating is well-known in sliding contacts but is also present in rolling contacts.

It is interesting to consider how the temperature influences the contact solution and, specifically, when temperature heating can be neglected or, on the contrary, must be accounted for. The simplest way to approach this problem is to focus on the mean temperature in the contact area since this may enable us to understand when thermal effects become prominent. For instance, Fig. 7.21 shows the mean temperature in the contact area as a function of the rolling speed for fixed a normal load $P = 20$ N. Here, two regimes are clearly distinguished: for low-speed values ($v < 100$ mm/s), temperature increase is negligible, but for higher speeds, there is a rapid growth in the temperature curve. This is coherent with the results obtained in Fig. 7.19, where

[1] Here, scatter is defined as σ/μ, i.e., the ratio between the standard deviation σ and the mean measured value μ.

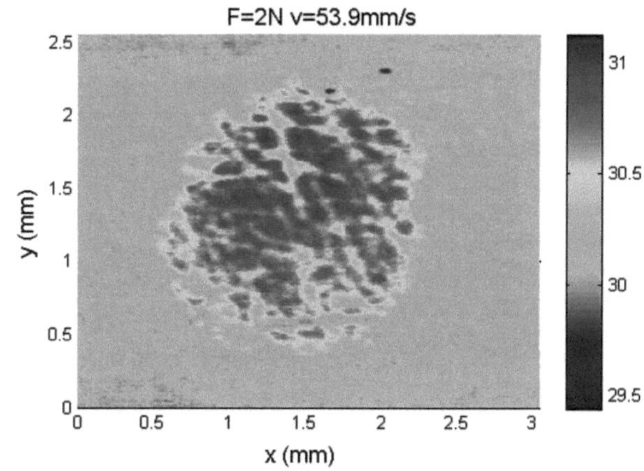

Fig. 7.20 Temperature map for a constant normal load of $F_N = 2$ N, a speed $v = 54$ mm/s and a room temperature $T = 30$ °C

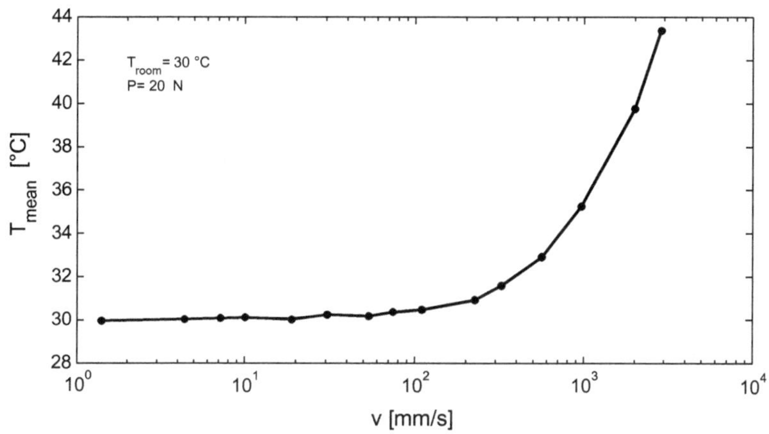

Fig. 7.21 Mean temperature in the contact area as a function of the rolling speed v for a constant normal load of $F_N = 20$ N and a room temperature $T = 30$ °C

the deviation from the athermal numerical model occurs for speeds larger than 100 mm/s.

We can now focus on the analysis of the contact area. In Fig. 7.22, the contact area is shown as a function of the rolling speed: a decreasing trend, due to the stiffening of the material, is evident. It is noteworthy to observe that, despite the simplicity of the proposed setup, it is possible to study a large speed range covering more than four orders of magnitude. Indeed, the test brings to light the viscoelastic effect, which in previous investigations (see, e.g., Johnson 1985; Krick et al. 2012; Lorenz et al. 2011) was partially obscured by, or was inseparable from, other phenomena. This is mainly due to the fact that these other studies either used only pure sliding conditions or were focused on the role of the roughness at low speeds. Actually, the approach

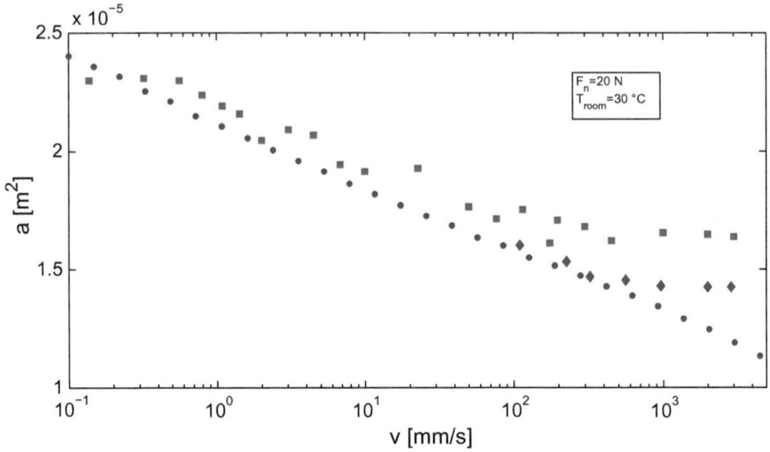

Fig. 7.22 Contact area as a function of the rolling speed v for a constant normal load of $F_N = 20$ N and a room temperature $T = 30$ °C. Red squares refer to experimental outcomes, blue circle to numerical results obtained without any thermal corrections, and green rhomboids to numerical simulations carried out with the viscoelastic response corrected with the temperature (color figure online)

taken in the present work, i.e., measuring contact area under pure rolling conditions, allows the heating and wear effects—that play a critical role also at very low sliding speeds—to be neglected over a larger rolling speed range. As a result, the significant decrease in contact area due to viscoelasticity can be clearly appreciated.

Turning now to the comparison with the numerical simulations, a good agreement is found for rolling speeds up to 100 mm/s, whereas, at higher speeds, numerical outputs are lower than experimental values. This can be explained by heating effects described in the preceding discussion regarding penetration. For this set of measurements, the scatter is less than 8%, which is quite low considering the convoluted nature of the acquisition and images processing procedures.

Viscoelastic Contacts Under Non-Steady-State Conditions: The Case of Reciprocating Motion

In the previous section, we have focused our attention on the case of steady-state relative motion between the viscoelastic solids. However, there may exist conditions more complicated, where contacts occur in viscoelastic regions that have been already deformed and are still relaxing. A case of particular practical interest is the case of reciprocating contact, where a punch oscillates periodically over a viscoelastic foundation. This Section focuses indeed on the Boundary Element formulation and the main results dealing with such loading and motion conditions.

Mathematical Formulation

The proposed formulation builds on the strengths of the boundary element method (BEM) in terms of accurately capturing interfacial stresses and displacements, and, as such, requires the determination of a viscoelastic reciprocating Green's function $G(\mathbf{x}, t)$.

To this end, let us first assume that the interfacial normal stress distribution obeys the law $\sigma(\mathbf{x}, t) = \sigma[\mathbf{x} - \xi_0 \sin(\omega t)]$, i.e., that the shape of normal stress distribution is fixed but moves on the viscoelastic half-space with a sinusoidal law of amplitude $|\xi_0|$ and angular frequency ω. The vector ξ_0 also identifies the direction of the reciprocating motion. Because of linearity and translational invariance, replacing $\mathbf{x} \to \mathbf{x} + \xi_0 \sin(\omega t)$ allows to write the relation between interfacial stresses and displacement as

$$u(\mathbf{x}, t) = \int d^2x' G(\mathbf{x} - \mathbf{x}', t)\, \sigma(\mathbf{x}') \tag{7.28}$$

In order to determine $G(\mathbf{x}, t)$, we recall that the general relation between stress and displacement fields is (Carbone and Putignano 2013):

$$\begin{aligned} u(\mathbf{x}, t) &= \mathcal{J}(0) \int d^2x' \mathcal{G}(\mathbf{x} - \mathbf{x}')\, \sigma(\mathbf{x}', t) \\ &\quad + \int_{-\infty}^{t} d\tau \dot{\mathcal{J}}(t - \tau) \int d^2x' \mathcal{G}(\mathbf{x} - \mathbf{x}')\, \sigma(\mathbf{x}', \tau) \end{aligned} \tag{7.29}$$

where $\mathcal{G}(\mathbf{x})$ and $\mathcal{J}(t)$ are the elastic Green's function and the creep material function, respectively. The symbol "$\cdot$" stands for the time derivative. The creep function is easily linked to the viscoelastic modulus $E(\omega)$ of the material by means of the relation $1/E(\omega) = i\omega\mathcal{J}(\omega)$ (Christensen 1982), where i is the imaginary unit and the Fourier transform of a function $f(t)$ is $f(\omega) = \int dt \exp(-i\omega t)\, f(t)$. The viscoelastic modulus has the general expression $1/E(\omega) = 1/E_\infty + \int_0^\infty d\tau \mathcal{C}(\tau) / (1 + i\omega\tau)$, where E_∞ is a real quantity corresponding to the elastic modulus of the material at very large excitation frequencies. $\mathcal{C}(\tau) > 0$ is usually defined as the creep spectrum, and τ is the relaxation time (Christensen 1982). In order to find $G(\mathbf{x}, t)$ we choose $\sigma(\mathbf{x}, t) = \delta[\mathbf{x} - \xi_0 \sin(\omega t)]$ and, after substituting in Eq. (7.29), we obtain

$$\begin{aligned} G(\mathbf{x}, t) &= \mathcal{J}(0)\, \mathcal{G}[\mathbf{x} - \xi_0 \sin(\omega\tau)] \\ &\quad + \int_{-\infty}^{t} d\tau \dot{\mathcal{J}}(t - \tau)\, \mathcal{G}[\mathbf{x} - \xi_0 \sin(\omega\tau)]\,. \end{aligned} \tag{7.30}$$

The term $\mathcal{G}[\mathbf{x}-\xi_0\sin(\omega t)]$ can be rewritten as

$$\mathcal{G}[\mathbf{x}-\xi_0\sin(\omega t)] = (2\pi)^{-2}\int d^2q\mathcal{G}(\mathbf{q})\,e^{-i\mathbf{q}\cdot[\mathbf{x}-\xi_0\sin(\omega t)]}, \qquad (7.31)$$

where $\mathcal{G}(\mathbf{q})$ is the Fourier transform of the function‘ $\mathcal{G}(\mathbf{x})$. Now, let us observe that $\int d\theta e^{ir\sin\theta}e^{-i\alpha\theta}$ is equal to

$$\int d\theta e^{ir\sin\theta}e^{-i\alpha\theta} = 2\pi\sum_{k=-\infty}^{+\infty}\delta(\alpha-k)\,J_k(r) \qquad (7.32)$$

where $J_k(r)$ is the kth-order Bessel function of the first kind. Consequently, Eq. (7.31) can be cast as

$$\mathcal{G}[\mathbf{x}-\xi_0\sin(\omega t)] = \sum_{k=-\infty}^{+\infty}A_k(\mathbf{x})\,e^{ik\omega t}, \qquad (7.33)$$

In Eq. (7.33), $A_k(\mathbf{x})$ can be written as

$$A_k(\mathbf{x}) = (2\pi)^{-1}\int_{-1}^{1}ds\mathcal{G}(\mathbf{x}-s\xi_0)\,B_k(s) \qquad (7.34)$$

with $B_k(s)$ being equal to $B_k(s)=(-i)^k\,T_k(s)\,B_0(s)$. $T_k(s)$ is the Chebyshev polynomial of the first kind and $B_0(s)=2\left(1-s^2\right)^{-1/2}$, for $|s|\le 1$ and 0 otherwise. Substituting (7.33) in (7.30) we obtain:

$$G(\mathbf{x},t) = \sum_{k=-\infty}^{+\infty}\frac{A_k(\mathbf{x})}{E(k\omega)}e^{ik\omega t} \qquad (7.35)$$

As mentioned above, the function $G(\mathbf{x},t)$ has been obtained under the assumption that the shape of the stress field at the interface, whose general form is $\sigma(\mathbf{x},t)=\sigma[\mathbf{x}-\xi_0\sin(\omega t),t]$, does not change during the reciprocating motion, *i.e.*, $\sigma(\mathbf{x},t)=\sigma[\mathbf{x}-\xi_0\sin(\omega t)]$. Such a condition holds true whenever $a_0/|\xi_0|\ll 1$, where a_0 the characteristic dimension of the contact region, and is equivalent to require that $|\partial\sigma/\partial t|/(|\xi_0\cdot\nabla\sigma|\,\omega)\ll 1$ (see Appendix A for more details). This assumption is justified in the majority of cases of reciprocating contact and is satisfied point-wise almost everywhere within the contact area in the analyses presented in this work. Now, to invert the linear operator in Eq. (7.28) we need a numerical approach which consists in discretizing the contact domain in M square cells. Indeed, assuming that in each boundary element the normal stress σ is constant and equal to σ_j, the normal displacement $u_i = u(\mathbf{x}_i, t)$ at the center $\mathbf{x}_i$ of the ith square can be written as

$$u_i = \frac{1}{N}\sum_{j=1}^{M}\sigma_j\sum_{r=1}^{N} L\left[\mathbf{x}_i-\mathbf{x}_j-\cos\left(\frac{2r-1}{2N}\pi\right)\xi_0\right]\sum_{k=-\infty}^{+\infty}\frac{e^{ik\omega t}(-i)^k}{E(k\omega)}\cos\left[k\left(\frac{2r-1}{2N}\pi\right)\right] \quad (7.36)$$

where $L(\mathbf{x})$ is related to the Love's solution (Johnson 1961). It should be observed that Eq. (7.36) is obtained by applying the Chebyshev–Gauss quadrature rule to the integral term $\int_{-1}^{1} ds T_k(s) L\left[\mathbf{x}_i - \mathbf{x}'_j - s\xi_0\right]\left(1-s^2\right)^{-1/2}$ at M nodes, thus making it easier to achieve the numerical convergence of the problem.

Equation (7.29) can be solved by using the iterative technique developed in Putignano et al. (2012b) for elastic contacts, thus providing contact areas, stresses, and strains. It should be noticed that the method does not require any discretization of the time domain as the time t is treated as a parameter.

Once the solution is known in terms of stresses and strains, following the approach stated in Carbone and Putignano (2013), it is straightforward to calculate the viscoelastic friction force as

$$F_T = \int_D d^2x\sigma(\mathbf{x})\frac{\partial u}{\partial x} \quad (7.37)$$

The friction coefficient is then obtained as $\mu = F_T/F_N$ where F_N is the external applied load.

Finally, we conclude noticing that the formulation, in the current form, does not explicitly account for the role of the tangential tractions at the contact interface. Indeed, this is out of the scope of our work. The purpose here is to determine the normal stresses and the normal displacements distribution, and, on this basis, calculate the viscoelastic friction that is proportional to the volume of deformed material. It is well-known that the normal and tangential contact problems have only a very weak coupling, which is normally neglected. Furthermore, in the case of a rigid body in contact with a soft layer—which can be usually assumed incompressible—it is absolutely rigorous to assert that tractions have no influence on normal pressure and displacements, and, consequently, on the viscoelastic dissipation.

Numerical Results

We study the contact of a rigid sphere of radius R undergoing reciprocating sliding against a viscoelastic material characterized by one relaxation time (being the ratio between the high-frequency modulus and the low-frequency $E_\infty/E_0 = 11$ and the Poisson ratio $\nu = 0.5$). We assume that the center $\mathbf{x}(t)$ of the sphere moves on the viscoelastic half-space following the law $\mathbf{x}(t) = [\xi_0 \sin(\omega t), 0]$. The dimensionless

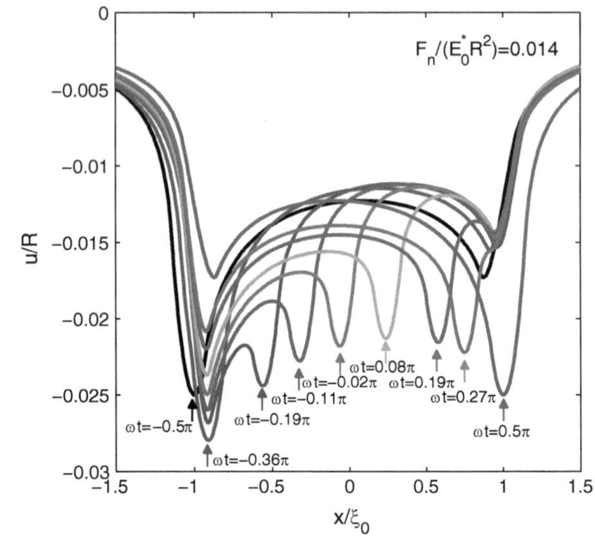

Fig. 7.23 The dimensionless normal displacements $u(x, y = 0)/R$ as a function of the dimensionless abscissa x/ξ_0 for a constant dimensionless normal force $F_n/R^2E_0^* = 0.014$, for an amplitude $\xi_0/R = 1$, and for several values of the dimensionless time $\omega t \in [-\pi/2, \pi/2]$

angular frequency of the reciprocating motion is $\omega\tau = 5$, being τ the relaxation time of the viscoelastic material.

Figure 7.23 shows the evolution of the dimensionless displacements, $u(x)/R$, at the center of the contact as a function of x/ξ_0 and for a specific dimensionless applied normal load $F_n/R^2E_0^* = 0.014$, and $\xi_0/R = 1$. Results are shown for different values of $\omega t \in [-\pi/2, \pi/2]$. An arrow refers, in each case, to the current position of the sphere. At $\omega t = -\pi/2$ the sphere has just reached the left dead-point and starts moving from left to right. Upon reversal of the sliding direction, and for $\omega t \leq -0.36\pi$, a marked increase in the dimensionless penetration at the center of the sphere is observed. This is due to the fact that, although the speed is increasing, it is still too low to cause a significant stiffening of the material, and the sphere is also moving over a portion of the viscoelastic half-space that has not yet had the time to relax after the previous contact of the rigid body. As the sliding speed increases, a non-negligible stiffening of the material and a marked decrease in the penetration are observed (see displacement in correspondence to the arrow). This is clearly shown by curves at $\omega t = -0.19\pi, -0.11\pi, -0.02\pi$, which also show additional deformation peaks, one at the left and one to the right of the arrow: this is the result of the interplay between the deformations, induced by the indenter as it moves to the right, and the original not yet fully relaxed footprints left by the sphere at preceding times. For $0 < \omega t < \pi/2$, the sliding speed begins to decrease and the material softens again, thus leading to an increase in penetration. It is now possible to justify the occurrence of three different deformations peaks within the track when the sphere is moving between the two dead-ends: one corresponds to the current position of the sphere and the other two are located close to the left and right dead-points, respectively, and are the result of the material inability to fully recover the viscoelastic deformations during a period of time comparable to the period $T = 2\pi/\omega = 6.28$ s of the reciprocating motion (recall that the relaxation time is $\tau = 5$ s).

The merging or separation of the previous and current sphere footprints, which takes place close to the dead-points of the reciprocating motion, has a significant effect on the interfacial normal stress distribution. This is clearly shown in Fig. 7.24, which depicts the evolution of the pressure distribution and shows the shape of the contact area. Let us first observe that at $\omega t = -\pi/2$, i.e., when the sliding speed goes to zero, the contact area as well as the interfacial normal stress distribution is characterized by an asymmetric shape. The observed asymmetry and, in particular, the presence of a peak on the left of the contact patch is a consequence of the viscoelastic time delay which prevents the material to relax immediately when the sliding speed vanishes. As the sphere starts moving to the right, such a peak cannot disappear suddenly but has to show a gradual decrease. At the same time, since the punch is traveling toward the right, as already observed in steady-state viscoelastic contacts moving at constant velocity (Carbone and Putignano 2013), a peak in the pressure distribution has to be originated also at the leading edge. Finally, at the center of the distribution, where we have the maximum of the displacement field in the contact area, the pressure must still resemble the classical elastic Hertzian solution. All this process strongly affects the evolution of the pressure distribution at the interface with the presence of multiple pressure peaks shown by the snapshots taken at $\omega t = -0.40\pi$, -0.38π, -0.36π (the reader may refer to Appendix B to appreciate the difference with steady-state conditions). A single peaked pressure distribution is later recovered: indeed, an asymmetric pressure profile marked by a peak closer to the contact leading edge is visible at $\omega t = -0.28\pi$.

We may observe that, for a single relaxation time material, in addition to the ratio E_∞/E_0, the behavior of the reciprocating contact is also governed by other two dimensionless parameters. The first dimensionless group is $\Pi = \tau/t_0$, where $t_0 = a_0/\omega\xi_0$ and a_0 is the Hertzian contact radius. This parameter can be also interpreted as a dimensionless sliding speed (Carbone and Putignano 2013) and compares the relaxation time τ with the time t_0 needed by the sphere to cover a distance a_0. The second group, $\Gamma = a_0/\xi_0 = \omega t_0 = 2\pi t_0/T$, compares, instead, the time t_0 with the period $T = 2\pi/\omega$ of the reciprocating motion. Since we have earlier assumed that in our problem $\Gamma = a_0/\xi_0 \ll 1$, we can focus on observing how the solution is affected by Π. For extremely small or extremely high values of Π, the response of the system is elastic (governed by either the high- or the low-frequency elastic limit of the material), and no tangential contact force will be generated. At intermediate values of Π, viscoelasticity will affect the solution leading to asymmetric contact areas and pressure distributions, and to the generation of tangential contact forces. In such a case, given the dimensionless parameter $\Xi = \Gamma\,\Pi = \omega\tau = 2\pi\tau/T$, if $\Xi < 1$, the reciprocating motion will occur on time scales longer than the relaxation time τ of the material and the system will resemble the steady-state behavior of the contact between a sphere moving on a viscoelastic half-space at constant speed (Carbone and Putignano 2013). If $\Xi \approx 1$, as in the case of Fig. 7.24, a strong interaction will be observed between different viscoelastic regions of the path covered by the sphere during the reciprocating motion. Note that, under the assumption small a_0/ξ_0 values

Fig. 7.24 The shape of the contact area and the contour plots of the normalized contact pressure distributions, p/E_0^*, for several values of ωt

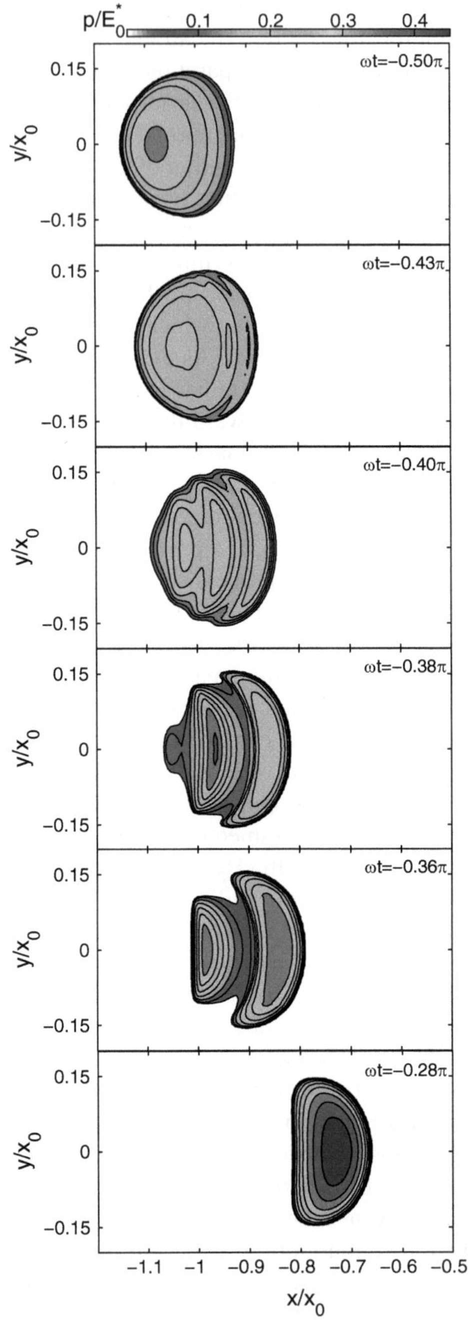

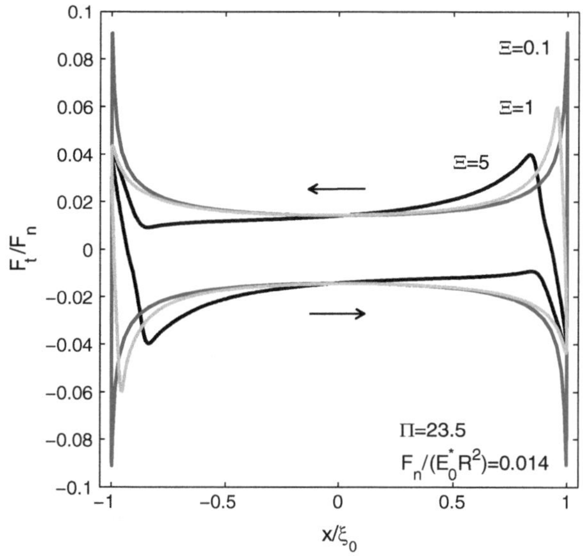

Fig. 7.25 The ratio between the tangential and the normal force F_t/F_n as a function of the dimensionless abscissa x/ξ_0 for different values of Ξ. Arrows refer to the hysteresis cycle direction

(always adopted in this Chapter), the condition $\Xi \gg 1$ implies $\Pi \gg 1$, and, in this case, the elastic response of the material will be recovered: the sphere will be just performing very fast oscillations, leading to a local stiffening and, ultimately, to a high-frequency elastic behavior.

In Fig. 7.25, the reduced tangential force, F_t/F_n, easily calculated once pressures and displacements are known (Carbone and Putignano 2013), is plotted as a function of the dimensionless abscissa x/ξ_0, which identifies the position of the sphere along the path, for different values of Ξ. For $\Xi = 0.1$ the material has the possibility to relax before a single reciprocating cycle is completed. In this case, as the solution resembles the steady-state viscoelastic sliding contact, the tangential force F_t/F_n always opposes the sphere speed at each point along the path. However, as Ξ is increased (see, e.g., results for $\Xi = 5$ presented in Fig. 7.25) the relaxation of the material involves time scales comparable to the time period of the reciprocating motion; in this case, there exist regions on the sphere track, specifically those close to the dead-points, where F_t/F_n has the same direction as the sliding speed. This is perfectly consistent with the results presented in Fig. 7.24.

Lubrication of Viscoelastic Solids: Fluid–Solid Interaction

Particularly interesting is the case of the lubrication between viscoelastic solids: in such a case, as shown in detail (Putignano and Dini 2017), a specific coupling between the solid viscoelasticity and the fluid viscosity is encountered and attention needs to be placed on the treatment of such problems using numerical tools. In this

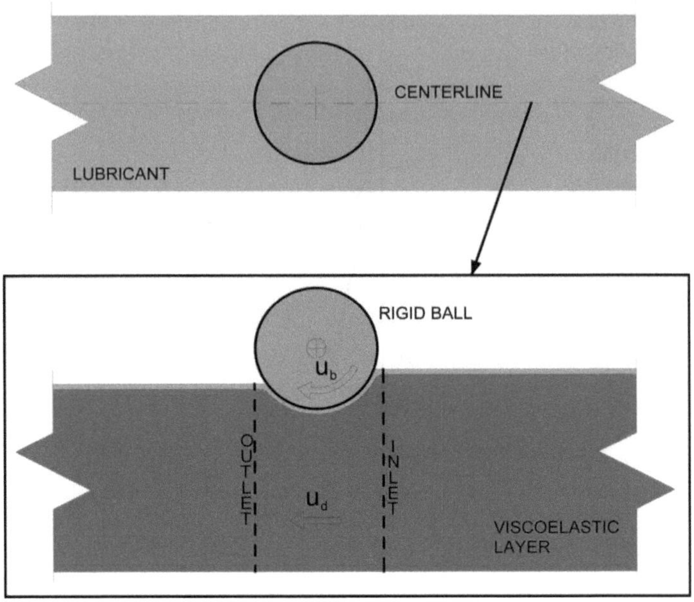

Fig. 7.26 Schematic of the model implemented in the numerical methodology. The spheres roll over a viscoelastic layer with a speed $\mathbf{u}_b = \Omega \times \mathbf{R}$ with Ω and $\mathbf{R}$ being, respectively, the constant angular velocity and the position vector. The viscoelastic layer slides with a constant speed $\mathbf{u}_d$

Section, we show the mathematical formulation and the numerical results, together with validation performed by employing ad hoc numerical procedures.

Mathematical Formulation

In order to deal with the lubrication in the presence of deformable bodies, as schematically reproduced in Fig. 7.26, let us focus on a scheme typical in tribology. A rigid spherical punch clamped in its center rolls, with a velocity $\mathbf{u}_b$ equal to $\mathbf{u}_b = \mathbf{\Omega} \times \mathbf{R}$ with $\mathbf{\Omega}$ and $\mathbf{R}$ being, respectively, the angular velocity and the position vector, over a viscoelastic layer sliding with a constant speed $\mathbf{u}_d$. We can assume that $\mathbf{u}_b$ and $\mathbf{u}_d$ are constant and have the same direction, i.e., $\mathbf{u}_b = u_b\mathbf{i}$ and $\mathbf{u}_d = u_d\mathbf{i}$ with $\mathbf{i}$ being the unit vector, identifying the common motion direction (corresponding to the horizontal direction in the figure). It is noteworthy to underline that the mathematical approach developed below is absolutely general and can be employed for any contact configuration, once the geometry of the contact and $\mathbf{u}_b$ and $\mathbf{u}_d$ are defined.

Now, the complete solution of the lubrication problem requires to determine two unknown distributions, i.e., the film thickness in the contact region and the normal interfacial stress. To this aim, it is necessary to couple a solver for the steady-state hydrodynamic lubrication equations with a methodology that, given the pressure distribution, provides the elastic (or viscoelastic) deformation experienced by the interacting pair. Indeed, the two aspects of the problem are coupled since the displacement of the solid surface influences the lubricating film and, consequently, the solution of the flow equations. Such coupling is particularly strong in soft materials since deformations can be significant when compared to the fluid film thickness.

As we have seen before, the solid problem can be formulated in terms of the following integral relation:

$$u\left(\mathbf{x}\right)=\int d^2x'G\left(\mathbf{x}-\mathbf{x}',\mathbf{u}_d\right)\sigma\left(\mathbf{x}'\right) \tag{7.38}$$

This can be properly discretized in the form of the following linear system:

$$u_i=L_{ik}\left(\mathbf{u}_d\right)\sigma_k \tag{7.39}$$

where the response matrix $L_{ik}\left(\mathbf{u}_d\right)$ parametrically depends on the velocity $\mathbf{u}_d$. We observe that the total load acting on the system F_n is equal to $F_n = D_k\, {}_{k=1}^{N}\, \sigma_k$ with the D_k being the area of each square cell.

Now, we can focus on the second equation, which is related to the fluid dynamics of the problem. We can assume valid all the assumptions commonly employed when dealing with soft lubrication and, in particular, we assume no-slip boundary conditions at both solids interface. We can, then, introduce the Reynolds equations, whose general form can be written as (see Hamrock et al. 2004; Venner and Lubrecht 2000):

$$\frac{\partial\rho h}{\partial t}+\nabla\cdot\left(\rho h\mathbf{U}\right)=\nabla\cdot\left(\frac{\rho h^3}{12\eta}\nabla\sigma\right) \tag{7.40}$$

where $\mathbf{U}$ is the entrainment speed, i.e., the mean surface velocity that for the system depicted in Fig. 7.26 is equal to $\mathbf{U}=\left(\mathbf{u}_b+\mathbf{u}_d\right)/2$, ρ is the density (which here is considered constant), η is the fluid viscosity and h is the film thickness. The latter quantity can be easily related to the normal displacement of the deformable surface u and, specifically, $h(x,y)=h_0+s(x,y)+u(x,y)$ with h_0 and $s(x,y)$ being, respectively, a rigid motion constant and the separation due to the undeformed geometry of the contacting surfaces. For the tribo-system sketched in Fig. 7.26, $s(x,y)$ is equal to $s(x,y)=R-\left(R-x^2-y^2\right)^{1/2}$ with R being the radius of the sphere.

Given the steady-state conditions of our study, Eq. (7.40) simplifies since the time derivative vanishes and is, then, solved by means of a finite difference scheme, whose nodes are equally spaced in the computational domain and correspond to the centers of the boundary elements previously defined for the solid problem. Indeed, such a procedure, where the differential terms in Eq. (7.40) are discretized with central finite differences (Hamrock et al. 2004; Venner and Lubrecht 2000), allows us to reduce Eq. (7.40) to the following linear system:

$$h_i=R_{ik}\left(\mathbf{U},\mu\right)\sigma_k \tag{7.41}$$

Ultimately, the problem consists in coupling the solid mechanics and fluid dynamics (Hamrock et al. 2004; Venner and Lubrecht 2000; Snoeijer et al. 2013) and, consequently, in finding the pressure distribution that satisfies, at the same time, both Eqs. (7.39) and (7.41). An iterative scheme is adopted to solve the system formed by these two equations provided in vector form. Basically, at each iteration, given the estimation of the film thickness $\tilde{h}_i$ computed at the previous iteration, Eq. (7.41) is inverted to calculate an estimated stress field $\tilde{\sigma}_k$, which then is inputted in Eq. (7.39) to obtain the new viscoelastic deformation field $\tilde{u}_i$ and, consequently, the film thickness to employ at the next iteration. The iterative procedure, properly underrelaxed by means of the Aitken acceleration approach (see, e.g., Venner and Lubrecht 2000), continues until film thickness and pressure distributions numerically converge in two consecutive iterations. Furthermore, with regard to the inversion of Eq. (7.39), in order to speed up the solving procedure, we may observe that the matrix R_{ik} is pentadiagonal thanks to the central finite differences discretization, and, consequently, we may implement a direct solver which requires to store only the nonzero elements of the matrix R_{ik} (Schenk and Gärtnerc 2004). The main advantage is the possibility of implementing a fine mesh and, at the same time, obtaining fast computational times. Once the problem is fully solved in terms of pressure distribution and deformations, it is straightforward to calculate the total friction as the sum of the viscoelastic hysteretic term (Carbone and Putignano 2013) and the contribution coming from the fluid losses (Hamrock et al. 2004).

Incidentally, we observe that, as usually done when dealing with numerical methods and, in particular, in lubrication problems, the outcomes of the methodology are reported in dimensionless form. To this aim, we note that the characteristic length of the problem is the radius R: consequently, when considering the film thickness h (and all deformations and quantities defined using units of length), we will look at the ratio h/R. Furthermore, when we have to analyze the stress distribution σ, it is convenient to make such a quantity dimensionless and write it as σ/E_0, by introducing the rubbery elastic modulus E_0 (Carbone and Putignano 2013). Such a modulus is, then, employed the reference for quantities characterized by units of stress. As a consequence, the normal dimensionless load will be $F_n/\left(R^2 E_0\right)$. Finally, to introduce a dimensionless speed ξ, we compare two time scales: the first one is a characteristic relaxation time τ of the material and the second one is the time employed by the fluid to cover the length R with the speed U (Carbone and Putignano 2013). Then, ξ is equal to $\xi = U\tau/R$. Finally, to deal with the viscosity η, we employ the dimensionless group $\eta U R/F_n$, that is the so-called Hersey number.

Film Thickness and Interfacial Pressure Distribution

We focus on the description of the physics governing the lubrication of a rigid sphere in pure rolling over a viscoelastic half-space. Consequently, the slide-roll ratio SRR, defined as $SRR = (u_b - u_d)/((u_b + u_d)/2)$ with u_b and u_d being the speed of the sphere and of the disk, respectively, is set to zero.

Although these two conditions, i.e., the layer assumed as an half-space and the SRR being equal to zero, can be seen as simplification with respect to the general formulation, they are only used here to reduce the number of parameters and show the effect of viscoelasticity in one of the simplest possible configurations. Furthermore, as a starting point and again in order to show the main peculiarities of lubricated viscoelastic response using a simplified material model, we employ a simple one relaxation time material with a ratio E_∞/E_0 equal to $E_\infty/E_0 = 100$ and several different values of the relaxation time τ. Without loss of generality and for illustration purposes, all the calculations are carried out for a constant dimensionless normal load $F_n/\left(R^2 E_0\right) = 8.5\ 10^2$.

Now, by employing the numerical methodology described above, we can calculate how the pressure distribution and the film thickness depends on the dimensionless speed $\xi = U\tau/R$, where the fluid entrainment speed equal U, in pure rolling conditions, to u_d. As shown in Fig. 7.27, for very low values of ξ, the deformable solid is in the elastic rubbery region and behaves, consequently, as a soft elastic body: no viscoelastic effect is present. When looking at the lubricating film, as expected (see, e.g., Hamrock et al. 2004; Bowden and Tabor 2001), we observe an almost perfectly circular shape and, due to the flow conservation, a minimum at the fluid outlet can be observed. This is particularly evident in Fig. 7.28, where both the fluid meatus and the normal stress distribution are plotted at the centerline of the contact depicted in Fig. 7.27. Indeed, for $\xi = 0.005$, we notice a Hertzian-like pressure distribution, typical of low-pressure contacts and iso-viscous fluids (Hamrock et al. 2004).

However, a very different story has to be told when the speed is increased. Indeed the contact zone (i.e., the region delimited by low film and high pressure) gradually decreases its size and, most importantly, evolves toward a shape that is increasingly far from a circle and is affected by a sharp shrinkage at the fluid outlet. In Fig. 7.27, the contour plots of the pressure clearly show this trend; as emphasized when focusing on the centerline in Fig. 7.28, there exists a stress peak that increases with the speed and produces, ultimately, a strong asymmetry toward the contact inlet. Interestingly, such changes produce an additional local minimum in the fluid film: by increasing the speed, this effect becomes predominant and the absolute film thickness minimum moves from the flow outlet to the inlet. This is in agreement with recent experimental evidence (Marx et al. 2016; Hutt and Persson 2016); however, this is certainly not intuitive and might come as a surprise for researchers familiar with classic lubrication models (Hamrock et al. 2004) and would be inadmissible if the rheology of the contacting solids were not to be considered. Hence, further considerations must be made to corroborate and explain these trends.

First of all, let us observe what happens in dry conditions when a rigid ball moves in rolling or sliding contact over a viscoelastic half-space. As shown in Carbone and Putignano (2013), the contact pressure has a peak at the leading edge, where the material is completely undeformed, and is smaller at the trailing edge, where the viscoelastic solid has been deformed and has not yet fully relaxed. Such a pressure field produces a contact area that is asymmetric and has a marked shrinkage at the trailing edge. When a fluid is inserted between the rigid punch and the viscoelastic layer, there occurs a similar mechanism: at the flow inlet, i.e., where the lubricant

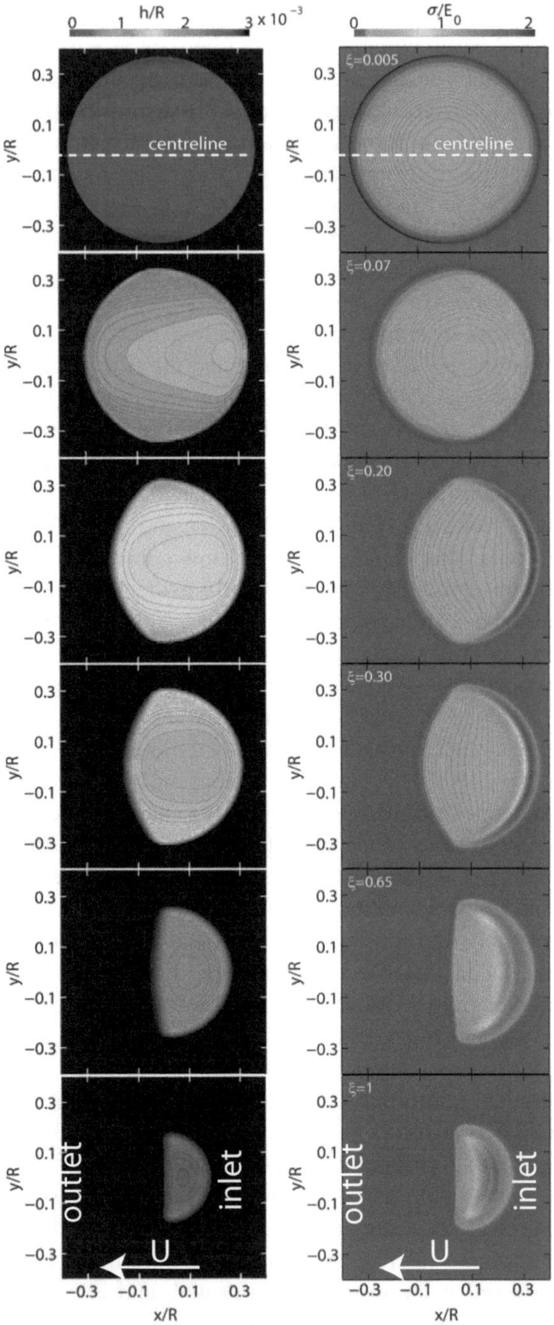

Fig. 7.27 Contour plots of the film thickness (*left*) and the pressure distribution (*right*) predicted for a simple viscoelastic material subjected to a normal load of $F_n/\left(R^2 E_0\right) = 8.5\ 10^2$ and different values of the dimensionless speed ξ. Calculations are carried out with a glass modulus $E_\infty = 10^8$ Pa, a ratio $E_\infty/E_0 = 100$, a relaxation time $\tau = 0.01$ s, a radius $R = 0.02$ m, and a viscosity equal to $\eta = 1$ Pa·s

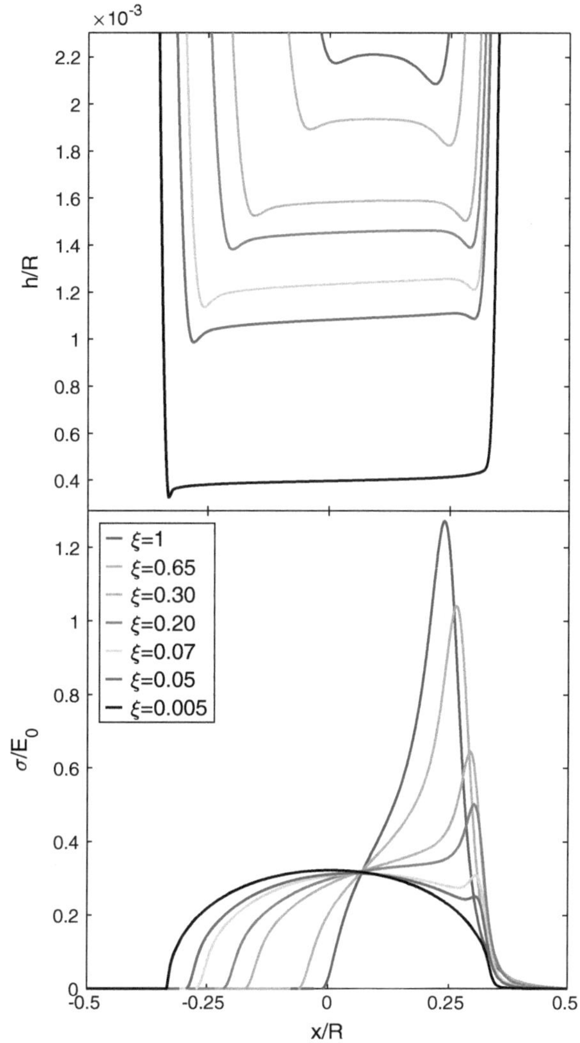

Fig. 7.28 Film thickness (*top*) and pressure distribution (*bottom*) measured at the centerline of the contact ($y/R = 0$) for a normal load of $F_n/\left(R^2 E_0\right) = 8.5\ 10^2$and different values of the dimensionless speed ξ

is "sucked in", the viscoelastic material is still undeformed; on the contrary, at the outlet, where the lubricant exits the contact region, the solid is still relaxing. This can be seen very clearly in Fig. 7.29, where the viscoelastic displacement $u\,(x, y)$ is plotted for different values of the dimensionless speed ξ. As a consequence, similarly to what happens in dry conditions, larger pressure values have to be expected toward the inlet rather than at the flow outlet. To such a pressure distribution corresponds a shrunk nonsymmetric film thickness and a minimum value at the inlet.

Furthermore, experimental comparisons with numerical results can be carried out by employing the same experimental setup shown in Marx et al. (2016). This is based on the optical interferometry. Such a technique is normally used to detect the fluid

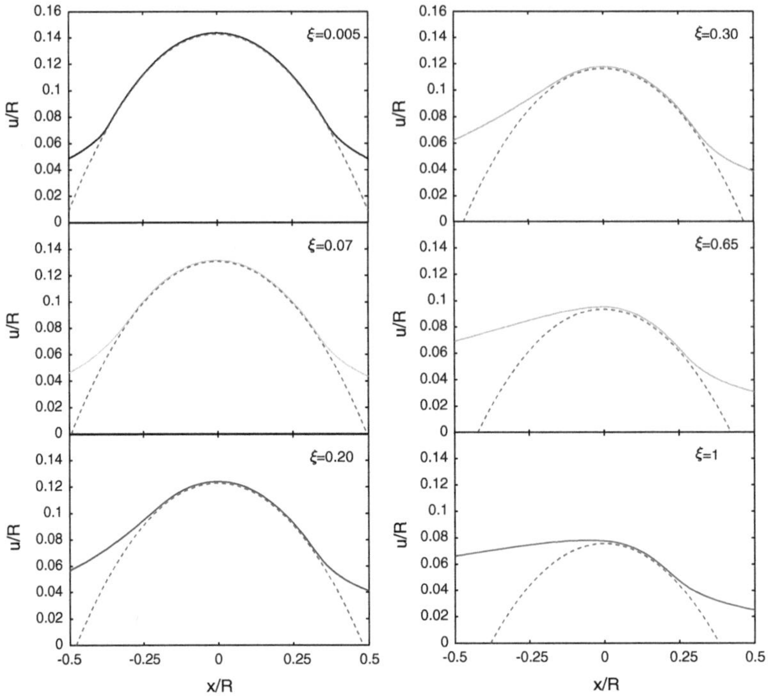

Fig. 7.29 Displacement distribution $u(x, y)$ for different values of the dimensionless speed ξ. Dotted lines refer to the position of the rigid punch

film between two lubricated bodies: fundamentally, light is shone into the lubricated contact through a transparent body that is usually glass or sapphire. Some of this light is reflected from the lower surface of the transparent disk while some passes through the lubricating film and is reflected on its turn from the reflective ball surface, which is usually steel. When the two light beams recombine, they interfere in a way that depends on the path difference between them and, consequently, on the lubricant film thickness. Such an experimental setup has been applied successfully to transparent polymethylmethacrylate (PMMA) disks by applying a semi-reflective chromium coating on the polymeric material. As for the lubricant, an additive-free base fluid, of a gas-to-liquid origin and corresponding broadly to API Group IIIÃ, is employed. More details, including the viscometric properties of the fluid, can be found in Marx et al. (2016). With regard to the disk material, PMMA is a viscoelastic polymer whose properties have been obtained by means of dynamic mechanical analysis performed on the Q800 Dynamic Mechanical Analyzer (DMA) manufactured by TA Instruments (the reader is referred to Appendix A for more details). In Fig. 7.30, results show a good qualitative and quantitative agreement between the contour map of the film thickness experimentally measured at the temperature of $T = 40$ °C and the equivalent contour plot obtained numerically. A marked deviation from the circular shape of the contact zone, which cannot be captured using classical elasto-

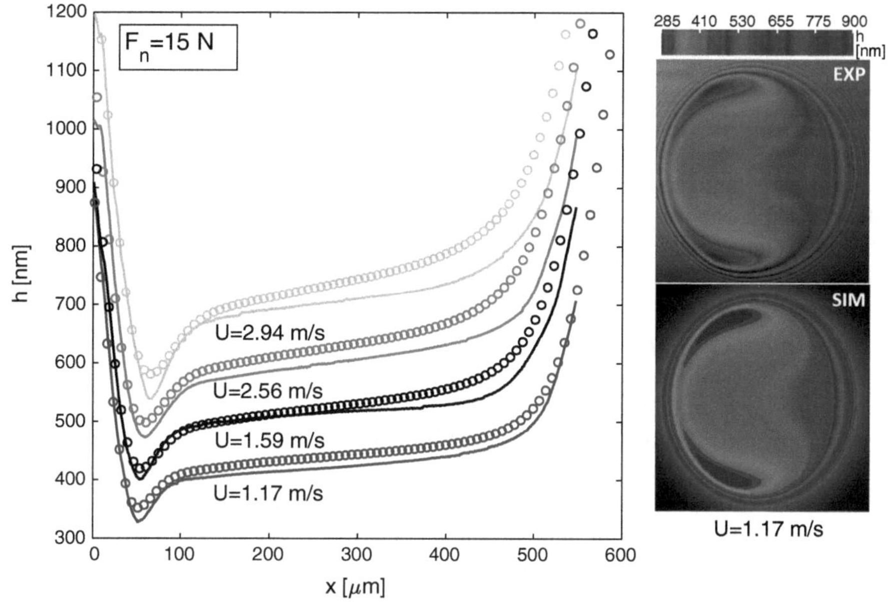

Fig. 7.30 Interferometer map (*top right*) and numerical contour plot (*bottom right*) of film thickness for a normal PMMA sample subjected to a normal load of $F_n = 15$ N and an entrainment speed $U = 1.17$ m/s. Comparison between numerical predictions and experimental outcomes at the centerline $(y/R = 0)$ for different values of the entrainment speed (*left*)

hydrodynamic lubrication and neglecting the effect of the solid rheology, is clearly shown. As shown also in Marx et al. (2016), and coherently with the theoretical framework so far developed, such an effect and, in particular, the shrinkage at the flow inlet are strongly dependent on the entrainment speed.

A more direct quantitative analysis of the film at the contact centerline, carried out for different values of entrainment speed and for a normal load of $F_n = 15$ N, shows a good agreement with the numerical outcomes with discrepancies always below 8%. The observation of the pronounced shrinkage of the lubricating film and the accuracy of the quantitative comparison between experimental and numerical results highlights the role that might be played by solid viscoelasticity. However, at the same time, in the current experimental setup, viscoelastic effects are not as strong as those observed, for example, in Figs. 7.27 and 7.28. Indeed, the experimental evidence and the numerical results presented so far suggest that viscoelastic effects can be more or less pronounced depending on the fluid viscosity and the frequency/speed range in which solid viscoelasticity is prominent for the specific material under investigation. In other words, the rheology of the fluid and of the solid undergo a complex interplay, with different levels of coupling between the fluid film and the deformation of the solid bodies, which leads to different lubrication scenarios and frictional response. Such interplay is discussed next.

Solid Viscoelasticity and Fluid Viscosity: A Coupling Criterion

To quantify the coupling and the interaction between viscoelastic deformations in the solid and fluid film, let us primarily focus on the two parameters that govern the phenomenon. Starting from the aspects linked to the dynamics of the fluid, all the properties and, in particular, friction, which will be used as a measure of the effect that the coupling has on dissipation, are determined by the Hersey number, $\eta U R/F_n$ (Hamrock et al. 2004). With regard to the solid, we have a frequency where viscoelastic losses and, specifically, the loss tangent reaches a maximum: in the very simple case of a one relaxation time material, such a frequency, which maximizes the ratio $\mathrm{Im}[E(\omega)]/|E(\omega)|$, can be estimated analytically and is equal to $\omega_{cr} \approx \pi^{-1}\sqrt{E_\infty/E_0}$ (Carbone and Putignano 2013). The critical disk speed associated to this frequency is equal to $u_d|_{cr} \approx \omega_{cr} R/\tau$ and, therefore, since in rolling conditions $U = u_d$, the critical entrainment speed can be estimated as $U_{cr} = (R/\pi\tau)\sqrt{E_\infty/E_0}$.

Now, it is straightforward to observe that, given a constant normal load, when we increase the speed, the minimum and the mean film thickness increase and, as expected from classical lubrication theory, once we pass from the elasto-hydrodynamic to the hydrodynamic regime, any deformation in the contacting solids tends to decrease. As shown also in Fig. 7.29, when the deformable layer is viscoelastic, the situation is similar with the only difference that, by increasing the speed, the solid becomes stiffer until we reach the elastic glassy regime where $E(\omega)$ is equal to E_∞. Therefore, the transition between a deformable regime that we could now define as visco-elasto-hydrodynamic (VEHL), to the hydrodynamic behavior can be even faster. Now, given a viscoelastic solid, we may wonder what happens if, for a given load, U_{cr} falls into the hydrodynamic region and, consequently, no deformation occurs at the critical speed: simply, no viscoelastic effect will be observed. We have an elasto-hydrodynamic regime—the solid behaves elastically with a rubber modulus E_0—followed by a hydrodynamic regime at larger speeds. It emerges that viscoelasticity is a necessary, but not a sufficient condition to see a marked deviation from classical EHL conditions. To observe the visco-elasto-hydrodynamic regime, the solid deformation at the critical speed U_{cr} has to be large and comparable with the film thickness. This observation leads to the introduction of the following coupling parameter:

$$\Gamma = \frac{h_{hydro}}{\delta_{cr}} \tag{7.42}$$

with h_{hydro} and δ_{cr} being, respectively, the minimum film thickness in hydrostatic conditions and the solid penetration at the critical speed. To have a VEHL regime, $\Gamma \approx 1$. The minimum hydrostatic film thickness can be found solving the Reynolds equations and, for our configuration, as shown in Esfahanian and Hamrock (1991), is equal to $h_{hydro} = RH_{hydro}$ with H_{hydro} being a dimensionless quantity equal to $H_{hydro} = \alpha(\eta U_{cr} R/F_n)^2$ with α a constant numerically found equal to $\alpha \approx 1.3 \cdot 10^2$. The penetration δ_{cr} can be approximately estimated using the Hertzian relations as $\delta_{cr} \approx \left(9F_n^2/16R\,(E^*(\omega_{cr}))^2\right)$.

This is a powerful tool, as evaluating Γ allows, given the viscoelastic modulus $E(\omega)$ and the fluid viscosity η, to determine if a visco-elasto-hydrodynamic behavior has to be expected in first-order approximation for materials governed by linear viscoelasticity. To show the effects induced by the VEHL regime, let us focus on the friction force, which is the sum of the fluid contribution and the viscoelastic dissipation in the solid. Incidentally, we observe that the role of the roughness may play a significant role in the so-called boundary and mixed regimes and may not influence the regimes in which the fluid film is fully formed, which are the subject of our investigation. Now, if viscoelastic effects are not significant we have a transition between EHL and hydrodynamic (HD) regimes regulated by the Stribeck curve: at the very low speed, the friction tends to zero, whereas, at larger speeds, we have a linear dependence between the friction and the logarithm of the speed. The situation is different when viscoelasticity is significant since viscoelastic hysteresis adds a source of dissipation and modifies the friction curve. This appears very clearly in Fig. 7.31 where the coefficient of friction is plotted for a material whose viscoelasticity is characterized by one relaxation time material, with a ratio $E_\infty/E_0 = 100$, and four different values of relaxation time. By changing the relaxation time, we are shifting the viscoelastic spectrum, thus obtaining four different values of Γ: $\Gamma_a = 0.065$, $\Gamma_b = 0.65$, $\Gamma_c = 6.5$ and $\Gamma_d = 65$. As shown in Fig. 7.31, when Γ is small, we have a deviation from the Stribeck curve with a marked "bump" due to the viscoelastic hysteretic contribution; for larger values of τ and, consequently, of Γ, the viscoelastic friction peak moves toward larger speeds and, then, as explained before, decreases its intensity until it disappears. Indeed, for $\Gamma = \Gamma_d$, we have a standard Stribeck curve: specifically, for low speeds, we have an EHL region, where the body deforms elastically with a modulus equal to the rubber modulus E_0, and, then, for larger values

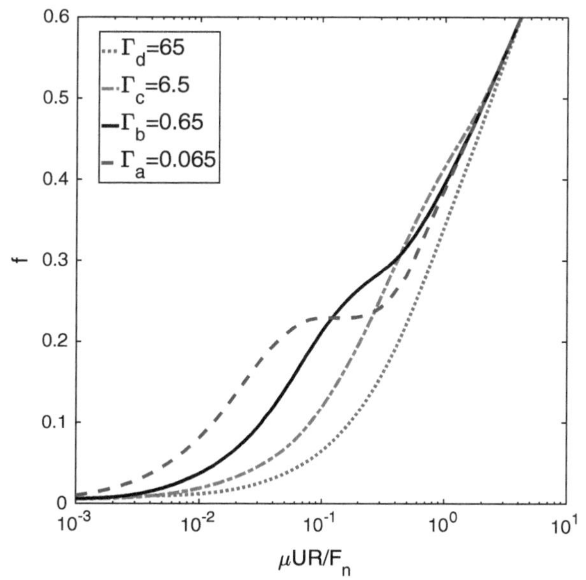

Fig. 7.31 Friction coefficient f as a function of the speed for a dimensionless load equal to $F_n/\left(R^2 E_0\right) = 8.5\ 10^2$ and different values of the parameter Γ: $\Gamma_a = 0.065$, $\Gamma_b = 0.65$, $\Gamma_c = 6.5$ and $\Gamma_d = 65$. They have been obtained by employing the following four different values of τ : $\tau_a = 10^{-2}$ s, $\tau_b = 3.5\ 10^{-2}$ s, $\tau_c = 10^{-3}$ s, and $\tau_d = 10^{-4}$ s

of the speed, there is the HD regime where the solid would be potentially able to show viscoelastic effects, but, since the pressure is not large enough to significantly deform the material, no energy dissipation occurs.

As expected and consistent with the proposed theoretical framework, all the four curves tend to collapse at low speed (low Hersey number) to a soft-EHL behavior and at large speeds (high Hersey number) we find the same hydrodynamic regime, which is dictated only by the ball geometry and the fluid viscosity.

Adhesion and Soft Contacts

Another important aspect to consider when dealing with soft materials in contact is that they can also be affected by strong adhesive interactions, something that must be carefully considered when tackling such problems. Although an extensive review of methods to deal with adhesion is outside the scope of this chapter, an overview of the main available methodologies is provided. For a more in-depth description of methodologies available to solve adhesive contacts and a historical summary of their development, the reader should refer to Vakis et al. (2018), Muser et al. (2017). The most frequently adopted adhesive models are the two analytical ones developed in the 1970s, the JKR model (Johnson et al. 1971) and the DMT model (Derjaguin et al. 1975). These considered adhesive contact between a smooth sphere and a flat body, but with different approaches and making significantly different assumptions. They were shown to apply equally well to different contact conditions by Tabor (1977), who identified a characteristic parameter, now known as the Tabor parameter, which can be systematically used to identify whether short-range or long-range adhesion dominates the contact interactions; in particular, the JKR model captures mainly short-range interactions, representative only for contacts with a large value for the Tabor parameter (>2, soft solids, small curvature, large adhesion), while the DMT model is valid for contacts with a small value (<0.01, rigid solids, large curvature, weak adhesion). Various later attempts were made later to bridge the two models and to achieve a higher level of accuracy using semi-analytical methodologies (see, e.g., Muller et al. 1980; Greenwood 1997; Maugis 1992). More recently, finite element models for adhesive contact problems have also been developed, where the contact description obtained using the Lennard-Jones potential is incorporated into the framework of nonlinear continuum mechanics (Sauer and Wriggers 2009; Eid et al. 2011). Alternative approaches have also been developed based on the BEM, which incorporates adhesion through energy minimization (Carbone and Mangialardi 2008a).

Most of the models discussed above were developed for or applied to smooth surface contact, nominally between a sphere and a flat. A common justification for neglecting adhesive forces is the existence of surface roughness and, starting from this point, an early and significant analysis was carried out by Fuller and Tabor (1975), who showed that the adhesive influence could be described by an adhesion parameter, which is, in effect, a ratio of the adhesive force of lower asperities to the elastic push of higher asperities. The theory was found to show reasonable agreement when fitted to experimental results. Fuller and Tabor had used the JKR model on an asperity level; Maugis repeated the analysis using the DMT model and found that an additional load would be caused by adhesive forces around each asperity (Maugis 1996). Looking at other theoretical and nondeterministic models of multi-asperity contacts, in some of the early contributions, Persson and Tosatti considered adhesion through a fractal representation of surface roughness and showed that adhesion dropped significantly at higher fractal dimensions (Persson and Tosatti 2001). More recently, Persson and Scaraggi (2014) used Persson's theory and a power spectrum representation of the contact roughness to introduce a Tabor number that depends on the length scale or magnification, and which gives information about the nature of the adhesion at different length scales. It was shown that adhesion problems that are JKR-like for large length scales and DMT-like for short length scales can be approximately treated using the theory with different levels of approximations, which depend on how quickly the behavior transitions between the two limits across the scales. While these rough surface models (or asperity models) are limited to a stochastic description of the surfaces and thus cannot provide a complete contact mechanics solution for all surfaces, they may constitute a good approximation and provide a useful design tool, especially when numerical simulations may struggle or fail to produce fast and reliable results. Deterministic adhesion models of contact in the presence of roughness are expected to provide an accurate representation of the response of real bodies in contact. Given the advent of new and improved numerical methodologies and increased computational power, there has been a recent resurgence in the development of contact mechanics models able to address contact between surfaces of arbitrary shape and roughness, of small and large scale, and capable of providing accurate information for contact forces, surface displacements and hysteretic effects (where present) throughout the contact. Many of these methodologies can be seen as boundary elements methods (BEM) relying on different discretization schemes and numerical techniques to solve the contact problem using brute force (Mser et al. 2017), and include GFMD (Pastewka and Robbins 2014), FFT-based (e.g., Bazrafshan et al. 2017), and Multi-level multi-integration (MLMI)-based techniques (Medina and Dini 2014). These methods have been shown to capture the response of rough contact surfaces in the presence of adhesion in a number of configurations and can be used successfully to predict the scales and regimes at which roughness will play a significant role in adhesive contacts, as well as computing hysteretic losses. An interesting extension of these methods would be to study the effects of adhesion in rough viscoelastic contacts.

Conclusion

In this chapter, we have reviewed and discussed recent advances in methodologies develop to study dry and lubricated contacts involving soft materials. Particular attention was focused on the treatment of viscoelastic solids and the effect that viscoelastic dissipation has on the frictional response of the components under investigation. Starting from the pioneering work by Hunter (1961), various formulations that can be used to model linear viscoelastic material have been presented in increased order of complexity. The effect of viscoelastic layer thickness as well as surface roughness is incorporated in the formulation and key findings are shown in terms of how the material properties as well as the loading conditions (sliding and reciprocating) affect the material response. An overview of how to extend the formulation to treat lubricated contacts has also been provided, together with examples of experimental validation conducted for a number of problems. These include the study of the evolution of contact area, thermal response, film thickness, and friction in various configurations. Interesting phenomena have been observed that can only be explained by invoking the specific material characteristics the availability of new modeling tools in this area of tribology opens new avenues to explore the mechanisms governing interactions between soft solids, which have become increasingly relevant for industrial applications.

References

Akarapu, S., Sharp, T., & Robbins, M. O. (2011). Stiffness of contacts between rough surfaces. *Physical Review Letters*, *106*, 204301.

Almqvist, A., Campañá, C., Prodanovb, N., & Persson, B. N. J. (2011). Interfacial separation between elastic solids with randomly rough surfaces: Comparison between theory and numerical techniques. *Journal of the Mechanics and Physics of Solids*, *59*, 11.

André, T., Lévesque, V., Hayward, V., Lefèvre, P., & Thonnard, J.-L. (2011). Effect of skin hydration on the dynamics of fingertip gripping contact. *Journal of the Royal Society Interface*, *8*, 64.

Archard, J. F., & Hirst, W. (1956). Proceedings of the Royal Society of London. *Series A: Mathematical and Physical Sciences*, *236*, 397.

Barber, J. R. (2003a). Bounds on the electrical resistance between contacting elastic rough bodies. *Proceedings of the Royal Society of London. Series A: Mathematical, Physical and Engineering Sciences, 459*, 53–66.

Barber, J. R. (2003b). Bounds on the electrical resistance between contacting elastic rough bodies. *Proceedings of the Royal Society of London. Series A: Mathematical and Physical Sciences*, *459*, 53–66.

Barber, J. R. (1974). Determining the contact area in elastic indentation problems. *Journal of Strain Analysis*, *9*, 230–232.

Barber, J. R., Davies, M., & Hills, D. A. (2011). Frictional elastic contact with periodic loading. *International Journal Solids Structures*, *48*, 2041–4047.

Barenblatt, G. I. (1996). *Scaling, self-similarity and intermediate asymptotics*. Cambridge: Cambridge University Press.

Bazrafshan, M., de Rooij, M., Valefi, M., & Schipper, D. (2017). Numerical method for the adhesive normal contact analysis based on a Dugdale approximation. *Tribology International*, *112*, 117–128.

Bhushan, B. (2004). *Springer handbook of nanotechnology*. Berlin, Heidelberg, New York: Springer.

Bjorkland, S. (1997). A random model for micro-slip between nominally flat surfaces. *ASME Journal of Tribology*, *119*, 726–732.

Borri Brunetto, M., Chiaia, B., & Ciavarella, M. (2001). Incipient sliding of rough surfaces in contact: A multi-scale numerical analysis. *Computer Methods in Applied Mechanics and Engineering*, *190*, 6053–6073.

Bottiglione, F., Carbone, G., & Mantriota, G. (2009a). Fluid leakage in seals: An approach based on percolation theory. *Tribology International*, *42*(5), 731–737.

Bottiglione, F., Carbone, G., Mangialardi, L., & Mantriota, G. (2009b). Leakage mechanism in flat seals. *Journal of Applied Physics*, *106*, 104902.

Bowden, F. P., & Tabor, D. (2001). *The friction and lubrication of solids*. Clarendon Press.

Bowden, F. P., & Tabor, D. (1939). The friction and lubrication of solids. *Proceedings of the Royal Society of London. Series A: Mathematical and Physical Sciences*, *169*, 391.

Bruggeman, D. A. G. (1935). Berechnung verschiedener physikalischer Konstanten von heterogenen Substanzen. *Ann. Phys. (Leipzig)*, *24*, 636–679.

Bureau, L., Caroli, C., & Baumberger, T. (2003). Elasticity and on set of frictional dissipation at a non-sliding multi-contact interface. Proceedings Royal Society (London). *A*, *459*, 2787–2805.

Bush, A. W., Gibson, R. D., & Thomas, T. R. (1975). The elastic contact of a rough surface. *Wear*, *35*, 87–111.

Campana, C., Mueser, M. H., & Robbins, M. O. (2008). Elastic contact between self-affine surfaces: Comparison of numerical stress and contact correlation functions with analytic predictions. *Journal of Physics-Condensed Matter*, *20*(35), 354013.

Campana, C., & Muser, M. H. (2007). Contact mechanics of real vs. randomly rough surfaces: A Green's function moleculardynamics study. *Europhysics Letters, 77*(3), 38005.

Campana, C., Persson, B. N. J., & Muser, M. H. (2011). Transverse and normal interfacial stiffness of solids with randomly rough surfaces. *Journal of Physics-Condensed Matter*, *23*(8), 085001.

Carbone, G., & Persson, B. N. J. (2005a). Crack motion in viscoelastic solids: The role of the flash temperature. *The European Physical Journal E-Soft Matter*, *17*(3), 261.

Carbone, G., & Persson, B. N. J. (2005b). Hot cracks in rubber: Origin of the giant toughness of rubber-like materials. *Physical Review Letters*, *95*, 114301.

Carbone, G., & Pierro, E. (2012a). Sticky bio-inspired micropillars: Finding the best shape. *Small*, *8*(9), 1449–1454.

Carbone, G., & Pierro, E. (2012b). Effect of interfacial air entrapment on the adhesion of bio-inspired mushroom-shaped micro-pillars. *Soft Matter*, *8*(30), 7904–7908.

Carbone, G., & Putignano, C. (2013). A novel methodology to predict sliding/rolling friction in viscoelastic materials: Theory and experiments. Accepted on *Journal of Mechanics and Physics of Solids*.

Carbone, G., Scaraggi, M., & Tartaglino, U. (2009). Adhesive contact of rough surfaces: Comparison between numerical calculations and analytical theories. *The European Physical Journal E-Soft Matter*, *30*(1), 65–74.

Carbone, G. (2009). A slightly corrected Greenwood and Williamson model predicts asymptotic linearity between contact area and load. *Journal of the Mechanics and Physics of Solids*, *57*(7), 1093–1102.

Carbone, G., & Bottiglione, F. (2008). Asperity contact theories: Do they predict linearity between contact area and load? *Journal of the Mechanics and Physics of Solids*, *56*, 2555–2572.

Carbone, G., & Mangialardi, L. (2004). Adhesion and friction of an elastic half-space in contact with a slightly wavy rigid surface. *Journal of the Mechanics and Physics of Solids*, *52*(6), 1267–1287.

Carbone, G., & Mangialardi, L. (2008a). Analysis of the adhesive contact of confined layers by using a Green's function approach. *Journal of the Mechanics and Physics of Solids, 56*, 684–706.

Carbone, G., & Mangialardi, L. (2008b). Analysis of adhesive contact of confined layers by using a Green's function approach. *The Journal of the Mechanics and Physics of Solids, 56*(2), 684–706.
Carbone, G., & Putignano, C. (2014). Rough viscoelastic sliding contact: Theory and experiments. *Physical Review E, 89*, 032408.
Carbone, G., Mangialardi, L., & Persson, B. N. J. (2004). Adhesion between a thin elastic plate, and a hard randomly rough substrate. *Physical Review B, 70*(12), 125–407.
Carbone, G., Lorenz, B., Persson, B. N. J., & Wohlers, A. (2009). Contact mechanics and rubber friction for randomly rough surfaces with anisotropic statistical properties. *European Physical Journal E, 29*, 275–284.
Carbone, G., Pierro, E., & Gorb, S. (2011). Origin of the superior adhesive performance of mushroom shaped microstructured surfaces. *Soft Matter, 7*(12), 5545–5552.
Cattaneo, C. (1938). Sul contatto di due corpi elastici: distribuzione locale degli sforzi. *Rend. Accad. Naz. Lincei, 27*, 342–348, 434–436, 474–478 (in Italian).
Christensen, R. M. (1982). *Theory of viscoelasticity*. New York: Academic Press.
Ciavarella, M. (1998a). The generalized Cattaneo partial slip plane contact problem I-Theory, II-Examples. *International Journal of Solids and Structures, 35*, 2349–2378.
Ciavarella, M. (1998b). Tangential loading of general three-dimensional contacts. *ASME Journal of Applied Mechanics, 65*, 998–1003.
Ciavarella, M., Delfine, V., & Demelio, G. (2006). A "re-vitalized" Greenwood and Williamson model of elastic contact between fractal surfaces. *Journal of the Mechanics and Physics of Solids, 54*, 2569–2591.
Ciavarella, M., Dibello, S., & Demelio, G. (2008). Conductance of rough random profiles. *International Journal of Solids and Structures, 45*, 879–893.
D'Amico, F., Carbone, G., Foglia, M. M., & Galietti, U. (2012). Moving cracks in viscoelastic materials: Temperature and energy-release-rate measurements. *Engineering Fracture Mechanics* (submitted).
Dapp, W. B., Lucke, A., Persson, B. N. J., & Muser, M. H. (2012). Self-affine elastic contacts: Percolation and leakage. *Physical Review Letter, 108*, 244301.
de Vicente, J., Stokes, J. R., & Spikes, H. R. (2006). Rolling and sliding friction in compliant, lubricated contact. *Proceedings Institutions of Mechanical Engineering Part J: Journal of Engineering Tribology, 220*.
Derjaguin, B. V., Muller, V. M., & Toporov, Y. P. (1975). Effect of contact deformations on the adhesion of particles. *Journal of Colloid and Interface Science, 53*, 314–326.
Dieker, A. B., & Mandjes, M. (2003). On spectral simulation of fractional Brownian motion. *Probability in the Engineering and Informational Sciences, 17*, 417–434.
Eid, H., Adams, G., McGruer, N., Fortini, A., Buldyrev, S., & Srolovitz, D. (2011). A combined molecular dynamics and finite element analysis of contact and adhesion of a rough sphere and a flat surface. *Tribology Transactions, 54*, 920–928.
Elsharkawy, A. A. (1996). Visco-elastohydrodynamic lubrication of line contacts. *Wear, 199*, 45–53.
Endlein, T., Barnes, W. J. P., Samuel, D. S., Crawford, N. A., Biaw, A. B., & Grafe, U. (2013). Sticking under wet conditions: The remarkable attachment abilities of the torrent frog. *Staurois guttatus, PLoS ONE, 8*, e73810.
Eriten, M., Polycarpou, A. A., & Bergman, L. A. (2011). Surface roughness effects on energy dissipation in fretting contact of nominally flat surfaces. *ASME Journal Applied Mechanics, 78*.
Esfahanian, M., & Hamrock, B. J. (1991). Fluid-film lubrication regimes revisited. *Tribology Transactions, 34*(4), 628–632.
Felhős, D., Xu, D., Schlarb, A. K., Váradi, K., & Goda, T. (2008). Viscoelastic characterization of an EPDM rubber and finite element simulation of its dry rolling friction. *Express Polymer Letters, 2*(3), 157–164.
Ferry, J. D. (1980). *Viscoelastic properties of polymers*. Wiley, Inc.
Fuller, K. N. G., & Tabor, D. (1975a). The effect of surface roughness on the adhesion of elastic solids. *Proceedings of the Royal Society of London. Series A: Mathematical, Physical and Engineering Sciences, 345*, 327–342.

Fuller, K. N. G., & Tabor, D. (1975b). The effect of surface roughness on the adhesion of elastic solids. *Proceedings of the Royal Society of London. Series A: Mathematical and Physical Sciences, 345*(1642), 327–342.

Geike, T., & Popov, V. L. (2007). Mapping of three-dimensional contact problems into one dimension. *Physical Review E, 76*, 036710.

Geim, A. K., Dubonos, S. V., Gricorieva, I. V., Novoselov, K. S., Zhukov, A. A., & Shapoval, S. Y. (2003a). Microfabricated adhesive mimicking gecko foot-hair. *Nature Materials, 2*, 461–463.

Geim, A. K., Dubonos, S. V., Gricorieva, I. V., Novoselov, K. S., Zhukov, A. A., & Shapoval, S. Y. (2003b). Microfabricated adhesive mimicking gecko foot-hair. *Nature Materials, 2*, 461–463.

Greenwood, J. (1997). Adhesion of elastic spheres. *Proceedings of the Royal Society of London. Series A: Mathematical, Physical and Engineering Sciences, 453*, 1277–1297.

Greenwood, J. A. (2006). A simplified elliptic model of rough surface contact. *Wear, 261*, 191–200.

Greenwood, J. A., & Williamson, J. B. P. (1966). Contact of nominally flat surfaces. *Proceedings of the Royal Society of London. Series A. Mathematical and Physical Sciences, 295*, 300–319.

Greenwood, J. A., Putignano, C., & Ciavarella, M. (2011). A Greenwood & Williamson theory for line contact. *Wear, 270*, 332–334.

Grosch K. A. (1963). The relation between the friction and visco-elastic properties of rubber. *Proceedings of the Royal Society of London. Series A: Mathematical and Physical, 274*(1356), 21–39.

Hamrock, B. J., Schmid, S. R., & Jacobson, B. O. (2004). *Fundamentals of fluid film lubrication*. CRC Press.

Harrass, M., Friedrich, K., & Almajid, A. A. (2010). Tribological behavior of selected engineering polymers under rolling contact. *Tribology International, 43*, 635–646.

Hooke, C. J., & Huang, P. (1997). Elastohydrodynamic lubrication of soft viscoelastic materials in line contact. *Proceedings Institutions of Mechanical Engineering Part J: Journal of Engineering Tribology, 211*, 185.

Hunter, S. C. (1961). The rolling contact of a rigid cylinder with a viscoelastic half space. *Transactions ASME, Series E, Journal of Applied Mechanics, 28*, 611–617.

Hutt, W., & Persson, B. N. J. (2016). Soft matter dynamics: Accelerated fluid squeeze-out during slip. *The Journal of Chemical Physics, 144*, 124903.

Hyun, S., & Robbins, M. O. (2007). Elastic contact between rough surfaces: Effect of roughness at large and small wavelengths. *Tribology International, 40*, 413–1422.

Hyun, S., Pei, L., Molinari, J.-F., & Robbins, M. O. (2004). Finite-element analysis of contact between elastic self-affine surfaces. *Physical Review E, 70*, 026117.

Jager, J. (1998). A new principle in contact mechanics. *ASME Journal of Tribology, 120*, 677–684.

Jang, H., & Barber, J. R. (2011). Effect of phase on the frictional dissipation in systems subjected to harmonically varying loads. *European Journal Mechanics A/Solids, 30*, 269–274.

Johnson, K. L. (1961). Energy dissipation at spherical surfaces in contact transmitting oscillating forces. *Journal of Mechanical Engineering Science, 3*, 362–368.

Johnson, K. L. J. (1985). *Contact mechanics*. Cambridge University Press.

Johnson, K. L., Kendall, K., & Roberts, A. D. (1971). Surface energy and the contact of elastic solids. *Proceedings of the Royal Society of London. Series A: Mathematical, Physical and Engineering Sciences, 324*, 301–313.

Kessler, M. (2004). *Advanced topics in characterization of composites*. Trafford Publishing.

Krick, B. A., Vail, J. R., Persson, B. N. J., & Sawyer, W. G. (2012). Optical in situ micro tribometer for analysis of real contact area for contact mechanics, adhesion, and sliding experiments. *Tribology Letters*.

Le Tallec, P., & Rahler, C. (1994). Numerical models of steady rolling for non-linear viscoelastic structures in finite deformations. *International Journal for Numerical Methods in Engineering, 37*, 1159–1186.

Lorenz, B., & Persson, B. N. J. (2010a). Leak rate of seals: Effective-medium theory and comparison with experiment. *European Physical Journal E, 31*(2), 159–167.

Lorenz, B., & Persson, B. N. J. (2010b). Time-dependent fluid squeeze-out between solids with rough surfaces. *European Physical Journal E*, *32*(3), 281–290.
Lorenz, B., & Persson, B. N. J. (2010c). On the dependence of the leak rate of seals on the skewness of the surface height probability distribution. *EPL*, *90*, 38002.
Lorenz, B., & Persson, B. N. J. (2009). Interfacial separation between elastic solids with randomly rough surfaces: comparison of experiment with theory. *Journal of Physics: Condensed Matter*, *21*(1), 015003.
Lorenz, B., Persson, B. N. J., Dieluweit, S., & Tada, T. (2011). Rubber friction: Comparison of theory with experiment. *European Physical Journal E*, *34*, 129.
Lorenz, B., Carbone, G., & Schulze, C. (2010). Average separation between solids in rough contact: Comparison between theoretical predictions and experiments. *Wear*, *268*(7–8), 984–990.
Luan, B., & Robbins, M. O. (2009). Hybrid atomistic/continuum study of contact and friction between rough solids. *Tribology Letters*, *36*, 1–16.
Luan, B. Q., Hyun, S., Molinari, J. F., Bernstein, N., & Robbins, M. O. (2006). Multiscale modeling of two-dimensional contacts. *Physical Review E*, *74*, 046710.
Mandelbrot, B. B. (1982). *The fractal geometry of nature*. New York: W. H. Freeman and company.
Manners, W., & Greenwood, J. A. (2006). Some observations on Persson's diffusion theory of elastic contact. *Wear*, *261*, 600–610.
Martina, D., Creton, C., Damman, P. M., Jeusette, & Lindner. A., (2012). Adhesion of soft viscoelastic adhesives on periodic rough surfaces. *Soft Matter, 8*(19), 5350–5357.
Marx, N., Guegan, J., & Spikes, H. A. (2016). Elastohydrodynamic film thickness of soft EHL contacts using optical interferometry. *Tribology International*, *99*, 267–277.
Maugis, D. (1992). Adhesion of spheres: The JKR-DMT transition using a dugdale model. *Journal of Colloid and Interface Science*, *150*, 243–269.
Maugis, D. (1996). On the contact and adhesion of rough surfaces. *Journal of Adhesion Science and Technology*, *10*, 161–175.
Medina, S., & Dini, D. (2014). A numerical model for the deterministic analysis of adhesive rough contacts down to the nano-scale. *International Journal of Solids and Structures*, *51*, 2620–2632.
Mindlin, R. D. (1949). Compliance of elastic bodies in contact. *ASME Journal of Applied Mechanics*, *16*, 259–268.
Mser, M. H., Dapp, W. B., Bugnicourt, R., Sainsot, P., Lesaffre, N., Lubrecht, A. A., et al. (2017). Meeting the contact-mechanics challenge. *Tribology Letters*, *65*, 118.
Muller, V., Yushchenko, V., & Derjaguin, B. (1980). On the influence of molecular forces on the deformation of an elastic sphere and its sticking to a rigid plane. *Journal of Colloid and Interface Science*, *77*, 91–101.
Munisamy, R. L., Hills, D. A., & Nowell, D. (1994). Static axisymmetrical Hertzian contacts subject to shearing forces. *ASME Journal of Applied Mechanics*, *61*, 278–283.
Nackenhorst, U. (2004). The ALE-formulation of bodies in rolling contact Theoretical foundations and finite element approach. *Computer Methods in Applied Mechanics and Engineering*, *193*, 4299–4322.
Nasdala, L., Kaliske, M., Becker, A., & Rothert, H. (1998). An efficient viscoelastic formulation for steady-state rolling structures. *Computational Mechanics*, *22*, 395–403.
Nowell, D., Dini, D., & Hills, D. A. (2006). Recent developments in the understanding of fretting fatigue. *Engineering Fracture Mechanics*, *73*, 207–222.
O'Boy, D. J., & Dowling, A. P. (2009). Tyre/road interaction noise–A 3D viscoelastic multilayer model of a tyre belt. *Journal of Sound and Vibration*, *322*(4–5), 829–850.
Olaru, D. N., Stamate, C., & Prisacaru, G. (2009). Rolling friction in a micro tribosystem. *Tribology Letters*, *35*, 205–210.
Padovan, J. (1987). Finite element analysis of steady and transiently moving/rolling nonlinear viscoelastic structure-I. *Theory. Computers & Structures*, *27*(2), 249–257.
Padovan, J., & Paramadilok, O. (1984). Transient and steady state viscoelastic rolling contact. *Computers & Structures*, *20*, 545–553.

Padovan, J., Kazempour, A., Tabaddor, F., & Brockman, B. (1992). Alternative formulations of rolling contact problems. *Finite Elements in Analysis and Design, 11*, 275–284.

Paggi, M., & Barber, J. R. (2011). Contact conductance of rough surfaces composed of modified RMD patches. *International Journal of Heat and Mass Transfer, 54*(21–22), 4664–4672.

Paggi, M., & Ciaveralla, M. (2010). The coefficient of proportionality k between real contact area and load, with new asperity models. *Wear, 268*, 1020–1029.

Pandey, A., Karpitschka, S., Venner, C. H., & Snoeijer, J. H. (2016). Lubrication of soft viscoelastic solids. *Journal of Fluid Mechanics, 799*, 433–447.

Panek, C., & Kalker, J. J. (1980). Three-dimensional contact of a rigid roller traversing a viscoelastic half space. *IMA Journal of Applied Mathematics, 26*, 299–313.

Pastewka, L., & Robbins, M. O. (2014). Contact between rough surfaces and a criterion for macroscopic adhesion. *Proceedings of the National Academy of Sciences, 111*, 3298–3303.

Persson, B. N. J. (2006a). Contact mechanics for randomly rough surfaces. *Surface Science Reports, 61*, 201–227.

Persson, B. N. J. (2006b). Rubber friction: Role of the flash temperature. *Journal of Physics Condensed Matter, 18*(32), 7789–7823.

Persson, B. N. J., & Brener, E. A. (2005). Crack propagation in viscoelastic solids. *Physical Review E, 71*(3), 036123.

Persson, B. N. J., Albohr, O., Heinrich, G., & Ueba, H. (2005). Crack propagation in rubber-like materials. *Journal of Physics-Condensed Matter, 17*(44), R1071–R114.

Persson, B. N. J. (2001). Theory of rubber friction and contact mechanics. *Journal of Chemical Physics, 115*, 3840–3861.

Persson, B. N. J. (2010). Rolling friction for hard cylinder and sphere on viscoelastic solid. *European Physical Journal E, 33*, 327–333.

Persson, B. N., & Scaraggi, M. (2014). Theory of adhesion: Role of surface roughness. *The Journal of Chemical Physics, 141*, 124701.

Persson, B. N., & Tosatti, E. (2001). The effect of surface roughness on the adhesion of elastic solids. *The Journal of Chemical Physics, 115*, 5597–5610.

Persson, B. N. J., Bucher, F., & Chiaia, B. (2002). Elastic contact between randomly rough surfaces: Comparison of theory with numerical results. *Physical Review B, 65*, 184106.

Persson, B. N. J., Lorenz, B., & Volokitin, A. I. (2010). Heat transfer between elastic solids with randomly rough surfaces. *European Physics Journal E, 31*, 3–24.

Putignano, C., Afferrante, L., Carbone, G., & Demelio, G. (2012a). The influence of the statistical properties of self-affine surfaces in elastic contact: A numerical investigation. *Journal of Mechanics and Physics of Solids, 60*(5), 973–982.

Putignano, C., Afferrante, L., Carbone, G., & Demelio, G. (2012b). A new efficient numerical method for contact mechanics of rough surfaces. *International Journal of Solids and Structures, 49*(2), 338–343.

Putignano, C., & Dini, D. (2017). Soft matter lubrication: Does solid viscoelasticity matter? *ACS Applied Materials & Interfaces, 9*(48), 42287–42295.

Putignano, C., Ciavarella, M., & Barber, J. R. (2011). Frictional energy dissipation in contact of nominally flat rough surfaces under harmonically varying load. *Journal of Mechanics and Physics of Solids, 59*(12), 2442–2454.

Putignano, C., Reddyhoff, T., Dini, D., & Carbone, G. (2013). The effect of viscoelasticity in rolling contacts: A combined experimental and numerical investigation. *Tribology Letters, 51*(1), 105–113.

Putignano, C., Le Rouzic, J., Reddyhoff, T., Carbone, G., & Dini, D. (2014). A theoretical and experimental study of viscoelastic rolling contacts incorporating thermal effects. *Proceedings of the Institution of Mechanical Engineers, Part J: Journal of Engineering Tribology, 228*(10), 1112–1121.

Putignano, C., Carbone, G., & Dini, D. (2015). Mechanics of rough contacts in elastic and viscoelastic thin layers. *International Journal of Solids and Structures, 69*, 507–517.

Rao, J. S. (2011). Finite element methods, history of mechanism and machine. *Science, 20*, 141–183.

MIX
Papier aus verantwortungsvollen Quellen
Paper from responsible sources
FSC® C105338

FSC
www.fsc.org

If you have any concerns about our products,
you can contact us on
ProductSafety@springernature.com

In case Publisher is established outside the EU,
the EU authorized representative is:
Springer Nature Customer Service Center GmbH
Europaplatz 3, 69115 Heidelberg, Germany

Printed by Libri Plureos GmbH
in Hamburg, Germany